Published August 1974

Registered at National Library Canberra, A.C.T. Australia
No. – ISBN 0 909969 98 1

Published by arrangement with and special thanks to
HONDA MOTOR Co. Ltd.
Foreign Service Division
18–1 Honcho, Wako-shi, Saitama-ken.
JAPAN, 351

CONTENTS

Australia Type

	Item			English	Metric	Note
Dimension	Overall length			139.76 in.	3,550 mm	
	Overall width			59.25 in.	1,505 mm	
	Overall height			52.95 in.	1,345 mm	
	Wheelbase			86.61 in.	2,200 mm	
	Track F/R			51.18/50.39 in.	1,300/1,280 mm	
	Ground clearance			7.09 in.	180 mm	
	Curb weight	2 Door type	MT	1,448 lbs	657 Kg	MT : manual shift transmission HM : Honda-matic
			HM	1,470 lbs	667 Kg	
		3 Door type	MT	1,475 lbs	669 Kg	
			HM	1,497 lbs	679 Kg	
	Gross vehicle weight rating	2D oor type	MT			
			HM			
		3 Door type	MT			
			HM			
	Curb weight distribu-tion F/R	2 Door type	MT	879/569 lbs	399/258 Kg	
			HM	893/577 lbs	405/262 Kg	
		3 Door type	MT	897/578 lbs	407/262 Kg	
			HM	911/586 lbs	413/266 Kg	
	Seating capacity			4		
	Overhang F/R			25.98/27.17 in.	660/690 mm	
Engine	Type			Water cooled 4 cycle, O.H.C.		
	Cylinder arrangement			4 cylinder in line, transverse		
	Bore and stroke			2.76×2.99 in.	70×76 mm	
	Displacement			71.3 cu. in.	1,170 cc	
	Compression ratio			8.1 : 1		
	Carburetor venturi dia. P/S			Down draft, 0.79/1.02 in.	20/26 mm	P : Primary S : Secondary
	Valve train			Timing belt driven single overhead camshaft		
	Lubrication system			Trochoid pump		
	Fuel required			Low lead or regular gasoline with 91 research octane number or higher		
Transmission	Clutch			Single plate dry type, diaphragm spring		
	Transmission		MT	Synchronized 4 forward 1 reverse		
			HM	2 speed with torque converter		
	Primary reduction			Direct 1 : 1		
	Gear ratio (MT)		I	3.000		HM 1 gear 1.521 2 gear 1.034 Reverse 2.916
			II	1.789		
			III	1.182		
			IV	0.846		
			Reverse	2.916		
	Final reduction		MT	Single helical gear, 4.933		
			HM	Single helical gear, 4.117		
	Clutch facing area			22.63 sq. in.	146 cm²	
Steering system	Type			Rack and pinion		
	Overall ratio			16.91		
	Turns, lock-to-lock			3.1		
	Steering wheel dia.			14.96 in.	380 mm	
Suspension system	Type, F			Independent, Macpherson strut, coil spring		
	Type, R			Independent, Macpherson strut, coil spring		
	Shock absorber F/R			Telescopic, hydralic		
Front wheel alignment	Front wheel alignment		Camber	30′		
			Caster	1°45′		
			Toe-out	0.04 in.	1 mm	
	Kingpin inclination			8°55′		
Rear wheel alignment			Toe-in	0		
			Camber	30′ positive		

	Item		English	Metric	Note
Brake system	Type, F		Disk brake		
	Type, R		Leading-trailing shoe type		
	Swept area, F		3.1 sq. in.	20 cm^2	
	Swept area, R		10.54 sq. in.	68 cm^2	
Tires	F/R		600 S 12-4PR		
Electrical system	Battery		12V-32AH		
	Starting motor		12V-0.7KW		
	Alternator		12V-35A (40A)		
	Fuses		15A×4 10A×4		
	Main fuse		45A		
	Headlight		12V-50/40 W		
	Front turn signal light/ position light		12V-21/5 W		
	Front side marker light		12V-4 cp		
	Gauge, indicator light		12V-1 W		
	Room light		12V-5 W		
	Rear turn/stop/taillight		12V-21/21/5 W		
	Backup light		12V-21 W		
	Licence light		12V-4 cp		
Service data (Engine)	Ignition timing		5° BTDC		
	Valve timing (Valve clearance 0, valve lift 0.043 in. (1.1 mm))	In. Open	10° ATDC		
		In. Close	20° ABDC		
		Ex. Open	30° BBDC		
		Ex. Close	10° BTDC		
	Spark plug		(NGK) BP-5ES BP-6ES		
			(Denso) W-20EP W-22EP		
	Spark plug gap		0.027 in.-0.031 in.	0.7-0.8 mm	
	Idle speed		750±50 rpm		With head light on
	Engine oil capacity		2.6 Imp. qt.	3.0 lit.	
	Transmission oil capacity		2.2 Imp. qt.	2.5 lit.	
	Fuel tank capacity		8.4 Imp. gal.	38.0 lit.	
	Radiator capacity		3.5 Imp. qt.	4.0 lit.	
	Honda-matic fluid capacity		2.2 Imp. qt.	2.5 lit.	
	AC generator belt tension		0.47~0.67 in.	12~17 mm	Applied load: 24.3 lbs (11 Kg)
	Recommended speed range: Up to 600 miles (1,000 Km)				
	Sepedometer range	I	0~15 mph	0~25 Km/h	
		II	~35 mph	~55 Km/h	
		III	~50 mph	~80 Km/h	
		IV	Over 15 mph	Over 25 Km/h	
	After 600 miles (1,000 Km)				
	Speedometer range	I	0~20 mph	0~ 35 Km/h	
		II	~40 mph	~ 65 Km/h	
		III	~60 mph	~100 Km/h	
		IV	Over 15 mph	Over 25 Km/h	
	Valve clearance In./Ex. cold		0.005±0.01 in.	0.12±0.02 mm	
	Breaker point gap		0.018~0.022 in.	0.45~0.55 mm	
	Compression pressure		170±28 psi	12±2 Kg/cm^2	
	Engine oil		API service SE (SD)		
	Transmission oil (MT)		API service SE (SD)		
	Honda-matic fluid (HM)		DEXRON		
(Chassis)	Tire pressure F/R		21 psi	1.5 Kg/cm^2	
	Brake fluid		SAE J1703 or DOT 3, 4 type		
	Brake pedal free play		0.04~0.2 in.	1~5 mm	
	Brake pedal to floor clearance		5.32 in.	135 mm	
	Brake pad wearing limit		0.06 in.	1.6 mm	
	Brake shoe wearing limit		0.08 in.	2.0 mm	

France type

	Item			English	Metric	Nole
Dimension	Overall length			139.76 in.	3,550 mm	
	Overall width			59.25 in.	1,505 mm	
	Overall height			52.76 in.	1,340 mm	
	Wheelbase			86.61 in.	2,200 mm	
	Track F/R			51.18/50.39 in.	1,300/1,280 mm	
	Ground clearance			6.89 in.	175 mm	
	Curb weight	2 Door type	MT	1,499 lbs	680 Kg	MT: manual shift transmission HM: Honda-matic
			HM	1,521 lbs	690 Kg	
		3 Door type	MT	1,521 lbs	690 Kg	
			HM	1,543 lbs	700 Kg	
	Gross vehicle weight rating	2 Door type	MT			
			HM			
		3 Door type	MT			
			HM			
	Curb weight distribution F/R	2 Door type	MT	915/584 lbs	415/265 Kg	
			HM	926.10/595.35 lbs	420/270 Kg	
		3 Door type	MT	904/617 lbs	410/280 Kg	
			HM	926/617 lbs	420/280 Kg	
	Seating capacity			4		
	Overhang F/R			25.98/27.17 in.	660/690 mm	
Engine	Type			Water cooled 4 cycle, O.H.C.		
	Cylinder arrangement			4 cylinder in line, transverse		
	Bore and stroke			2.76×2.99 in.	70×76 mm	
	Displacement			71.39 cu. in.	1,170 cc	
	Compression ratio			8.1 : 1		
	Carburetor venturi dia. P/S			Down draft, 0.79/1.02 in.	20/26 mm	S: Primary S: Secondary
	Valve train			Timing belt driven single overhead camshaft		
	Lubrication system			Trochoid pump		
	Fuel required			Low lead or regular gasoline with 91 research octane number or higher		
Transmission	Clutch			Single plate dry type, diaphragm spring		
	Transmission		MT	Synchronized 4 forward 1 reverse		
			HM	2 speed with torque converter		
	Primary reduction			Direct 1 : 1		
	Gear ratio (MT)		I	3.000		HM 1 gear 1.521 2 gear 1.034 Reverse 2.916
			II	1.789		
			III	1.182		
			IV	0.846		
			Reverse	2.916		
	Final reduction		MT	Single helical gear, 4.933		
			HM	Single helical gear, 4.117		
	Clutch facing area			22.63 sq. in.	146 cm²	
Steering system	Type			Rack and pinion		
	Overall ratio			16.91		
	Turns, lock-to-lock			3.1		
	Steering wheel dia.			15.16 in.	385 mm	
Suspension system	Type, F			Independent, Macpherson strut, coil spring		
	Type, R			Independent, Macpherson strut, coil spring		
	Shock absorber F/R			Telescopic, hydraulic		
Front wheel alignment	Front wheel alignment		Camber	30′		
			Caster	1°45′		
			Toe-out	0.04 in.	1 mm	
	Kingpin inclination			8°55′		
Rear wheel alignment			Toe-in	0		
			Camber	30′ positive		

	Item			English	Metric	Note
Brake system	Type, F			Disk brake		
	Type, R			Leading-trailing shoe type		
	Swept area, F			3.1 sq. in.	20 cm²	
	Swept area, R			10.54 sq. in.	68 cm²	
Tires	F/R			155 SR 12		
Electrical system	Battery			12V–45AH		
	Starting motor			12V–0.7 KW		
	Alternator			12V–35A (40A)		
	Fuses			15A×4 10A×4		
	Main fuse			45A		
	Headlight			12V–45/40W		
	Front turn signal light/ position light			12V–21/5W		
	Front side marker light			12V–4 cp		
	Gauge, indicator light			12V–1W		
	Room light			12V–5W		
	Rear turn/stop/taillight			12V–21/21/5W		
	Backup light			12V–10W		
	Licence light			12V–10W		
Service data (Engine)	Ignition timing			5°BTDC		
	Valve timing (Valve clearance 0, valve lift 0.043 in. (1.1 mm))		In. Open	10° ATDC		
			In. Close	20° ABDC		
			Ex. Open	30° BBDC		
			Ex. Close	10° BTDC		
	Spark plug			(NGK) BPR-6ES		
	Spark plug gap			0.027 in.–0.031 in.	0.7–0.8 mm	
	Idle speed			750±50 rpm		With headlight on
	Engine oil capacity			2.6 Imp. qt.	3.0 lit.	
	Transmission oil capacity			2.2 Imp. qt.	2.5 lit.	
	Fuel tank capacity			8.4 Imp. gal.	38 lit.	
	Radiator capacity			3.5 Imp. qt.	4.0 lit.	
	Honda-matic fluid capacity			2.2 Imp. qt.	2.5 lit.	
	AC generator belt tension			0.47∼0.67 in.	12∼17 mm	Applied load: 24.3 lbs (11 Kg)
	Recommended speed range	Up to 600 miles (1,000 Km)				
		Speedometer range	I	0∼15 mph	0∼25 Km/h	
			II	∼35 mph	∼55 Km/h	
			III	∼50 mph	∼80 Km/h	
			IV	Over 15 mph	Over 25 Km/h	
		After 600 miles (1,000 Km)				
		Speedometer range	I	0∼20 mph	0∼ 35 Km/h	
			II	∼40 mph	∼ 65 Km/h	
			III	∼60 mph	∼100 Km/h	
			IV	Over 15 mph	Over 25 Km/h	
	Valve clearance In./Ex. cold			0.005±0.001 in.	0.12±0.02 mm	
	Breaker point gap			0.018∼0.022 in.	0.45∼0.55 mm	
	Compression pressure			170±28 psi	12±2 Kg/cm²	
	Engine oil			API service SE (SD)		
	Transmission oil (MT)			API service SE (SD)		
	Honda-matic fluid (HM)			DEXRON		
(Chassis)	Tire pressure F/R			24 psi	1.7 Kg/cm²	
	Brake fluid			SAE J1703 or DOT 3, 4 type		
	Brake pedal free play			0.04∼0.2 in.	1∼5 mm	
	Brake pedal to floor clearance			5.32 in.	135 mm	
	Brake pad wearing limit			0.06 in.	1.6 mm	
	Brake shoe wearing limit			0.08 in.	2.0 mm	

Specifications

U.S.A. type

	Item			English	Metric	Note
Dimension	Overall length			139.76 in.	3,550 mm	
	Overall width			59.25 in.	1,505 mm	
	Overall height			52.95 in.	1,345 mm	
	Wheelbase			86.61 in.	2,200 mm	
	Track F/R			51.18/50.39 in.	1,300/1,280 mm	
	Ground clearance			7.09 in.	180 mm	
	Curb weight	2 Door type	MT	1,536 lbs	697 Kg	MT: manual shift transmission HM: Honda-matic
			HM	1,552 lbs	704 Kg	
		3 Door type	MT	1,552 lbs	704 Kg	
			HM	1,565 lbs	710 Kg	
	Gross vehicle weight rating	2 Door type	MT	2,400 lbs	1,089 Kg	
			HM	2,400 lbs	1,089 Kg	
		3 Door type	MT	2,400 lbs	1,089 Kg	
			HM	2,400 lbs	1,089 Kg	
	Curb weight distribu-tion F/R	2 Door type	MT	930/606 lbs	422/275 Kg	
			HM	946/606 lbs	429/275 Kg	
		3 Door type	MT	924/628 lbs	419/285 Kg	
			HM	946/619 lbs	429/281 Kg	
	Seating capacity			4		
	Overhang F/R			25.98/27.17 in.	660/690 mm	
Engine	Type			Water cooled 4 cycle, O.H.C.		
	Cylinder arrangement			4 cylinder in line, transverse		
	Bore and stroke			2.76×2.99 in.	70×76 mm	
	Displacement			71.39 cu. in.	1,170 cc	
	Compression ratio			8.3 : 1		
	Carburetor venturi dia. P/S			Down draft, 0.79/1.02 in.	20/26 mm	P: Primary S: Secondary
	Valve train			Timing belt driven single overhead camshaft		
	Lubrication system			Trochoid pump		
	Fuel required			Low lead or regular gasoline with 91 research octane number or higher		
Transmission	Clutch			Single plate dry type, diaphragm spring		
	Transmission		MT	Synchronized 4 forward 1 reverse		
			HM	2 speed with torque converter		
	Primary reduction			Direct 1 : 1		
	Gear ratio (MT)		I	3.000		HM 1 gear 1.521 2 gear 1.034 Reverse 2.916
			II	1.789		
			III	1.182		
			IV	0.846		
			Reverse	2.916		
	Final reduction		MT	Single helical gear, 4.933		
			HM	Single helical gear, 4.117		
	Clutch facing area			22.63 sq. in.	146 cm^2	
Steering system	Type			Rack and pinion		
	Overall ratio			16.91		
	Turns, lock-to-lock			3.1		
	Steering wheel dia.			15.16 in.	385 mm	
Suspension system	Type, F			Independent, Macpherson strut, coil spring		
	Type, R			Indedendent, Macpherson strut, coil spring		
	Shock absorber F/R			Telescopic, hydraulic		
Front wheel alignment	Front wheel alignment		Camber	30′		
			Caster	1°45′		
			Toe-out	0.04 in.	1 mm	
	Kingpin inclination			8°55′		

	Item		English	Metric	Note
Rear wheel alignment	Toe-in		0		
	Camber		30′ positive		
Brake system	Type, F		Power-assisted disk brake		
	Type, R		Leading-trailing shoe type		
	Swept area, F		3.1 sq. in.	20 cm^2	
	Swept area, R		10.54 sq. in.	68 cm^2	
Tires	F/R		600 S 12-4PR		
Electrical system	Battery		12V-45AH		
	Starting motor		12V-0.7KW		
	Alternator		12V-35A (40A)		
	Fuses		15A×4 10A×4		
	Main fuse		45A		
	Headlight		12V-50/40W		
	Front turn signal light/ parking light		12V-32/3 cp		
	Side marker light F and R		12V-4 cp		
	Gauge, indicator light		12V1 W		
	Room light		12V-5 W		
	Rear turn/stop/taillight		12V-32/32/3 cp		
	Backup light		12V-32 cp		
	Licence light		12V-4 cp		
Service data (Engine)	Ignition timing		5° BTDC. static 0° (TDC) at idle		
	Valve timing (Valve clearance 0, valve lift 0.43 in. (1.1 mm))	In. Open	10° ATDC		
		In. Close	20° ABDC		
		Ex. Open	30° BBDC		
		Ex. Close	10° BTDC		
	Spark plug		(NGK) BP-5ES BP-6ES (Denso) W-20EP W-22EP		
	Spark plug gap		0.028 in.~0.031 in.	0.7~0.8 mm	
	Idle speed	MT	800±50 rpm		With headlight on
		HM	750±50 rpm (1 range)		
	Engine oil capacity		3.2 US. qt.	3.0 lit.	
	Transmission oil capacity		2.6 US. qt.	2.5 lit.	
	Fuel tank capacity		10 US. gal.	38 lit.	
	Radiator capacity		4.2 US. qt.	4.0 lit.	
	Honda-matic fluid capacity		2.6 US. qt.	2.5 lit.	
	AC generator belt tension		0.47~0.67 in.	12~17 mm	Applied load: 24.3 lbs (11 Kg)
	Recommended speed range: Up to 600 miles (1,000 Km)				
	Speedometer range	I	0~15 mph	0~25 Km/h	
		II	~35 mph	~55 Km/h	
		III	~50 mph	~80 Km/h	
		IV	Over 15 mph	Over 25 Km/h	
	After 600 miles (1000 Km)				
	Speedometer range	I	0~20 mph	0~ 35 Km/h	
		II	~40 mph	~ 65 Km/h	
		III	~60 mph	~100 Km/h	
		IV	Over 15 mph	Over 25 Km/h	
	Valve clearance In./Ex. cold		0.006±0.001 in.	0.15±0.02 mm	
	Breaker point gap		0.018~0.022 in.	0.45~0.55 mm	
	Compression pressure		155 psi	11 Kg/cm^2	
	Engine oil		API service SE (SD)		
	Transmission oil (MT)		API service SE (SD)		
	Honda-matic fluid (HM)		DEXRON		
(Chassis)	Tire pressure F/R		24 psi	1.7 Kg/cm^2	
	Brake fluid		SAE J1703 or DOT 3, 4 type		
	Brake pedal free play		0.04~0.2 in.	1~5 mm	
	Brake pedal to floor clearance		5.32 in.	135 mm	
	Brake pad wearing limit		0.06 in.	1.6 mm	
	Brake shoe wearing limit		0.08 in.	2.0 mm	

Lubrications

Lubrication point

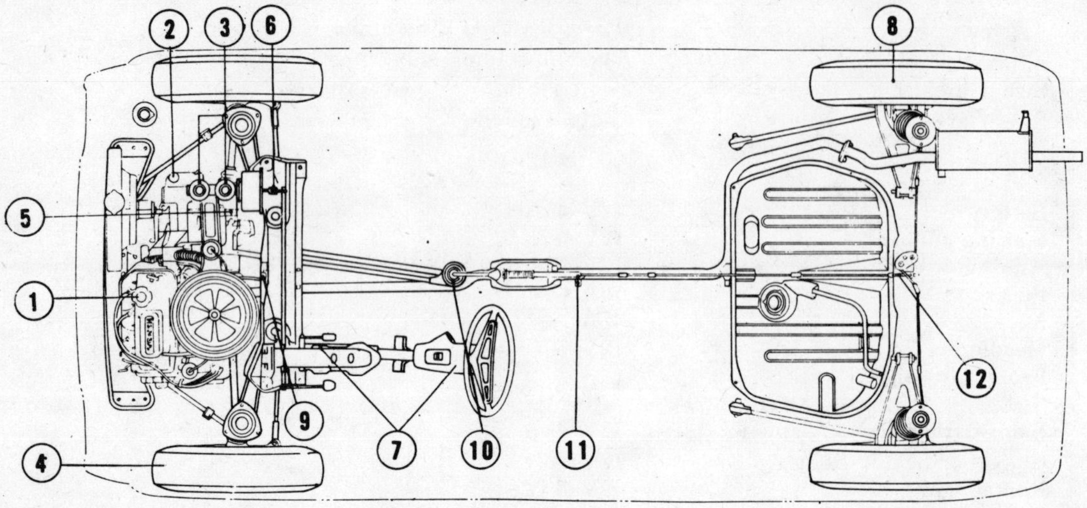

Ref. No.	Lubrication Point	Type of Oil and Grease	Remarks
1	Engine	Engine oil (see below)	Oil replace first 3,000 miles (5,000 Km) For U.S.A. type every 6,000 miles (10,000 Km) For the other type every 3,000 miles (5,000 Km)
2	Transmission Differential gear	Engine oil	Oil replace first 3,000 miles (5,000 Km) Every 24,000 miles (40,000 Km)
3	Brake master cylinder	DOT 3 or 4 SAE J, 1703	Brake fluid replace Every 24,000 miles (40,000 Km) or 24 months whichever comes first
4	Front wheel bearing	Grease	Factory sealed-lubrication system, grease supplied during overhaul
5	Drive shaft joint	Grease	
6	Pedal linkage	Grease	
7	Steering shaft bushing	Grease	
8	Rear wheel bearing	Grease	
9	Steering gear box	Grease	
10	Gearshift rod linkage	Grease	
11	Parking brake linkage	Grease	
12	Parking brake equalizer	Grease	

Recommended lubricant

	Temperature	Grade	Class
Engine Oil			
Multigrade	Above 5°F (−15°C)	SAE 10 W/40	Certified to meet or exceed US car manufacture's requirements for service SE (SD)
Multigrade	5°F 85°F (−15°C 30°C)	SAE 10 W/30	
Multigrade	Above 32°F (0°C)	SAE 20 W/40 SAE 20 W/50	
Grease			
	Multipurpose	NLGI......... 2	Multipurpose type

MAINTENANCE STANDARD

Maintenance Standard (Except Honda-matic Transmission)

Unit: mm (in.)

Group	Part and item		Assembly standard	Repair limit
Engine	**Cylinder head**			
	Gasketed surface warpage		0.03 (0.0012)	0.10 (0.0039)
	Cylinder			
	Cylinder block gasketed surface warpage		0.03 (0.0012)	0.10 (0.0039)
	I.D.		70.00~70.015 (2.7559~2.7565)	70.10 (2.7598)
	Difference in I.D.s between top and bottom			0.05 (0.0020)
	Difference in I.D.s between cylinders		0.01 (0.0004), max.	0.05 (0.0020)
	Wear			1.00 (0.03937) max.
	Piston			
	O.D. at skirt		69.98~70.00 (2.7551~2.7559)	69.91 (2.7524)
	Min. clearance in cylinder		0.03 (0.0012)	0.10 (0.0039)
	Piston ring groove width	Top	1.510~1.520 (0.0594~0.0598)	1.55 (0.0610)
		Second	1.510~1.520 (0.0594~0.0598)	1.55 (0.0610)
		Oil	2.805~2.820 (0.1104~0.1110)	2.85 (0.1122)
	Piston ring			
	Travel in groove	Top	0.020~0.045 (0.0008~0.0018)	0.13 (0.0051)
		Second	0.020~0.045 (0.0008~0.0018)	0.13 (0.0051)
	Ring gap	Top	0.2~0.4 (0.0079~0.0157)	0.6 (0.0236)
		Second	0.2~0.4 (0.0079~0.0157)	0.6 (0.0236)
		Oil	0.2~0.9 (0.0079~0.0354)	1.1 (0.0433)
	Connecting rod			
	Press-fitting interference with piston pin		0.016~0.039 (0.0006~0.0015)	0.016 (0.0006)
	Big end axial play		0.15~0.30 (0.0059~0.0118)	0.4 (0.0157)
	Bend between big and small ends		0.08/100 (3/100000), max.	
	Crankpin bearing			
	Axial play		0.2~0.45 (0.0079~0.0177)	
	Oil clearance		0.020~0.038 (0.0008~0.0015)	0.07 (0.0028)
	Crankshaft			
	Dia.	Journal	50.0~49.97 ( )	——
		Crankpin	40.0~39.97 ( )	——
	Out of round	Journal	0.005 (0.0002), max.	0.010 (0.0004)
		Crankpin	0.005 (0.0002), max.	0.010 (0.0004)

Group	Part and item		Assembly standard	Repair limit
Engine	Axial play		0.10～0.35 (0.0039～0.0138)	0.45 (0.0177)
	Bend		0.03 (0.0013), max.	0.05 (0.0020)
	Main bearing			
	Oil clearance		0.024～0.042 (0.0009～0.0017)	0.07 (0.0028)
	Lug axial play		0.2～0.45 (0.0079～0.0177)	
	Camshaft			
	Oil clearance		0.05～0.09 (0.0020～0.0035)	——
	Axial play		——	0.5 (0.0197)
	Cam height	IN	36.77～36.48 (1.4477～1.4363)	0.3 (0.0118)
		EX	36.65～36.35 (1.4429～1.4311)	0.3 (0.0118)
	Bend		0.03 (0.0012)	0.05 (0.0020)
	Valve			
	Clearance (in cold) IN. EX.	U.S.A. type	0.13～0.17 (0.005～0.007)	——
		The other type	0.10～0.14 (0.004～0.006)	
	Stem O.D.	IN	6.580～6.590 (0.2591～0.2594)	6.55 (0.2579)
		EX	6.55～6.56 (0.2579～0.2583)	6.52 (0.2567)
	Valve guide to sem clearance	IN	0.01～0.04 (0.0004～0.0016)	0.08 (0.0031)
		EX	0.05～0.08 (0.0020～0.0031)	0.11 (0.0043)
	Valve seat			
	Width	IN	1.4 (0.0551)	——
		EX	1.4 (0.0551)	——
	Sinkage		0	0.2 (0.0079)
	Valve spring			
	Free length	Inner	42.0 (1.6535)	41.0 (1.6142)
		Outer	39.95 (1.5728)	38.9 (1.5315)
	Squareness, deg.	Inner	±2°	1.5°
		Outer	±2°	1.5°
	Valve guide			
	I.D.	IN	6.61～6.63 (0.2602～0.2610)	6.65 (0.2618)
		EX	6.61～6.63 (0.2602～0.2610)	6.65 (0.2618)

Group	Part and item	Assembly standard	Repair limit
Lubricating system	**Oil pump**		
	Displacement, ℓ (US. gal. / Imp.)/min	32 (8.4 / 7.0)	—
	Inner-to-ouer rotor radial clearance	0.15 (0.0059)	0.2 (0.0079)
	Outer rotor-to-body clearance	0.1～0.18 (0.0039～0.0071)	0.2 (0.0079)
	Rotor-to-body clearance	0.03～0.10 (0.0012～0.0039)	0.15 (0.0059)
	Oil regulator		
	Pressure setting, kg/cm² (lb/in²)	3.5～4.1 (49.77～58.30)	1.5 (21.33), min. (at 80°C or 176°F and at idle speed)
Fuel system	**Fuel pump**		
	Delivery pressure, kg/cm² (lb/in²)	0.18 (2.56)	—
	Displacement, cc (cu. in.)/min. — Cam rpm	Displacement	
	300	450 (27.45), min.	
	1500	700 (42.71), min.	
	3000	750 (45.76), min.	
	Carburetor		
	Float level (when float rises)	11 (0.4331)	
	Needle valve stroke	1.3～1.7 (0.0512～0.0669)	
	Fast idle (throttle valve-to-body clearance)		
	DCG306-45, 46	1.48 (0.058) 19.5°	
	DCG306-41, 42	1.22 (0.048) 17°	
	Throttle valve opening	5.83 (0.2295) (51°)	
	Secondary fuel supply orifice	1.51～1.99 (0.0594～0.0783)	
	Power valve opening, mmHg	−156	
	Accelerator pump displacement, cc (qt)/stroke	0.6 (0.0006)	
	No. of idle screw returns	1 3/4/4±1/4	
Cooling system	**V-belt**		
	Deflection at center/load, mm/kg (in./lb)	12～17 (0.4724～0.6693)/ 9～11 (19.84～24.25)	
	Radiator		
	Cooling water (incl. heater), ℓ (US qt. / Imp qt.)	4.2 (4.4 / 3.7)	
	Filler cap opening pressure, kg/cm² (lb/in²)	0.75～1.05 (10.67～14.93)	
	Thermostat		
	Valve starts opening at, °C (°F)	80～84 (176～183)	
	Valve fully opens at, °C (°F)	95 (203)	
	Valve lift	8 (0.3150), min.	
	Water pump		
	Pump-to-crankshaft speed ratio	1.2	
	Displacement, ℓ (US. gal / Imp.)/min./rpm	70 (18.5 / 15.4)/5500	

Group	Part and item	Assembly standard	Repair limit
Cooling system	**Fan**		
	Fan-to-core clearance	8 (0.31)	
	Operating temperature, °C (°F)	88.5∼91.5 (191∼197)	
Electrical system	**Battery**		
	Lighting capacity (20 hr. rating), AH	32 or 45	
	Starting capacity (5 sec. rating), V	8, min.	
	Ignition coil		
	Specified voltage, V	12	
	Test spark gap at 8V	8 (0.3150)	
	Insulation resistance, MΩ	10, min.	
	Spark plug characteristics (Measure by means of (+) sparks.)	Cam rpm / Spark gap / Voltage 150 / 8 (0.3150), min. / 8V 3000 / 7 (0.2756), min. / 12V	
	Spark advancer		
	Centrifugal		
	Beginning of advancement, rpm	575	
	Maximum advance angle, deg/rpm	14.5/2000	
	Vacuum		
	Beginning of advancement, mmHg	120	
	Maximum advance angle, deg/mmHg	12/275	
	Contact breaker		
	Dwell angle, deg	49∼55	
	Point gap	0.45∼0.00 (0.0177∼0.0217)	
	Point pressure, g (oz)	500∼600 (17.64∼21.16)	
	Capacitor		
	Capacity, μF	0.22∼0.24	
	Insulation resistance, MΩ	10, min.	
	Spark plug		
	Plug gap	0.7∼0.8 (0.0276∼0.0315)	
	AC generator (Values in ( ) are for a car equipped with an air conditioner.)		
	Voltage, V	12	
	Speed under no load, V/rpm	14/1200 (820)	
	Speed under load, V/A/rpm	14/35 (40)/5000	
	Coil resistance (rotor), Ω	4.2 (4.5)	±0.1
	Generator-to-crankshaft speed ratio	2.24	
	O.D., slip ring without air conditioner	32.5 (1.2795)	32.1 (1.2638)
	O.D., slip ring with air conditioner	32.0 (1.2598)	29.6 (1.1654)
	Brush wear	12.5 (0.4921)	5.5 (0.2165)
	Brush spring tension, g (oz)	300∼500 (10.58∼17.64)	

Group	Part and item	Assembly standard	Repair limit
Electrical system	**Regulator**		
	Voltage relay		
	Operating voltage, V	4.5~5.8	
	Relay point gap	0.4~1.2 (0.0016~0.0472)	
	Contact spring deflection (when attracted)	0.2~0.45 (0.0079~0.0177)	
	Voltage regulator		
	Regulated voltage, V	13.5~14.5	
	Armature gap	0.3 (0.0118), min.	
	Point gap	0.25~0.45 (0.0098~0.0177)	
	Contact spring deflection (when attracted)	0.2~0.6 (0.0079~0.0236)	
	Angle gap	0.2 (0.0079), min.	
	Starting motor		
	Mica depth	0.5~0.8 (0.0197~0.0315)	0.2 (0.0079)
	Commutator	0	0.08 (0.0032)
	Out of round	33 (1.2992)	32 (1.2598)
	O.D.	0.03~0.1	0.2 (0.0079)
	Bushing-to-shaft clearance	(0.0012~0.0039)	
	Brush length	16 (0.6299)	4 (0.1575)
	Brush spring tension, kg (lb) (at new parts)	1.6 (3.528)	——
	Pinion-to-pinion stop collar clearance	0.3~1.5 (0.0118~0.0591)	——
Power transmission system	**Pressure plate**		
	Face runout	0.03 (0.0012)	0.05 (0.0020)
	Friction disc		
	Depth up to rivet head	1.2~1.5 (0.0472~0.0591)	
	Face runout	0.5 (0.0197)	0.6 (0.0236)
	Spline radial play	0.04~0.10 (0.0016~0.0039)	0.4 (0.0158)
	Thickness	8.3~9.0 (0.3268~0.3543)	5.9~6.6 (0.2322~0.2598)
	Clutch release holder		
	I.D.	29.00~29.03 (1.1417~1.1535)	29.06 (1.1440)
	Holder-to-case clearance	0.02~0.08 (0.0008~0.0032)	0.2 (0.0079)
	Clutch pedal		
	Entire pedal stroke	145 (5.7087)	
	Pedal lash	15~20 (0.5906~0.7874)	20 (0.5906)
	Floor clearance	——	30 (1.1811)
	Main shaft		
	Axial play	0.09 (0.0035)	0.3 (0.0118)
	Needle roller bearing journal O.D.	20~19.98 (0.7874~0.7866)	
	6304 ball bearing journal O.D.	22~21.98 (0.8661~0.8654)	

Group	Part and item	Assembly standard	Repair limit
Power transmission system	**Countershaft**		
	Axial play	0.09 (0.0035)	0.3 (0.0118)
	Runout	0.03 (0.0012)	0.07 (0.0028)
	Needle roller bearing journal O.D.	31.98～32 (1.2591～1.2598)	31.97 (1.2587)
	Countershaft low and top gears		
	I.D.	37.00～37.02 (1.4567～1.4575)	37.05 (1.4587)
	Axial play	0.03 (0.0019)	0.18 (0.0071)
	Countershaft second and third gears		
	I.D.	37.00～37.02 (1.4567～1.4575)	37.05 (1.4589)
	Axial play	0.03 (0.0019)	0.18 (0.0071)
	Spacers between gears		
	O.D.	32～31.98 (1.2598～1.2590)	31.96 (1.2582)
	I.D.	25.98～25.99 (1.0228～1.0232)	——
	Width		——
	Reverse gear shaft		
	O.D.	14.98～14.97 (0.5898～0.5902)	19.94 (0.5882)
	Reverse idle gear		
	I.D.	15.04～15.01 (0.5921～0.5909)	15.08 (0.5937)
	Gear-to-shaft clearance	0.07～0.03 (0.0028～0.0012)	——
	Shift fork		
	Synchronizer sleeve sliding surface width	6.4～6.5 (0.2520～0.2559)	6.0 (0.2362)
	Fork-to-synchronizer sleeve clearance	0.45～0.65 (0.0177～0.0256)	1.0 (0.0394)
	Differential pinion shaft		
	O.D.	17.96～17.98 (0.7071～0.7079)	17.95 (0.7067)
	Final driven gear		
	Backlash	0.03 (0.0012)	0.18 (0.0071)
	Differential gear case		
	Differential pinion shaft hole dia.	18.04～18.01 (0.7104～0.7093)	18.1 (0.7126)
	Differential pinion shaft-to-hole clearance	0.077～0.032 (0.0030～0.0013)	0.1 (0.0039)

Group	Part and item		Assembly standard	Repair limit
Steering system	**Steering wheel**			
	Play (at periphery)		10 (0.3937), max.	
	Steering gear backlash		0.15 (0.0059) (measured when adjusting screw is turned back 45°)	0.3 (0.0118)
	Steering column			
	Column bushing I.D.	Upper	17.85~18.15 (0.7028~0.7146)	
		Lower	18.15~18.45 (0.7146~0.7264)	
Braking system	**Parking brake lever**			
	Effective stroke		1~5 notches (to lock)	
	Brake pedal			
	Height (above floor mat)		135 (5.3150)	
	Free play (at toe)		1~5 (0.0394~0.1969)	8 (0.3150)
	Master cylinder			
	Bore-to-piston clearance		0.02~0.105 (0.0008~0.0039)	0.13 (0.0051)
	Wheel cylinder			
	Bore to piston clearance	Front	0.02~0.105 (0.0008~0.0039)	0.15 (0.0059)
		Rear	0·016~0.086 (0.0006~0.0034)	0.13 (0.0051)
	Brake drum			
	Drum-to-lining clearance	Front	4 clicks (turn back)	
		Rear	2 clicks (turn back)	
	I.D.	Front	215 (8.4646)	216 (8.5039)
		Rear	180 (7.0866)	181 (7.1260)
	Brake lining			
	Thickness	Front	5.0 (0.1969)	2.0 (0.0787)
		Rear	5.0 (0.1969)	2.0 (0.0787)
	Wheel			
	Rim face runout		0.20 (0.0079)	3.0 (0.1181)
	Disc brake			
	Disc thickness		9.6 (0.3780)	9.0 (0.3543)
	Disc face runout		0.1 (0.0039)	0.15 (0.0059)
	Disc parallelism		0.03 (0.0012)	0.07 (0.0028)
	Pad thickness		10.3 (0.4055)	1.6 (0.0630)

Maintenance Standard (Honda-matic Transmission only)

Unit: mm (in.)

Group	Part and item	Assembly standard	Repair limit
	Hydraulic pressure		
	Line, kg/cm² (lb/in²) at 1000 rpm	5.0～8.0 (71.10～113.76)	4.0 (56.88)
	"D" clutch, kg/cm² (lb/in²) at 1000 rpm	5.0～8.0 (71.10～113.76)	4.0 (56.88)
	"L" clutch, kg/cm² (lb/in²) at 1000 rpm	5.0～8.0 (71.10～113.76)	4.0 (56.88)
	Stall rpm		
	Normal driving (Except driving on highlands)	2200～2600	2100
Torque converter	**Turbine**		
	Thickness of washer attaching surface	11.3～11.4 (0.4449～0.4488)	11.0 (0.4331)
	Stator cam		
	Roller attaching surface	——	Dents
	Stator hub		
	O. D.	40.066～40.091 (1.5774～1.5784)	Dents
	I. D.	31.000～31.039 (1.2205～1.2220)	Dents
	Stator side plate		
	Thickness	5.95～6.00 (0.2343～0.2362)	5.90 (0.2323)
	Cover		
	Bushing O. D.	13.000～13.018 (0.5118～0.5125)	13.1 (0.5157)
	Thrust washer (37×54×2)		
	Thickness	1.95～2.05 (0.0768～0.0807)	1.90 (0.0748)
	Thrust washer (17×45×2.5)		
	Thickness	2.45～2.55 (0.0965～0.1004)	2.40 (0.0945)
	Thrust washer (38×66×2)		
	Thickness	1.95～2.00 (0.0768～0.0787)	1.90 (0.0748)
Clutch	**"L" clutch assembly**		
	Initial clearance	0.5～0.8 (0.0197～0.0315)	
	"D" clutch assembly		
	Initial clearance	0.5～0.8 (0.0197～0.0315)	
	Clutch return spring		
	Free length	39.7 (1.5630)	36.0 (1.4173)
	Clutch disc		
	Thickness	1.95～2.05 (0.0768～0.0807)	Glazing (oil grooves have gone)
	Clutch plate		
	Thickness	1.95～2.05 (0.0768～0.0807)	Flaking and discoloration
	Clutch end plate		
	Thickness	1.7～1.8 (0.0669～0.0709)	Flaking and discoloration
		2.0～2.1 (0.0787～0.0827)	Flaking and discoloration

Group	Part and item	Assembly standard	Repair limit
Clutch	**Clutch end plate**		
	Thickness	2.3~2.4 (0.0906~0.0945)	Flaking and discoloration
		2.6~2.7 (0.1024~0.1063)	Flaking and discoloration
		2.9~3.0 (0.1142~0.1181)	Flaking and discoloration
		3.2~3.3 (0.1260~0.1299)	Flaking and discoloration
Transmission	**Main shaft**		
	Diameter of stator shaft needle roller bearing surface	17.983~27.994 (0.7080~1.1021)	Scores, scratches or excessive wear
	Diameter of "D" gear needle roller bearing surface	27.980~27.993 (1.1016~1.1021)	Scores, scratches or excessive wear
	Countershaft		
	Diameter of "L" side needle roller bearing surface	29.987~30.000 (1.1806~1.1811)	Scores, scratches or excessive wear
	Diameter of "D" gear needle roller bearing surface	25.980~25.993 (1.0228~1.0233)	Scores, scratches or excessive wear
	Reverse idle shaft		
	O. D.	13.983~13.994 (0.5505~0.5509)	Scores, scratches or excessive wear
	Reverse idle shaft holder		
	I. D.	14~14.027 (0.5512~0.5522)	Scores, scratches or excessive wear
	"L" gear		
	Axial play	0.1~0.276 (0.0040~0.0109)	0.6 (0.0197)
	I. D.	33~33.016 (1.2992~1.2998)	Scores, scratches or excessive wear
	"D" gear		
	Axipl play	0.3~0.956 (0.0118~0.0376)	1.0 (0.0394)
	I. D.	33~33.016 (1.2992~1.2998)	Scores, scratches or excessive wear
	Countershaft "D" gear		
	Axial play	0.05~0.35 (0.0020~0.0138)	0.5 (0.0197)
	I. D.	31~31.016 (1.2205~1.2211)	Scores, scratches or excessive wear
	Countershaft reverse gear		
	Axial play	0.1~0.2 (0.0040~0.0079)	0.5 (0.0197)
	I. D.	36~36.016 (1.4173~1.2211)	Scores, scratches or excessive wear
	Reverse idle gear		
	Axial play	0.1~0.4 (0.0040~0.0157)	0.6 (0.0236)
	I. D.	18~18.013 (0.7087~0.7092)	Scores, scratches or excessive wear
	Reverse gear hub		
	O. D.	51.89~51.92 (2.0429~2.0441)	Scores, scratches or excessive wear
	Thrust washer (22×41×5)		
	Thickness	2.0~2.1 (0.0787~0.0827)	1.95 (0.0768)
	Thrust washer (28×41×1.2)		
	Thickness (on right side of "D" gear)	1.15~1.25 (0.0453~0.0492)	1.10 (0.0433)
	Thrust washer (28×46.5×3)		
	Thickness (on right side of "D" clutch)	2.95~3.05 (0.1161~0.1201)	2.90 (0.1142)

Group	Part and item	Assembly standard	Repair limit
Transmission	**Thrust washer** (22×45×2)		
	Thickness (on left side of main shaft "L" gear)	1.95~2.05 (0.0768~0.0807)	1.90 (0.0748)
	Thrust washer (33×46×1.2)		
	Thickness (on left side of "L" gear)	1.15~1.25 (0.0453~0.0492)	1.10 (0.0433)
	Thrust washer (20×46.5×3)		
	Thickness (on left side of "L" clutch)	2.95~3.05 (0.1161~0.1201)	2.90 (0.1142)
Control system	**Reverse shift fork**		
	Thickness of pawl	5.8~5.9 (0.2283~0.2323)	5.2 (0.2047)
	Parking brake pawl		
	Pawl		Abrasive wear or galling
	Pawl release spring		
	Free length	33 (1.2992)	30 (1.1811)
	Parking brake shaft		
	Runout	0.1 (0.0040), max.	
	Parking gear		
	Teeth profile		Abrasive wear or galling
	Servo body		
	Shift fork shait bore diameter	14~14.018 (0.5512~0.5519)	14.025 (0.5522)
	Shift fork shaft valve bore diameter	37~37.039 (1.4567~1.4582)	37.045 (1.4585)
	Check valve spring		
	Free length	38 (1.4961)	34 (1.3386)
Hydraulic system			
	Oil pump gear		
	Gear-to-housing thrust clearance	0.03~0.06 (0.0012~0.0024)	0.08 (0.0031)
	Gear-to-housing clearance	0.05~0.09 (0.0020~0.0035)	0.15 (0.0059)
	Stator shaft		
	Bushing I. D.	17~17.018 (0.6693~0.6700)	17.1 (0.6732)
	Diameter of needle roller bearing surface	24~24.013 (0.9449~0.0454)	Scores, scratches or excessive wear
	Right side cover		
	"L" clutch bearing bore diameter	52.05~52.08 (2.0492~2.0504)	52.1 (2.0512)

Group	Part and item	Assembly ttandard	Repair limit
Selector lever	**Lock pin spring**		
	Free length	92.5 (3.6417)	90 (3.5433)
	Push knob spring		
	Free length	43.0 (1.6929)	41 (1.6142)
	Selector stopper spring		
	Free length	27.0 (1.0630)	28.5 (1.1220)
	Control cable		
	Pin hole diameter (on both engine and body sides)	7~7.036 (0.2756~0.2770)	7.096 (0.2794)
	Selector lever bushing		
	I. D.	9.75~9.9 (0.3839~0.3898)	9.98 (0.3929)
	Control rod pin A		
	O. D.	6.9~6.95 (0.2717~0.2736)	6.85 (0.2697)
	Selector lever stopper arm		
	I. D.	9.05~9.1 (0.3563~0.3583)	9.15 (0.3602)
	Selector stopper bolt		
	O. D.	8.9~9.0 (0.3504~0.3543)	8.85 (0.3484)

TORQUE SPECIFICATIONS

Engine

Unit: kg-m (lb-ft)

No.	Parts	Thread dia. (mm)	Torque
1	Crankshaft bearing cap	10×95	3.7~4.3 (27~31)
2	Connecting rod bearing cap	8	2.5~2.9 (18~21)
3	Cam holder	8×77	1.8~2.2 (13~16)
4	Flywheel and crankshaft	10	4.7~5.3 (34~38)
5	Clutch cover and flywheel	6×18	1.0~1.4 (7~10)
6	Cylinder head bolt	10	4.2~4.8 (30~35)
7	Main shaft (left-hand threads) lock nut	20	4.0~5.0 (29~36)
8	Countershaft lock nut	23	6.0~7.0 (43~51)
9	Final driven gear and differential case	10	9.0~9.5 (65~69)
10	Crankshaft pulley	12×25	4.7~5.3 (34~38)
11	Timing belt driven pulley	8×22	2.5~3.5 (18~25)
12	Engine mount bracket	10×27	3.1~3.9 (22~28)
13	Intake manifold	8	1.8~2.4 (13~17)
14	Carburetor assembly	6	0.8~1.2 (6~ 9)
15	Oil pan	6×18	0.3~0.7 (2~ 5)
16	Clutch case cover (oil pan)	6×18	0.3~0.7 (2~ 5)
17	Fuel pump	8	1.8~2.4 (13~17)
18	Tappet adjusting nut	7	1.8~2.2 (13~16)
19	Oil filter	20	1.5~2.0 (11~14)
20	Water pump	6	1.0~1.4 (7~10)
21	Exhaust manifold	8	1.8~2.4 (13~17)
22	Oil pump assembly	6	1.0~1.4 (7~10)
23	Clutch spring stopper	6×10	0.7~1.1 (5~ 8)
24	Oil pass block	6×32	1.0~1.4 (7~10)
25	Oil pump gear holder	6×28	1.0~1.4 (7~10)
26	Engine drain bolt	20	4.0~5.0 (29~36)
27	Cylinder head cover	6	0.8~1.2 (6~ 9)
28	Breather cover	6×12	0.7~1.1 (5~ 8)
29	Tachometer special bolt	6	1.0~1.4 (7~10)
30	Thermounit	16	2.0~2.5 (14~18)
31	Spark plug	14	1.3~1.7 (9~12)
32	Distributor fixing plate	6×14	1.0~1.4 (7~10)
33	Oil pressure unit	PT 1/8	1.4~1.8 (10~13)
34	AC generator	10×60	4.0~4.6 (29~33)
35	AC generator stay	8×22	2.0~2.5 (14~18)
36	Spring ball setscrew	12	1.0~2.4 (7~17)
37	Holder arm B complete	6×18	1.0~1.4 (7~10)
38	Low shift fork	6	1.0~1.4 (7~10)
39	Top shift fork	6	1.0~1.4 (7~10)
40	Reverse shift fork	6	1.0~1.4 (7~10)
41	Shift arm A	8×24	2.0~2.8 (14~20)
42	Clutch release arm A	8×24	2.0~2.8 (14~20)
43	Transmission drain bolt	14	3.5~4.5 (25~33)
44	Clutch case	10	4.0~5.0 (29~36)
45	Starting motor	10	4.0~5.0 (29~36)
46	Transmission case	8×40	2.3~3.1 (17~22)
47	Transmission cover	6×32	1.0~1.4 (7~10)
48	Back light switch	14	2.3~2.7 (17~20)
49	Timing belt upper cover	6×35	0.8~1.2 (6~ 9)
50	Timing belt lower cover	6×22	0.8~1.2 (6~ 9)
51	Breathing bolt	10	0.7~1.1 (5~ 8)
52	Hot Air cover	6×12	1.0~1.4 (7~10)
53	Speedometer gear	6	0.7~1.0 (5~ 7)
54	Clutch control cable stay	8×20	2.0~2.8 (14~20)

Frame

Unit: kg-m (lb-ft)

No.	Parts	Thread dia. (mm)	Torque
1	Gear box	8×16	1.9~2.5 (14~18)
2	Rack end	12	5.0~6.0 (36~43)
3	Tie lod end lock nut	12	4.0~4.8 (29~35)
4	Tie rod ball joint	10	4.0~4.8 (29~35)
5	Steering joint bolt	8	2.4~3.0 (17~22)
6	Steering wheel nut	12	3.0~4.5 (22~33)
7	Steering column	8×20	1.8~2.5 (13~18)
8	Rack guide lock nut	45	4.0~5.0 (29~36)
9	Pedal bracket	8×45	4.0~5.0 (29~36)
10	Front backing plate mounting bolts	8×18	1.9~2.5 (14~18)
11	Rear backing plate mounting bolts	8×12	1.9~2.5 (14~18)
12	Front spindle nut	32	12.0~18.0 (87~130)
13	Rear axle nut	16	10.0~13.0 (72~ 94)
14	Master cylinder	8	1.5~2.0 (11~14)
15	Parking brake lever	8×20	1.9~2.5 (14~18)
16	Caliper set bolt	10×26	5.0~6.0 (36~43)
17	Bleeder screw plug	8	0.5~0.7 (4~ 5)
18	Parking brake equalizer	8	1.9~2.5 (14~18)
19	Rear wheel bearing cap	6×12	0.4~0.7 (3~ 5)
20	Two-way joint	6	0.7~1.2 (5~ 9)
21	Three-way joint	6×25	0.7~1.2 (5~ 9)
22	Exhaust manifold and engine	8	1.9~2.5 (14~18)
23	Exhaust manifold and pipe	8	1.9~2.5 (14~18)
24	Exhaust pipe and silencer	8	1.9~2.5 (14~18)
25	Silencer mount	8×25	1.9~2.5 (14~18)
26	Fuel tank and body	8×20	1.9~2.5 (14~18)
27	Brake pedal mechanical stop switch	10	0.7~1.2 (5~ 9)
28	Brake pipe flare nut	10	1.3~1.5 (9~11)
29	Gearshift lever and rod	8	1.9~2.5 (14~18)
30	Gearshift extension mount	6×20	0.7~1.2 (5~ 9)
31	Front beam and body	10	3.5~4.3 (25~31)
32	Engine mount beam and body	10×70	3.5~4.3 (25~31)
33	Front beam, engine mount beam and center beam	10×25	3.5~4.3 (25~31)
34	Engine mount rubber A and center beam	8	1.0~1.6 (7~12)
35	Engine mount rubber A and engine	8×25	1.9~2.5 (14~18)
36	Engine mount rubber B and engine	8×16	1.9~2.5 (14~18)
37	Torque rod and body	10×40	3.5~4.3 (25~31)
38	Torque rod and engine	12	3.5~4.3 (25~31)
39	Radiator and body	6×10	0.7~1.2 (5~ 9)
40	Lower arm ball joint	10×40	3.0~4.0 (22~29)
41	Radius rod	10×60	3.5~5.0 (25~36)
42	Lower arm	10	3.5~5.0 (25~36)
43	Front shock absorber and knuckle	10	5.0~6.0 (36~43)
44	Front shock absorber and body	8	1.0~1.6 (7~12)
45	Stabilizer bracket	6×16	0.7~1.2 (5~ 9)
46	Front shock absorber center nut	12	5.5~7.0 (39.8~50.7)
47	Rear shock absorber center nut (upper)	12	5.5~7.0 (39.8~50.7)
48	Rear shock absorber center nut (lower)	12	5.5~7.0 (39.8~50.7)
49	Rear shock absorber and body	8	1.0~1.6 (7~12)
50	Rear lower arm and boey	10	3.5~5.0 (25~36)
51	Rear lower arm and shock absorber	12	5.0~6.5 (36~47)
52	Hub carrier and shock absorber	10×60	3.5~5.0 (25~36)
53	Rear radius rod and hub carrier	14	5.5~7.5 (40~54)
54	Rear radius rod and body	10×60	3.5~5.0 (25~36)

No.	Parts	Thread dia. (mm)	Torque
55	Front wheel clip nut	10	7.0~9.0 (51~65)
56	Rear wheel clip nut	10	7.0~9.0 (51~65)
57	Door striker	8	1.5~2.0 (11~14)
58	Door lock	6×25	0.4~0.7 (3~ 5)
59	Door inside cable	6×12	0.4~0.7 (3~ 5)
60	Front seat	8×22	1.9~2.5 (14~18)
61	Front seat belt		3.0~4.5 (22~33)
62	Seat belt anchor bolt	10×25	3.0~4.5 (22~33)
63	Heater blower	8	1.8~2.6 (13~19)
64	Front and rear bumpers	8	1.8~2.6 (13~19)

V–3. Honda-Matic

Unit: kg-m (lb-ft)

No.	Parts	Thread dia. (mm)	Torque
1	Drive plate (Crankshaft)	10	4.7~5.3 (34~38)
2	Drive plate (Torque converter)	6×12	1.0~1.4 (7~10)
3	Torque converter	6×40	1.0~1.4 (7~10)
4	Main shaft	16	4.7~5.3 (34~38)
5	Countershaft	20	4.7~5.3 (34~38)
6	Final driven gear	10	9.0~9.5 (65~69)
7	Bearing retainer (Pan head screw)	6×16	0.8~1.2 (6~ 9)
8	Bearing retainer (Hex bolt)	6×18	1.0~1.4 (7~10)
9	Main valve body	6	1.0~1.4 (7~10)
	Servo valve body	6	
	Seal ring guide	6	
	Oil strainer	6	
10	Inhibitor switch	6×12	0.8~1.2 (6~9)
11	Turnbuckle	6	0.8~1.2 (6~9)
12	Selector lever	6	0.8~1.2 (6~9)
13	Control cable clamp	6	0.3~0.5 (2~4)

Physical Dimension

U.S.A. type

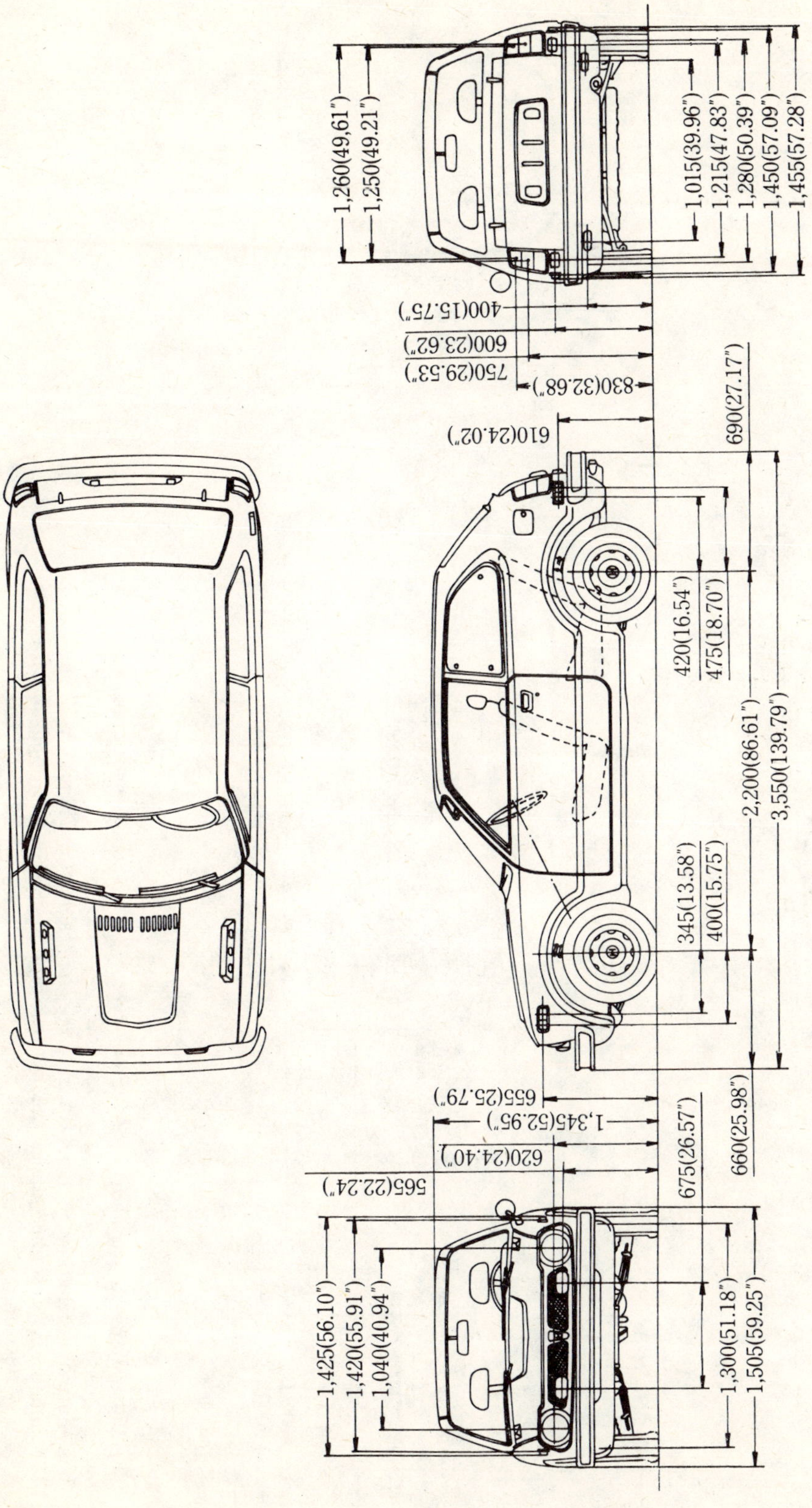

Remark:

The illustrations shown above are 2 Door type. 3 Door type has also the same dimensions.

Australia and France type

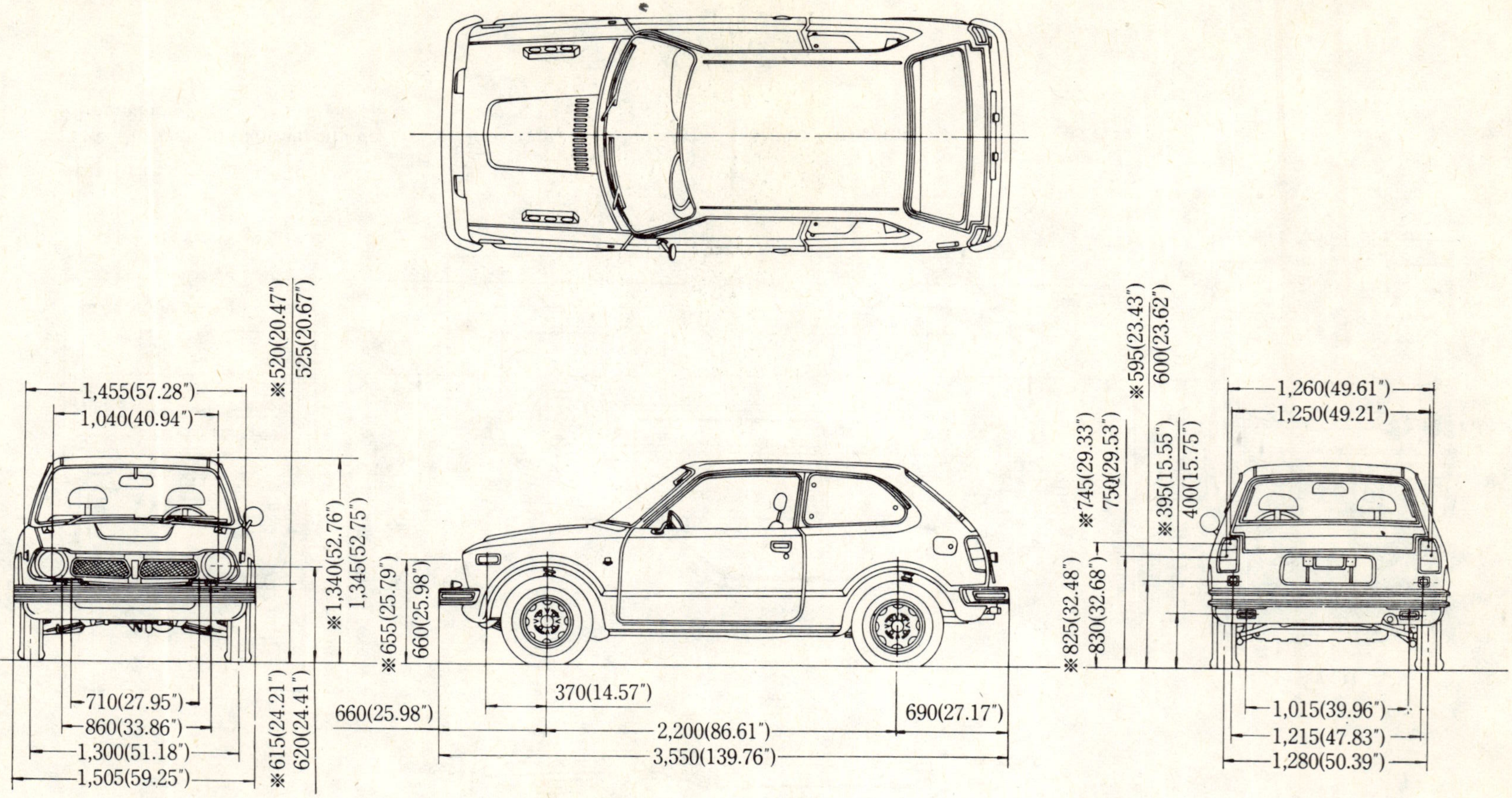

Remarks:

1. The illustrations shown above are 3 Door type. 2 Door type has also the same dimensions.
2. The dimensions marked with an asterisk are applicable for France type only. The dimensions described under the figures for France type are applicable for Australia type only.
3. Other dimensions are for both types.

Location of the Serial Numbers

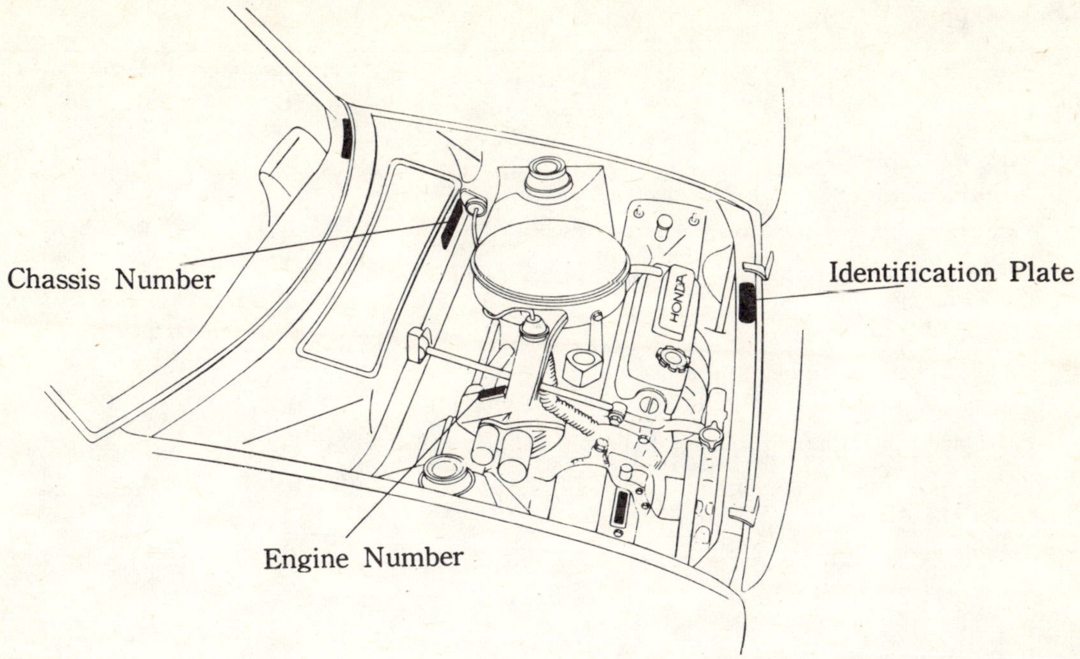

Identification plate

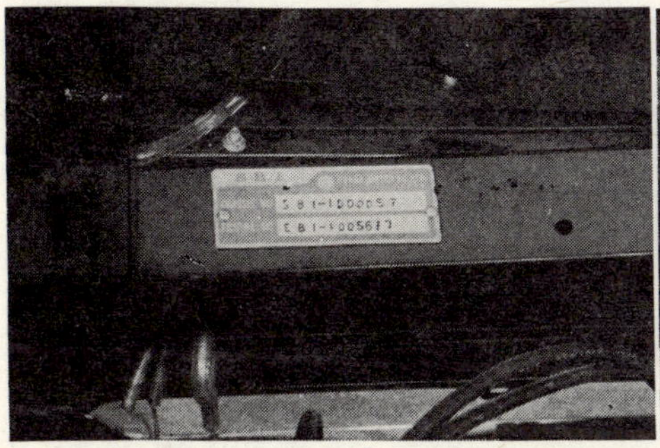

Chassis number

Engine number

• Always order replacement parts by the number stamped on the identification number plate.

Standard Tightening Torque

Unless otherwise specified, the bolts and nuts should be tightened to the torque value listed in the following table.

Standard Torque Table

Bolt/Nut nominal diameter (mm)	Thread pitch (mm)	Torque (kgm (lbft)
4	0.7	0.2 ~0.3 (1.45~2.15)
5	0.9	0.45~0.7 (3.25~5.05)
6	1.0	0.8 ~1.2 (5.80~8.70)
8	1.25	2.0 ~2.8 (14.5~20.0)
9	1.25	3.0 ~4.0 (22.0~29.0)
10	1.25	4.0 ~5.0 (29.0~39.0)
12	1.25	7.0 ~8.5 (50.5~61.5)

Maintenance Schedule

For U.S.A. type

ITEMS		Mileage Interval × 1000 Miles (Km)								
		3 (5)	6 (10)	12 (20)	18 (30)	24 (40)	30 (50)	36 (60)	42 (70)	48 (80)
Engine	Valve clearance	A		A		A		A		A
	Alternator drive belt	A		I		I		I		I
	Vacuum fittings, hoses and connections			I		I		I		I
	Engine oil	R	R	R	R	R	R	R	R	R
	Engine oil filter	R	R	R	R	R	R	R	R	R
	Radiator coolant					R				R
	Cooling system, hoses and connections			I		I		I		I
	Exterior of radiator core			C		C		C		C
	Intake and exhaust manifold tightness check	A								
Fuel System	Idle speed and idle CO	A		A		A		A		A
	Choke Mechanism	I		I		I		I		I
	Throttle opener and control valve	I		I		I		I		I
	Air cleaner element			R		R		R		R
	Fuel filter					R				R
	Intake air temp. control system			I		I		I		I
	Tank, fuel lines and connections	I				I				I
Evaporative Emission System	Carbon canister and idle cut-off valve			I		R		I		R
	One way valve			I		I		I		I
	Fuel filler cap			I		I		I		I
Ignition Components	Ignition timing			A		A		A		A
	Contact breaker points			R		R		R		R
	Spark plugs			R		R		R		R
	Distributor cap and rotor			I		R		I		R
	Ignition wiring					I				I
	Idle retarder (spark control)			I		I		I		I
	Advance mechanism in distributor					I				I
	Crankcase emission system			I		I		I		I
4 Speed	Transmission controlled spark advance					I				I
2 Speed	Temperature controlled spark advance					I				I
Rear brake		A	A	A	A	A	A	A	A	A
Parking brake		A								
Brake hoses, lines, master cylinder fluid level			I	I	I	I	I	I	I	I
Brake fluid *2						R				R
Front brake pads, yoke performance			I	I	I	I	I	I	I	I
Rear brake linings				I		I		I		I
Air filter of master power *1				C		C		C		C
Front wheel alignment				I		I		I		I
Clutch pedal travel			I	I	I	I	I	I	I	I
Transmission oil		R				R				R
Radiator fan			I	I	I	I	I	I	I	I
Engine exhaust silencer, susp. mounting bolts			I	I	I	I	I	I	I	I
Steering operation, tie rod ends, rack guide, grease steering gear box		I		I		I		I		I

A: Adjust R: Replace C: Clean

I: Inspect, adjust or replace if necessary

*1 Every 12,000 miles (20,000 Km) or 12 months whichever comes first

*2 Every 24,000 miles (40,000 Km) or 24 months whichever comes first

Australian Type

ITEMS		Mileage Interval × 1000 Miles (km)									
		3 (5)	6 (10)	9 (15)	12 (20)	15 (25)	18 (30)	21 (35)	24 (40)	27 (45)	30 (50)
Engine oil	Every 3,000 miles (5,000 km)	R	R	R	R	R	R	R	R	R	R
Engine oil filter	Every 6,000 miles (10,000 km)		R		R		R		R		R
Front disc pad wear and yoke function			I		I		I		I		I
Front brake lining adjustment			A		A		A		A		A
Brake system failure			I		I		I		I		I
Clutch pedal travel			I		I		I		I		I
Ignition timing	Every 12,000 miles (20,000 km)	A			A				A		
Contact breaker point					R				R		
Valve clearance		A			A				A		
Alternator drive belt		A			I				I		
Air cleaner element					R				R		
Spark plug					R				R		
Front wheel alignment					I				I		
Front/rear brake lining wear					I				I		
Steering wheel play and operation Grease replenishment		I			I				I		
Crankcase emission system					I				I		
Cooling system, hoses and connections					I				I		
Radiator core					C				C		
Automatic intake air control system					I				I		
Fuel filler cap					I				I		
Air filter of master power *1					C				C		
Radiator fluid	Every 24,000 miles (40,000 km)								R		
Fuel strainer									R		
Transmission oil (incl. ATF)		R							R		
Brake fluid *2									R		
Parking brake lever effective stroke	Initial time	A									
Engine, engine mount, exhaust pipe and suspension parts tightness		A									

A: Adjust R: Replace C: Clean

I: Inspect, adjust or replace if necessary

*1 Every 12,000 miles (20,000 km) or 12 months whichever comes first

*2 Every 24,000 miles (40,000 km) or 24 months whichever comes first

Vehicle Supporting Position

The garage jack and safety stands must be used as shown below.

- When jacking up the front end, put the parking brake on, and for rear end, put the gearshift lever in reverse.

NOTES:

- Use wheel chocks to prevent the car from moving.
- Be sure to support both sides of the car with safety stands. It is dangerous to do the jobs with the car supported by the garage jack only, even for a short period of time.

a. Where to use garage jack

Front

Front jack-up bracket

NOTE:

Take care not to damage the front skirt.

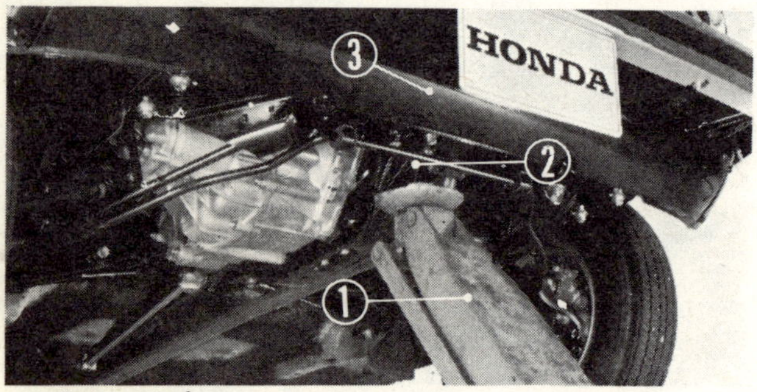

① Garage jack
② Jack-up bracket
③ Front skirt

Rear

Rear jack-up bracket

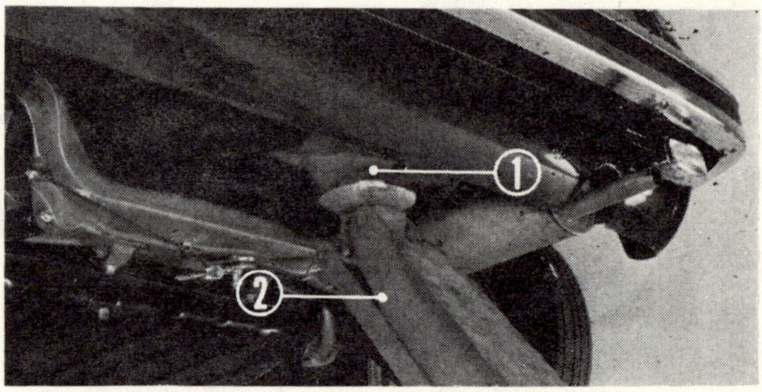

① Jack-up bracket
② Garage jack

b. Where to use safety stands

Front

Front side sills

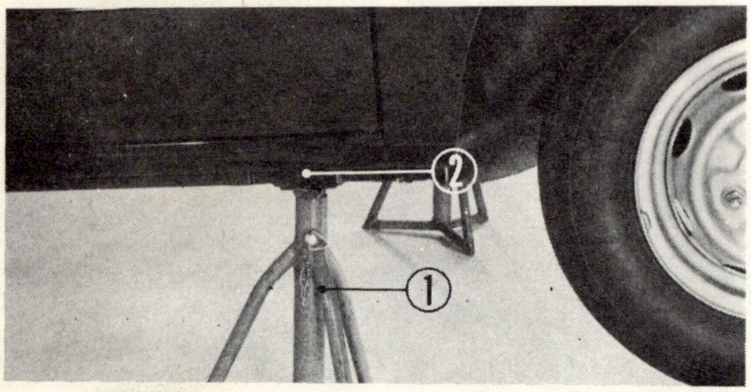

① Safety stand
② Rear side sill

Rear

Rear side sills

① Safety stand
② Front side sill

INSPECTION AND MAINTENANCE

Steering System

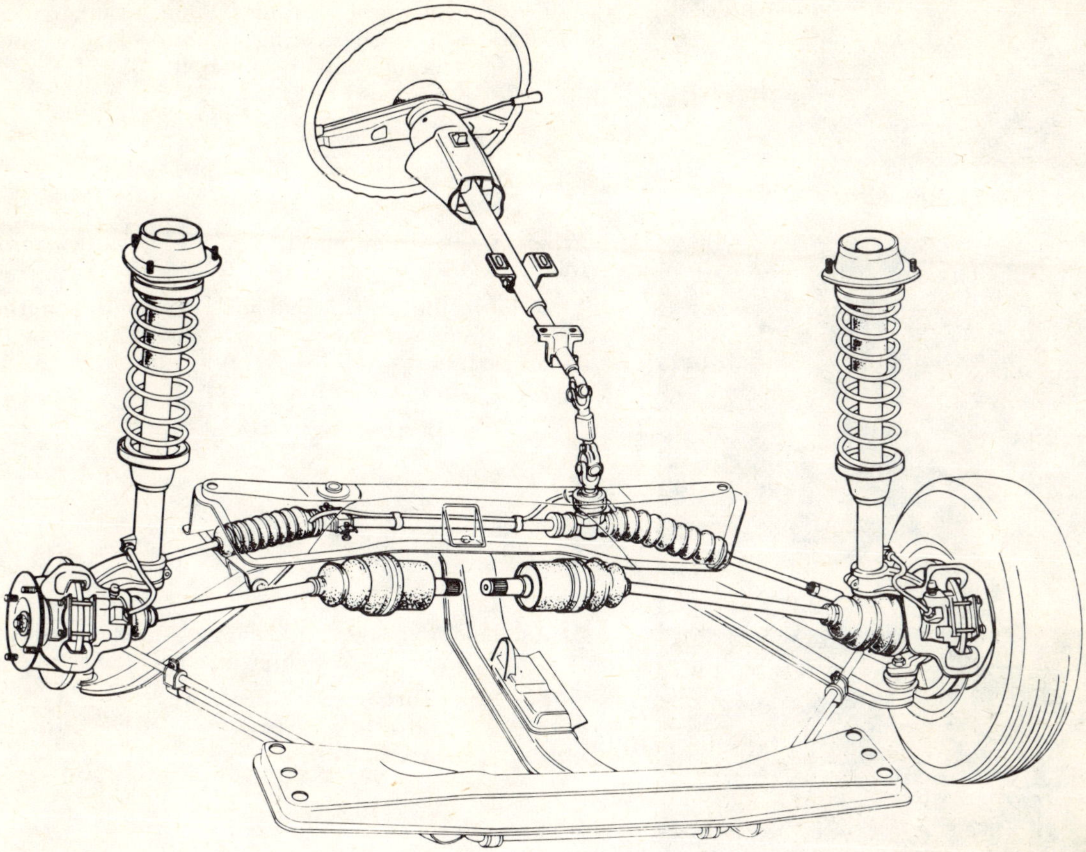

Fig. B-1

a. Steering wheel

a-1. Play and looseness

To check the steering wheel play, first place the front wheels in the straight-ahead position.

Then, turn the steering wheel and right and left and measure the play at the steering wheel rim.

Unit: mm (in.)

	Specification
Steering wheel	0~10 (0.394)

If the play exceeds the specification, check for:

- Excessively worn steering rack and pinion.
- Loose or excessively worn link connections.
- Excessively worn steering pinion guide.

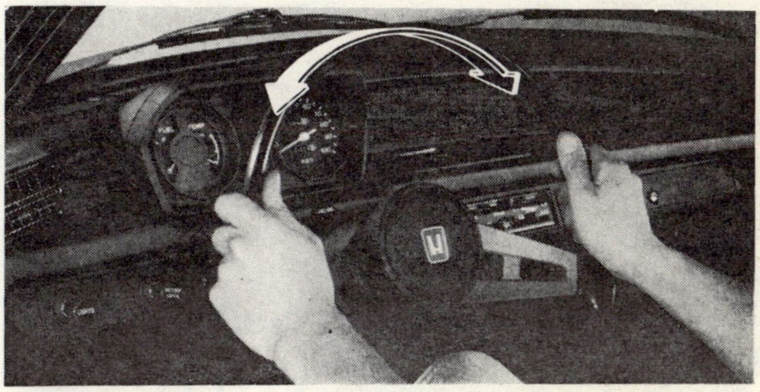

Fig. B-2
Checking play

Fig. B-3
Checking rattle

To check the steering wheel for rattle, hold the steering wheel as shown in Fig. B-3, and push or pull it in axial direction. If unusual noise is heard, it is an indication that the nut securing to the steering shaft is loose. The following failures are also suspected:

- Excessively worn steering column bushing.
- Excessively worn steering column joint.
- Loose steering column joint bolt.

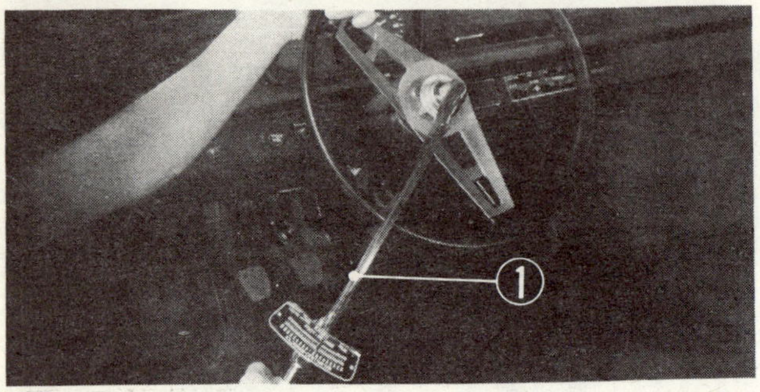

Fig. B-4
① Torque wrench

Where the steering wheel is loose or rattles, proceed as follows:

1. Remove the steering pad and examine the steering wheel nut torque with a torque wrench. Retighten the nut if necessary.

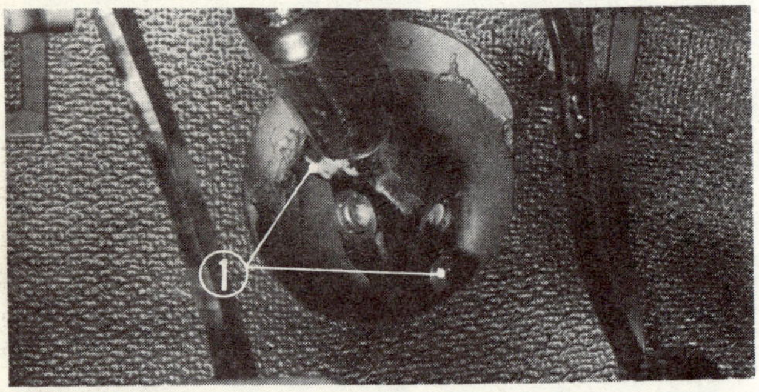

Fig. B-5
① Steering column joint bolt

2. Examine the steering column joint bolt torque with a torque wrench. Retighten the bolt if necessary.

   See the Torque Specifications.

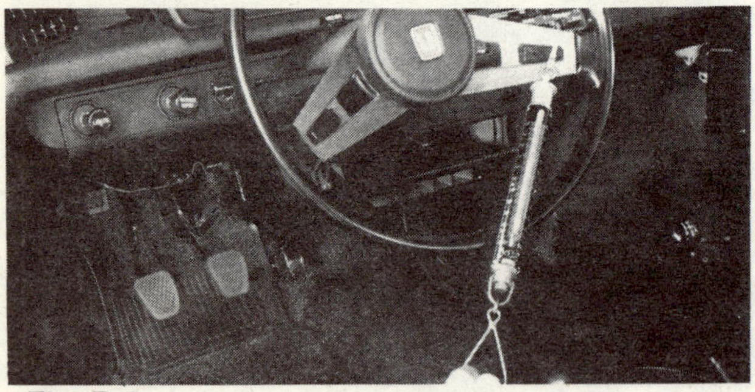

Fig. B-6
Checking movement of steering wheel

a-2. Inspect the operation

Drive the car for a certain distance and check the steering wheel for any sign of shimmy, tramp, noise, hard steering or any other abnormal condition.

Raise the front wheels off the ground and place them in the straight-ahead position.

Then turn the steering wheel from that position with a spring scale, and see if it can be turned with a force of below **1.5 kg (3.31 lb)**.

Unusual vibration, rattle, pull to one side, hard steering—these are suspected to be caused by the following failure:

1. Insufficiet grease in steering gear box.

   Excessively worn steering rack and pinion.
2. Insufficiently lubricated front suspension and tie rod end ball joints.
3. Too low or uneven tire inflation pressures.
4. Insufficiently lubricated, excessively worn or loose steering linkage connections.
5. Improper wheel alignment.
6. Maladjusted steering gear box.

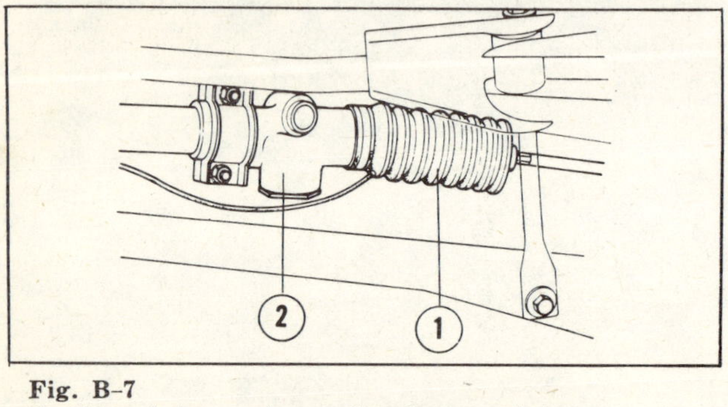

Fig. B-7
① Tie rod dust seal ② Steering gear box

b. Steering gear box

b–1. Inspect the gear box

Check the steering gear box bolts for looseness, and retighten them if necessary.

See the Torque Specifications.

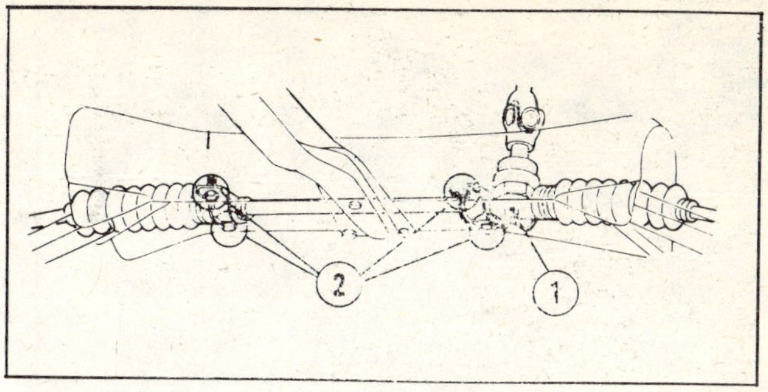

Fig. B–8
① Steering gear box ② Attaching bolts

b–2. Inspect the steering rack

Jack up the front of the car and support it with safety stands. Remove the dust seal, and hand-move the steering rack up and down to check the clearance between the rack and gear box.

- If the clearance is found to be excessive, see Par. c below. Replace the gear box if necessary.

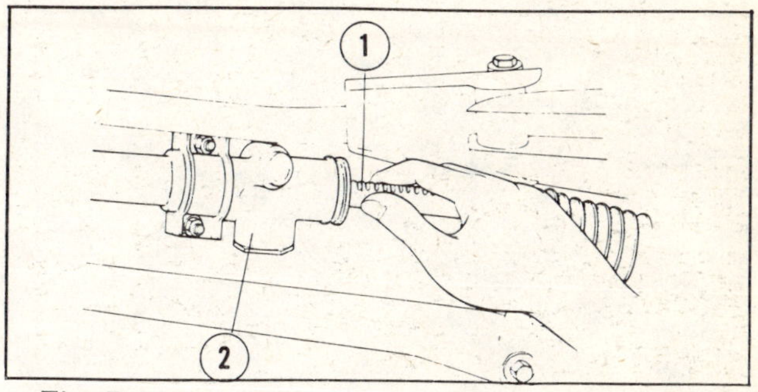

Fig. B–9
① Steering rack ② Gear box

b–3. Rack guide screw

Drive the car and if the steering wheel still rattles, or if abnormal noise is heard, check the rack guide screw and rack guide screw lock nut for looseness.

1. Jack up the front of the car with the wheels placed in the straight-ahead position and support it with safety stands.
2. Loosen the rack guide screw lock nut.
3. Tighten then pinion guide screw. Then turn it back 45° from the position where the rack is locked, and tighten the lock nut.
4. After tightening, drive the car and check for hard steering and/or excessive play.

   See the Torque Specifications.

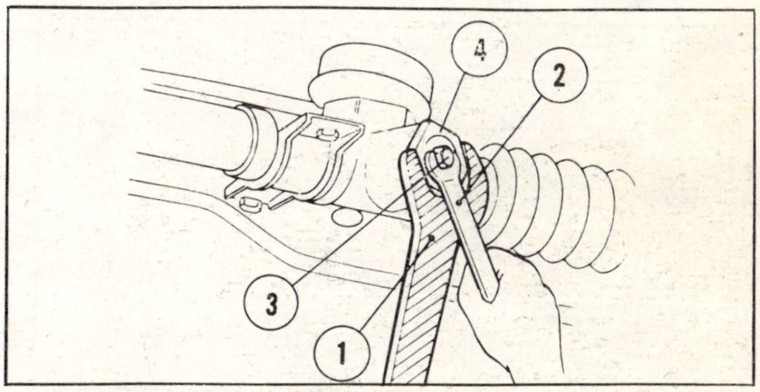

Fig. B–10
① Special tool ② 14 mm wrench
③ Rack guide screw ④ Rack guide screw lock nut

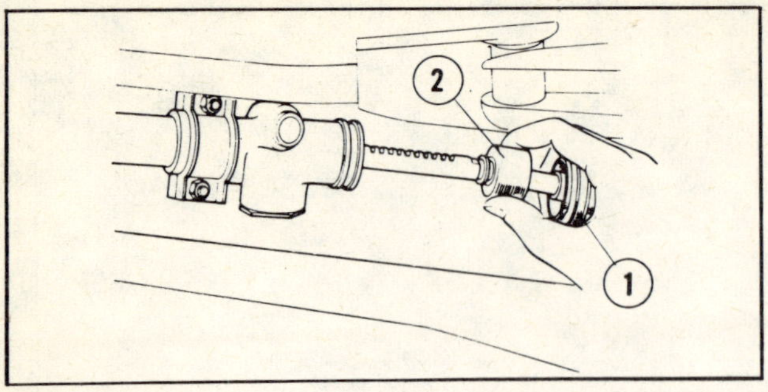

Fig. B-11
① Tie rod dust seal ② Tie rod ball joint

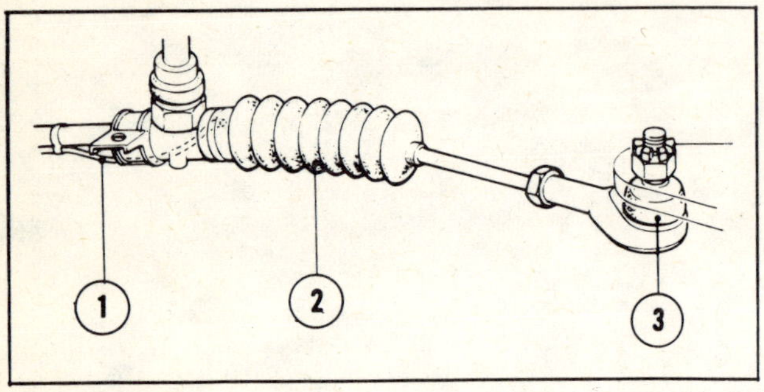

Fig. B-12
① Air tube ② Tie rod dust seal ③ Tie rod end boot

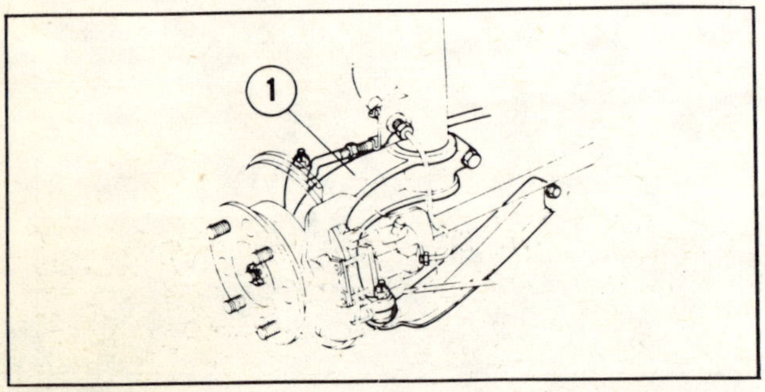

Fig. B-13
① Knuckle

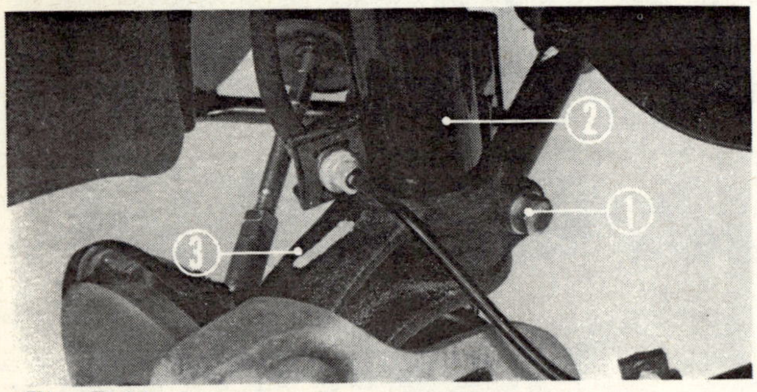

Fig. B-14
① Front shock absorber lock bolt
② Front shock absorber ③ Knuckle

c. Tie rods

c-1. Inspect the steering column joints and tie rod ball joints

Let an assistant turn the steering wheel and check the moving parts for rattle, and the connections and joints for looseness. Also, check for any other abnormal condition. Pay special attention to the following points:

1. Excessively loose or damaged steering column joint.
2. Excessively loose rack end ball joints. Check it by removing the tie rod dust seal.
3. Excessively loose or damaged tie rod end ball joints.

c-2. Inspect the air tube, tie rod dust seal, etc.

1. Check the tie rod dust seal for breakage.
2. Check the air tube for breakage and loose connection.
3. Check the tie rod end boot for breakage.
4. Check the rack end and tie rod end for wear, bend or damage. Replace a defective part, if any.

c-3. Inspect the knuckle arms

Check the knuckle arms for cracks and the tie rod end ball joints for cracks and looseness.

d. Knuckles

d-1. Rattle at the joint

Check the front shock absorber lock bolt and ball joints for looseness.

See the Torque Specifications.

d-2. Inspect for cracks

Check the knuckle for cracks.

e. Front wheels

e-1. Inspect the wheel alignment

Before attempting to measure the wheel alignment, it is advisable to inflate the tires to the specified pressures, check each part of the steering and suspension systems for looseness or wear, and repair if necessary.

Slippage indicated on the side-slip tester is the result of maladjusted toe-in and camber.

If the steering wheel vibrates or unusually rattles, even when the alignment is proper, check the wheel balance.

Side-slip measurment

Procedure	Remarks
1. Set testers: • On level floor. • With gauges at outside. • In parallel to center line of car. • So that right and left wheels pass through center lines on testers at the same time.	• It is desirabe to set the testers so that their upper surfaces are flush with the floor surface.
2. Set the gauge indicators to zero (0).	• The indicators are moved by the pointer and do not return.
3. Pass the car through tester: • Slowly (at a speed of about 3 km/h or 1.86 mph). • Straight ahead (without turning in steering wheel in either direction). • Until front wheels step off testers.	• Without passengers and/or load.
4. Read values registered by indicators on gauges. • 1/2 of the sum of readings of right and left scales must be within specification.	• Specified slippage: Inside: 2 mm (0.079 in.) per 1 m (3 ft. 3.37 in.) Outside: 3 mm (0.118 in.) per 1 m (3 ft. 3.37 in.)
5. If slippage is beyond specification. • Check and adjust toe-in. • Measure side-slip again.	

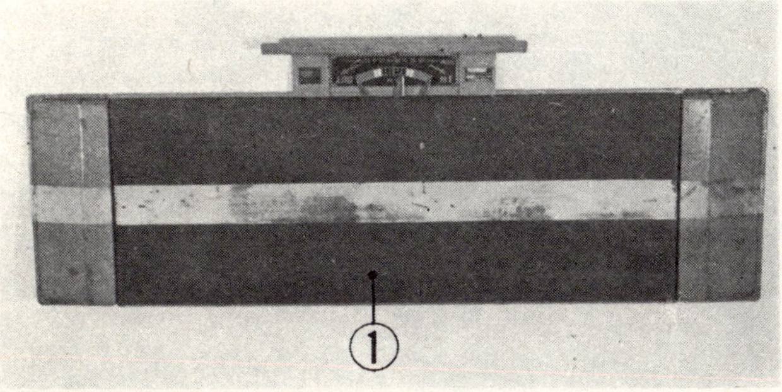

Fig. B-15
① Side-slip tester

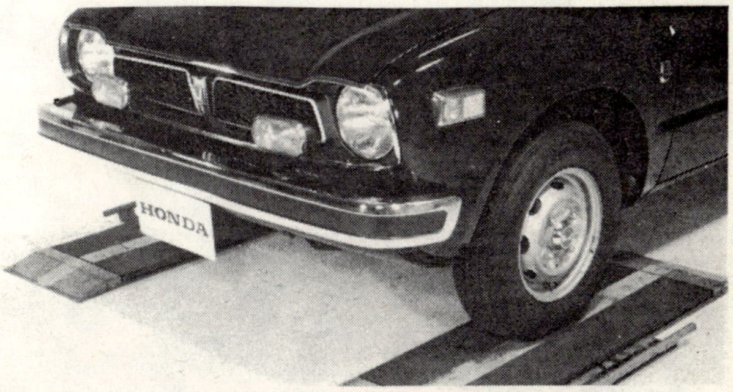

Fig. B-16
Measuring side-slip

● Adjust the toe-in

Check and adjust the toe-in without a driver or any load.

1. Place the car on level ground with front wheels in the straight-ahead position.
2. Chalk marks on the center of the tires at the height of the axle center.
3. Align the toe-in gauge with the marks on the tires as shown.
4. Read the scales of the gauges.
5. Slowly back up the car and rotate the wheels 180° until the marks on the tires are aligned with the height of the gauges on the rear side.
6. Measure the toe-in on the rear part of the tires a tthe same point.

Fig. B-17
① Toe-in gauge

Unit: mm (in.)

	Specification
Toe-in	1 (0.039) out

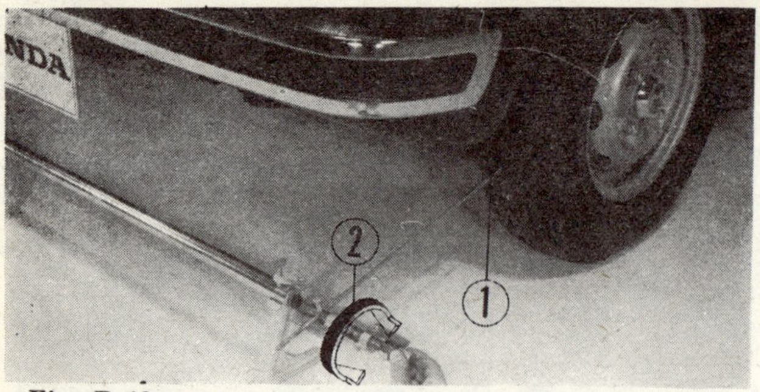

Fig. B-18
① Tire center ② Measure

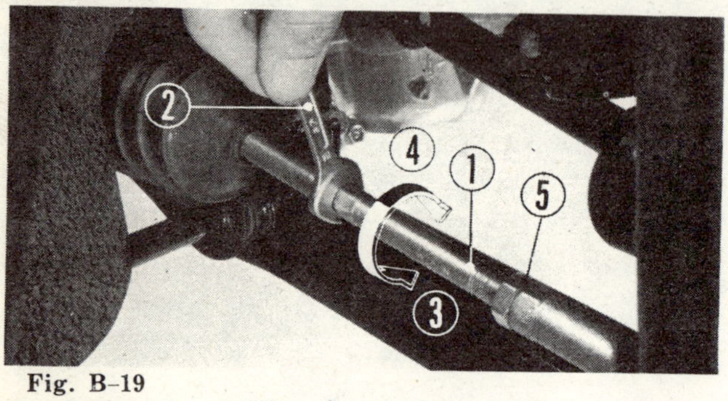

Fig. B-19
① Rack end ② Wrench ③ Increase
④ Decrease ⑤ Lock nut

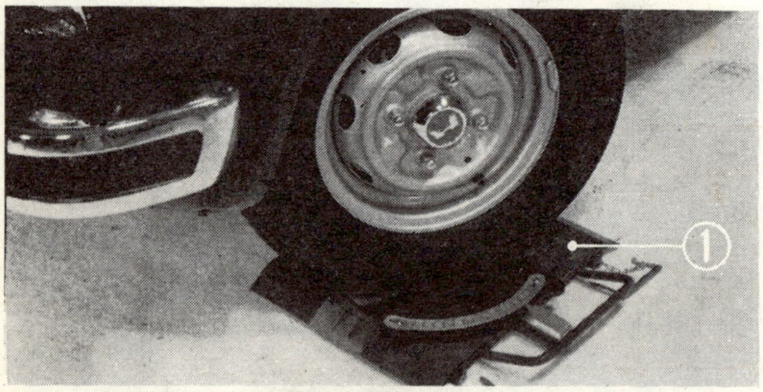

Fig. B-20
① Turning radius gauge

7. When the toe-in is out of specification, adjust it by changing the length of the tie rods while measuring the toe-in.

Procedures

1. Raise the wheels off the ground.
2. Loosen the tie rod lock nuts and turn the tie rods.

   The right and left tie rods must be turned equally, since it may vary the steering angles.

   - After adjusting, retighten the lock nuts.
   - If the side-slip is still incorrect even after the toe-in is adjusted, measure the camber and caster with the wheel alignment tester.

   Camber: 30′
   Caster: 1°45′
   King pin angle: 8°55′

   If the camber and caster are not correct, damage of the lower arm, front beam and center beam or misalignment of the car body is suspected to be the cause. Check and replace if necessary.

e-2. Adjust the steering angles (Without passengers or load)

1. Set the turning radius gauges in position.
2. Carefully place the front wheels on the centers of the gauges in the straight-ahead position as shown.
3. Calibrate the scales of the right and left gauges to zero (0), and pull out the dowel pins. Where it is difficult to pull out the dowel pins, it is an indication that the gauges are set in wrong position. Reset them in correct position.
4. With the brakes applied, fully turn the steering wheel to right and then to left and read the scales of the gauges.

   Inner wheel: 36°
   Outer wheel: 30°

5. Where the steering angles of the inner and outer wheels are not equal or are out of specification, check for toe-in and readjust.

a. Brake pedal

a–1. Inspect the pedal play and height

● Pedal play

Slightly depress the brake pedal, and check for the pedal play at the center of the pedal pad, that is, the distance the pedal moves until the brake starts working.

Unit : mm (in)

	Specification
Pedal play	1～5 (0.039～0.197)

NOTE :

To obtain the proper pedal play, adjust the stop switch position by screwing it in or out.

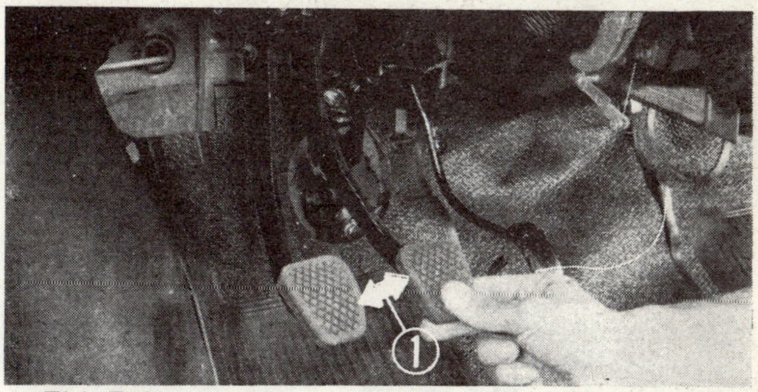

Fig. B–21
① Brake pedal play

● Pedal height

1. Check the pedal height above the floor mat with the pedal fully depressed.

Unit : mm (in)

	Specification
Pedal height	Over 90 (3.54) above floor mat
	Depressing force : 20 kg (44 lb)

2. Check the brake lines for entrapped air whenever occasioned as follows :
   - When the pedal is spongy.
   - When the pedal falls away gradually if depressed.
   - When the effective stroke decreases after the pedal is depressed several times.

In any of the above cases, bleed the brake lines.

- If the pedal height is out of specification even after the lines are bled completely, adjust the clearance between the brake drum and linings.

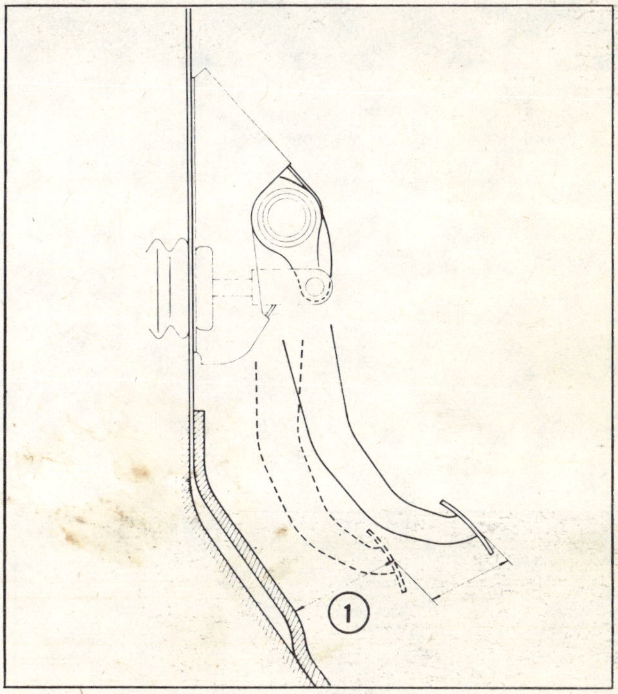

Fig. B–22
① Pedal height

a–2. Inspect the braking effect

Using a tester or driving the car, check for proper braking effect. If necessary, check the following items :

1. Adjustment of brakes
2. Tire inflation pressure
3. Check the conditions of brake drum and linings.

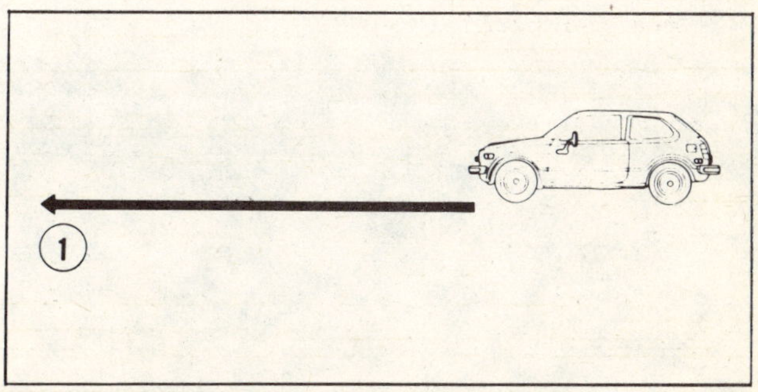

Fig. B–23 ① Brake distance
France type 14.5 m (47.6 ft)
Australia type 13.5 m (44.3 ft)
(Initial speed : 50 km/h or 31.08 mph)

b. Parking brake lever

b-1. Inspect the effective stroke

Count the ratchet teeth pass through when the brake lever is fully pulled up: effective stroke.

	Specification
Effective stroke	1~5 teeth

Full stroke: 7 teeth

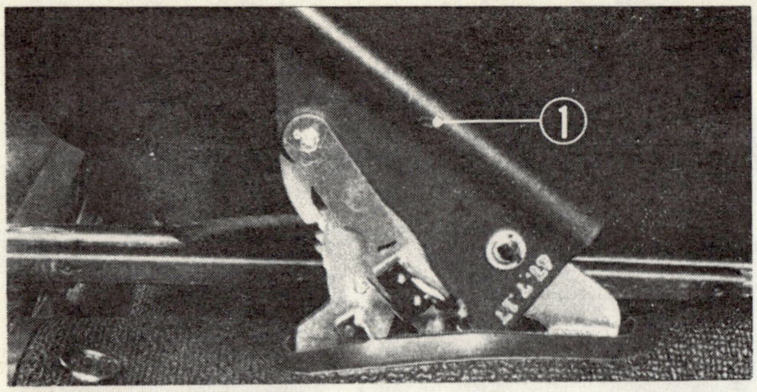

Fig. B-24
① Parking brake lever

b-2. Adjust proceedures

1. Check the service brakes for proper braking effect.
2. Fully release the parking brake lever.
3. Adjust the stroke by turning the equalizer adjusting nut, as shown Fig. B-25.
4. After adjusting, check, while rotating the rear wheels, to see if the wheels are dragged when the lever is released.

NOTE:

Make sure to adjust service brake first, then parking brake.

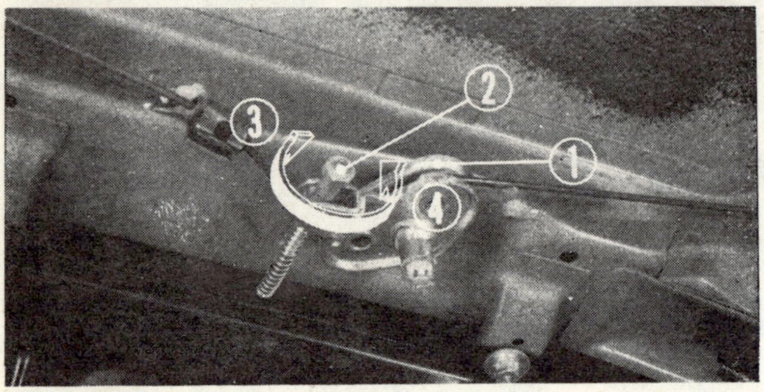

Fig. B-25
① Equalizer ② Adjusting nut
③ Effective stroke will be decreased
④ Effective stroke will be increased

b-3. Inspect the braking effect

Using a tester or stopping the car on grade, check the braking effect. Or drive the car at low speeds and apply the parking brake lever to check the braking effect. If the effect is insufficient, check the lever control cable, brake drum and linings.

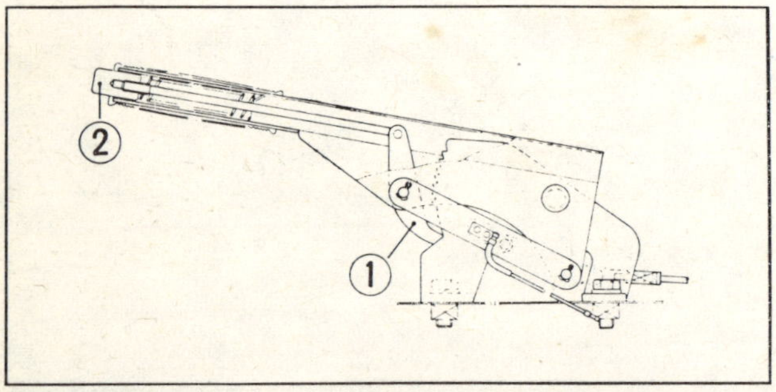

Fig. B-26
① Ratchet ② Push rod knob

b-4. Inspect the ratchet

Check the ratchet teeth and pawl for wear or othe rdamage. Replace the ratchet if necessary.

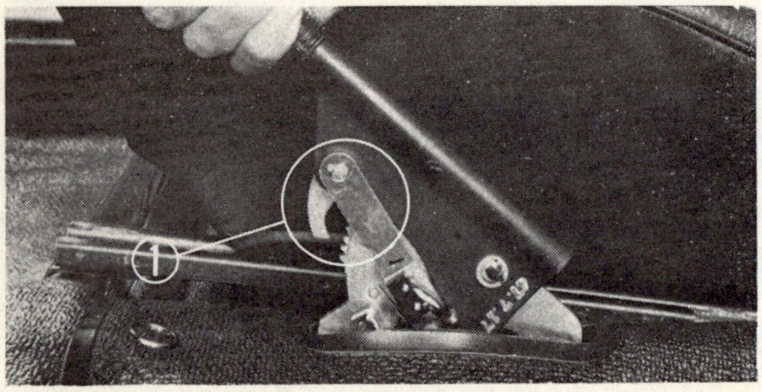

Fig. B-27
① Checking wear or damage to ratchet

c. Parking brake rod and cables

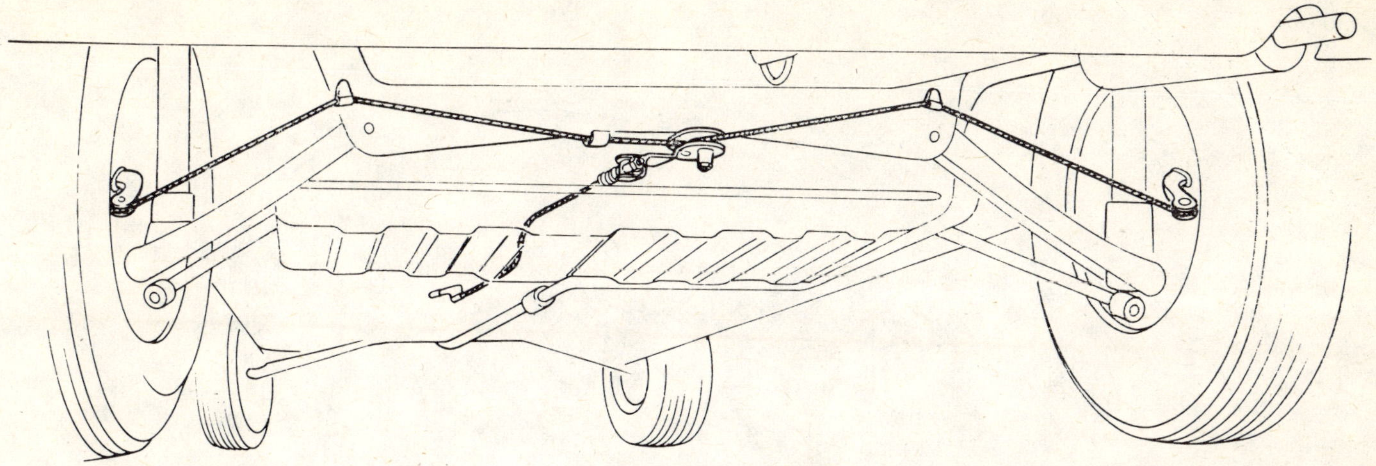

Fig. B-28

c-1. Inspect the control cables and equalizer

Apply the parking brake and check the control cables for damage, the equalizer connection for looseness or damage.

- Check the sliding surfaces of the equalizer and control cables for condition. If necessary, apply a coat of grease to the sliding surfaces.

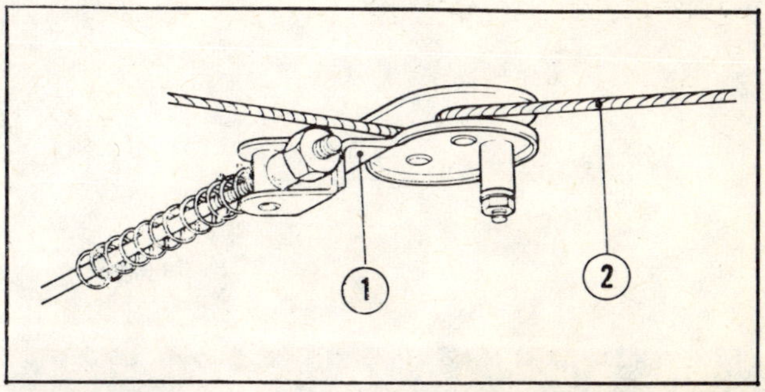

Fig. B-29
① Equalizer ② Parking brake control cable

- Check the cable pins for proper installation.
- Check the sliding surface of the brake lever for condition, and apply a coat of grease if necessary.

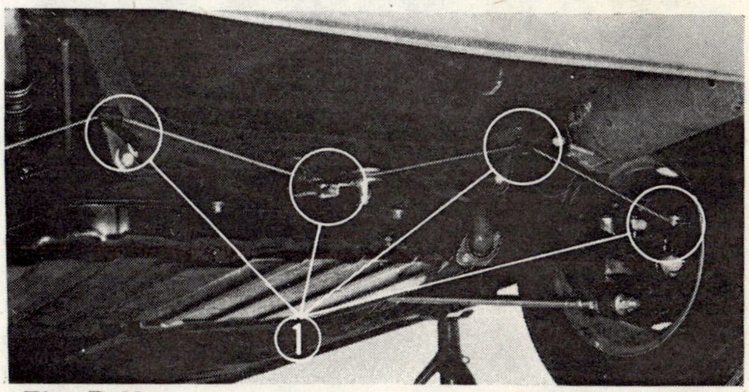

Fig. B-30
① Apply a coat of grease to cable connections

d. Brake hoses and pipes

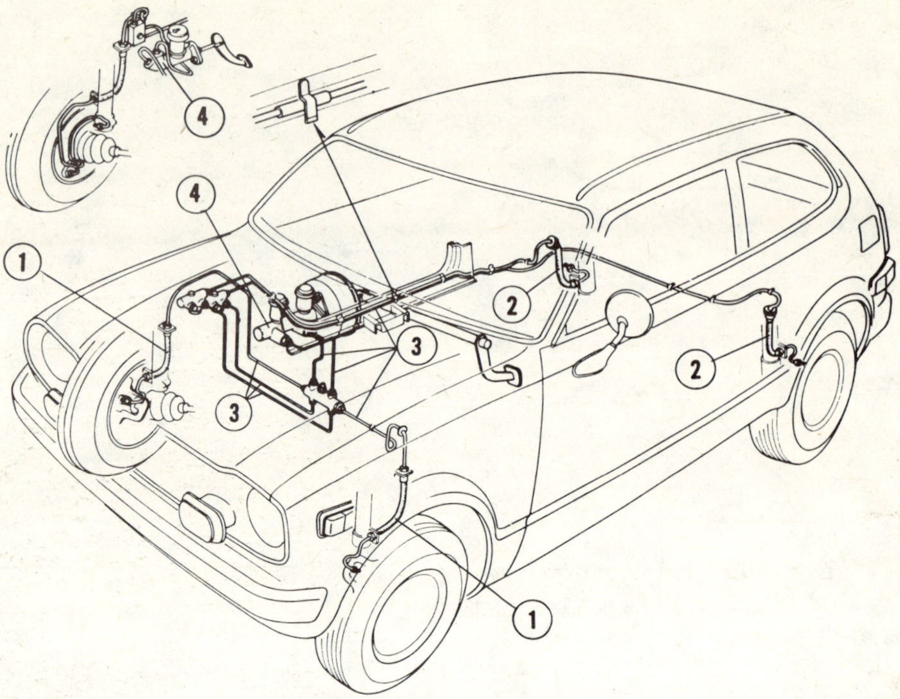

Fig. B–31

① Front brake hose
② Rear brake hose
③ Brake pipe
④ Master cylinder

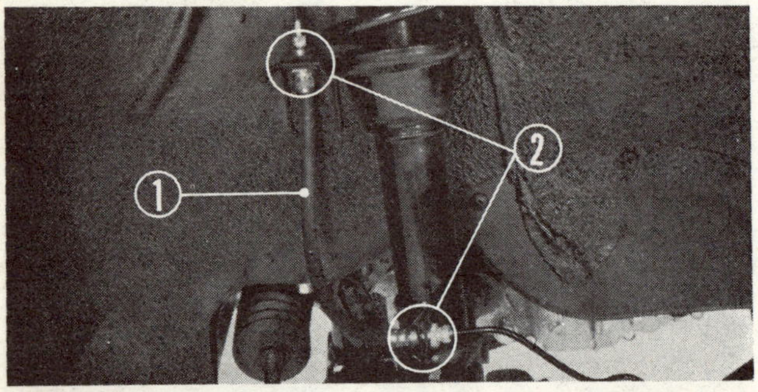

Fig. B–32
① Brake hose ② Checking fluid leak at hose joints

d–1. Inspect for fluid leak and damage

1. Check the brake hoses and pipes for damage, deterioration, twist or fluid leak.
2. Check the brake hoses for any sign of binding. Even when the steering wheel is fully turned to right and left, the hose-to-wheel rim clearance must be more than 40 mm (1.57 in.).
3. Check the connections and joints of hoses, pipes, master cylinder and wheel cylinders for fluid leak.
4. Check the brake pipes for collapse, bend or corrosion.

● When fluid leaks

- Retighten all joints.
- Replace a cracked or damaged hose or pipe, if any.
- After repairing a damaged hose or pipe, check the oil level in the master cylinder oil reservoir. Also, check for trapped air in the brake lines.
- See the Torque Specifications.

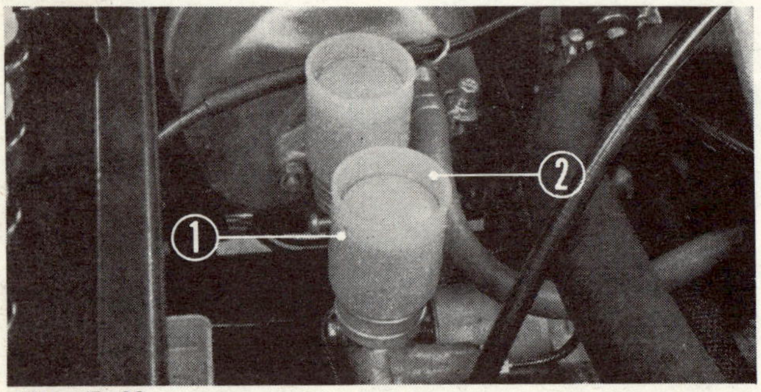

Fig. B–33
① Level line ② Reservoir tank

e. Hydraulic brakes

e–1. Inspect the brake fluid

Check the fluid level in the master cylinder reservoir tank, and add brake fluid up to the specified level line if necessary. If the fluid is excessively dirty, change it.

- Use SAE J 1703 or (DOT 3, 4 type).
- Do not mix brake fluid of different brands.
- Never use any fluid other than brake fluid in the master cylinder oil reservoir.
- Take care not to drop brake fluid on the painted surfaces of the car.

e–2. Change the brake fluid

Brake fluid is, regardless of the type of brake system, always subject to temperature changes. This causes water vapor in the air to condense in the brake lines, resulting in the drop of boiling point and poor braking effect. Therefore, brake fluid must be changed periodically.

- Change brake fluid, beginning on a wheel cylinder (rear left and right wheel cylinders and front left and right wheel cylinders in this order).

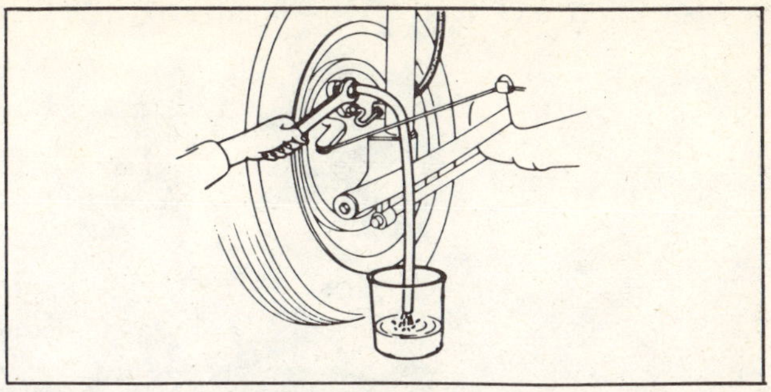

Fig. B–34 Changing brake fluid

1. Remove the bleeder screw rubber cap and connect a 6 mm (0.236 in.) vinyl tube to the bleeder screw. Then, receive the fluid flowing out of the hose in a container.
2. Loosen the bleeder screw and depress the brake pedal several times until the fluid no longer flows out.
3. After draining out brake fluid completely, refill the master cylinder oil reservoir with brake fluid. Then, bleed the brake lines.

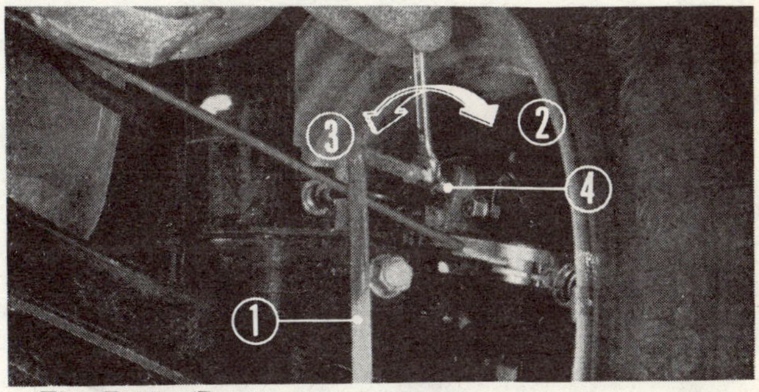

Fig. B–35 (Rear)
① Vinyl tube ③ Loosen
② Tighten ④ Bleeder screw

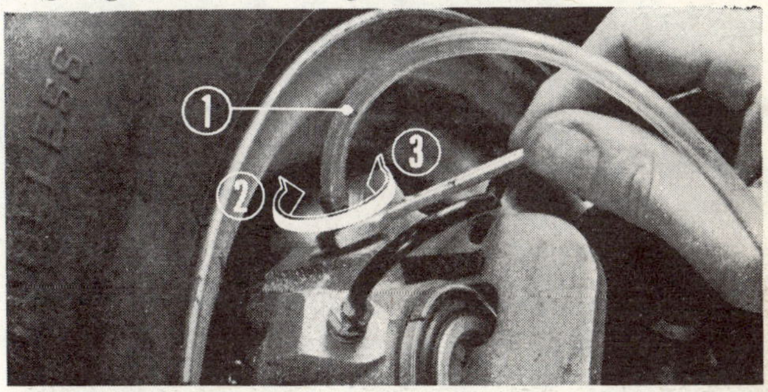

Fig. B–36 (Front, disc type)
① Vinyl tube ② Tighten ③ Loosen

e–3. Bleed the brake lines

If air is in the brake lines, bleed the lines in the following procedures:

1. Remove the bleeder screw rubber cap on the rear left wheel cylinder and connect a 6 mm (0.236 in.) vinyl tube to the bleeder screw.
2. Depress the brake pedal several times and loosen the bleeder screw to allow brake fluid to flow out.
   - The bleeder screw must be loosened just before the brake pedal reaches the floor mat.
   - Continue this operation until no air bubbles are contained in the fluid.
- Do this operation with an assistant.

3. Take care not to stop refilling the master cylinder oil reservoir to prevent air from getting into the inlet and return ports in the master cylinder.
4. Add brake fiuid up to the level line on the reservoir. Tighten the bleeder screw cap with the " F " mark facing the front.

NOTES:

1) Bleed the rear left and right wheel cylinders, and the front left and right wheel cylinders in this order.
2) If the brake pedal is felt spongy when depressed or if the effective stroke is decreased as the pedal is decreased several times, it is an indication that the brake lines are not bleed completely

e–4. Inspect the master cylinder and wheel cylinders

1. By using a tester or by driving the car, check to see if the master cylinder and wheel cylinders operate properly, that is, the brakes are applied positively and evenly, or if the brake pedal returns smoothly.
2. If fluid leaks from the master cylinder or wheel cylinders of any abnormal condition is noted during step 1, disassemble the cylinder and check the cylinder, piston, piston cup and check valve for wear, damage or corrosion.

f. Inspection and test of the master power

Fig. B-37

f-1. Instant method

To judge the performance of the master power accurately, a proper instrument is necessary. The instant method outlined below may also be used for testing the master power.

1. Test the function

   With the engine stopped, depress the brake pedal several times with normal pressure and make sure that the pedal height does not vary. Then start the engine with the pedal depressed. If the pedal sinks slightly, the master power is in good condition. If the pedal height does not vary, the master power is defective.

2. Test the load and airtightness

   Depress the brake pedal with the engine running and then stop the engine. If the pedal height does not vary when it is kept depressed for about 30 seconds, the master power is in good condition. If the pedal rises, the master power is defective.

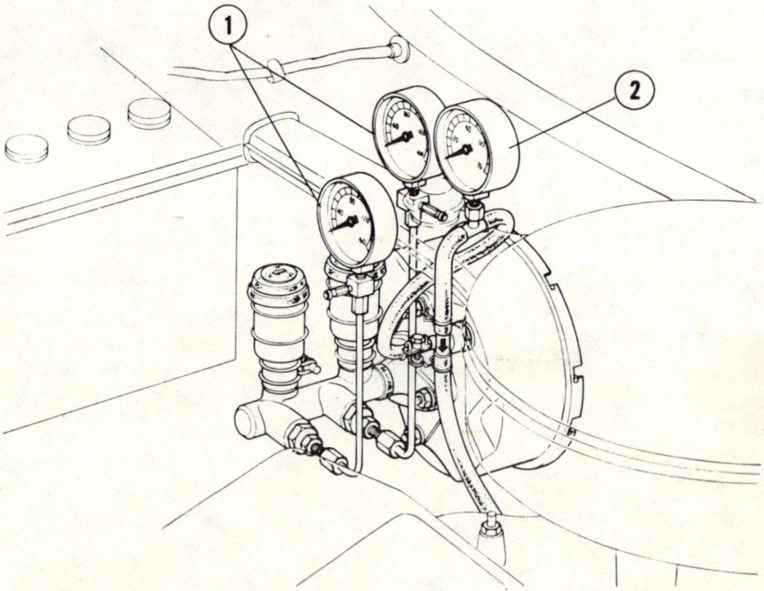

Fig. B-38 ① Hydralic pressure gauges ② Vacuum gauge

f-2. Test with the use of the approved instrument

- To test, install two hydraulic pressure gauges, a vacuum gauge, a vacuum hose and a check valve as shown.

1. Disconnect the brake pipes at the master cylinder and install the hydraulic pressure gauges. After installing them, be sure to bleed the master cylinder.

   NOTE:

   When disconnecting the brake pipes, place a waste cloth under the pipes to prevent brake fluid from coming in contact with other parts.

2. Disconnect the vacuum hose at the master power and install the check valve and vacuum gauge as shown. Install the check valve into between the vacuum gauge and intake manifold.

   NOTE:

   Be careful of the installation direction of the check valve.

Test the airtightness

Start the engine and depress the accelerator pedal until a reading of 300 or 500 mmHg is indicated on the vacuum gauge. Then stop the engine. If vaccum drop is 0 to 20 mmHg for 30 seconds, the master power is in good condition. If it is more than 20 mmHg, any one of the following failures may be the cause.

1. Defective•check valve.
2. Leak from vacuum hose.
3. Poor airtightness of poppet valve.
4. Broken diaphragm.
5. Poor airtightness of piston seel.
6. Poor airtightness of rod seal.

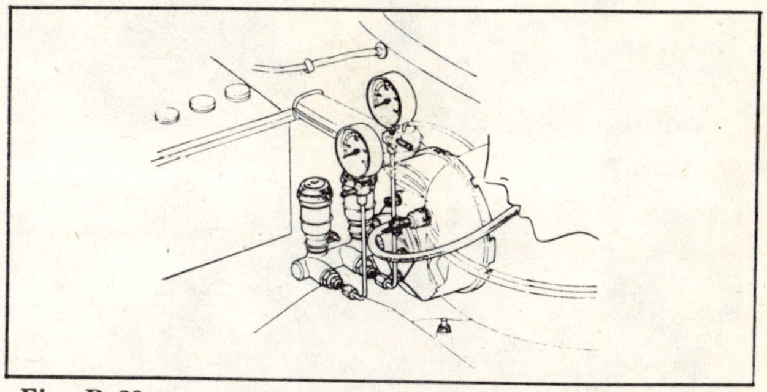

Fig. B-39

Test the function of the check valve

If you can suck the vacuum hose on the intake manifold side (engine side) and you cannot blow it on the same side, the check valve is in good condition. Replace the valve if defective. (Similarly test the check valve of the attachment furnished as an accessary of the instrument.)

Test the function of the master power

Start the engine and depress the accelerator pedal until a reading of 300 or 500 mmHg is indicated on the vacuum gauge. If the pedal depressing force and hydraulic pressure are within the specification below when the brake pedal is depressed with a force of 20 kg (44 lb), the master power is in good condition. If the outlet side pressure is out of the specification, the hydraulic circuit of the master cylinder or the vacuum circuit of the master power may be defective.

Negative pressure, mmHg	Depressing force kg, (lb)	Outlet side pressure, kg/cm^2 (lb/in^2)
300	20 (44)	40 (569), min.
500	20 (44)	55 (782), min.

If 300 or 500 mmHg cannot be obtained or if the said vacuum cannot be maintained, inspect or adjust the following items. Again measure the outlet side pressure.

- Adjust the idle speed by means of the idle adjusting screw and throttle stop screw of the carburetor.
- Tappet clearance.
- Ignition timing.
- Condition of air cleaner element.

After testing, be sure to bleed the brake system.

g. Brake discs and pads

g-1. Inspect the pads

Remove the front wheel and measure the thickness of the pad at its most worn point with a caliper.

Unit: mm (in.)

	Specification	Service limit
Pad thickness	10.3 (0.406)	1.6 (0.063)

Fig. B-40
① Pad ② Caliper

- Replace a pad when it wears down to the service limit.
- If the brakes squeak after replacing the pad, apply a thin coat (0.3~0.5 mm or 0.012~0.020 in.) of grease to the rear part of the pad. Be sure not to put grease to the pad surface.
- Check for smooth movement of the caliper. If not, apply a thin coat of grease to the sliding surfaces.
- Rotate the front wheel and check the wheel cylinder for jamming.
- Improper clearance between the disc and pad is due to a defective caliper or wheel cylinder, or excessive face runout of the disc. Check and repair if necessary.

NOTE:
The front disc brakes require no adjustment. Therefore, it is normal that there is no clearance between the disc and pad.

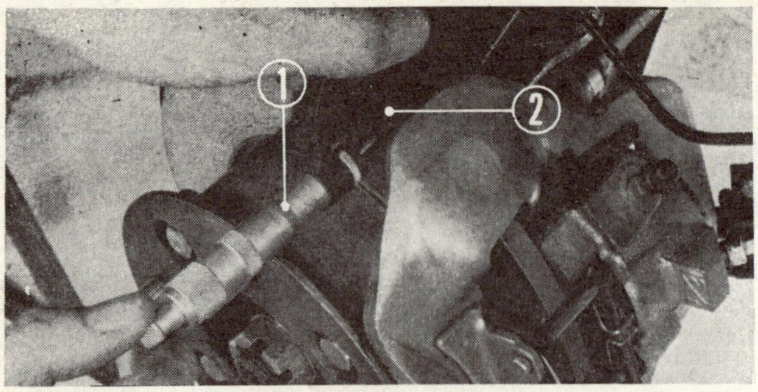

Fig. B-41
① Micrometer ② Disc

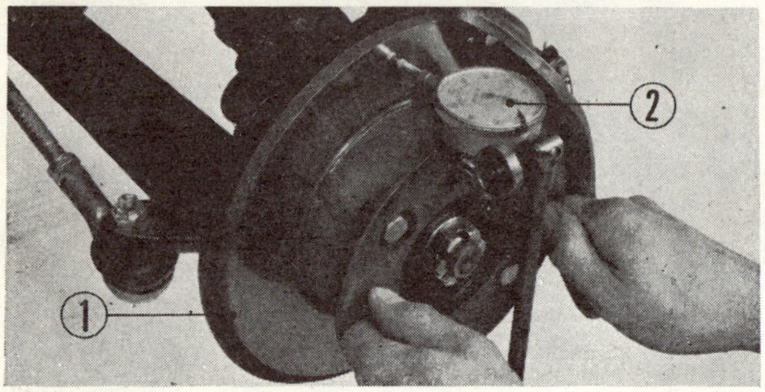

Fig. B-42
① Disc ② Dial gauge

g-2. Inspect the discs

1. Remove the front wheel and measure the disc thickness with a micrometer.

Unit: mm (in.)

	Specification	Service limit
Disc thickness	9.6 (0.378)	Replace when it is below 9.0 (0.354).

To check for surface squareness of the disc, measure the thickness of the sliding surfaces on the way around circumferences of the disc with a micrometer. The desired difference in thickness is below 0.07 mm (0.003 in.)

2. Check the discs for cracks or any other damage.

3. Check the face runout of the disc with a dial gauge.

Unit: mm (in.)

	Service limit
Disc face runout	Replace when it exceeds 0.15 (0.006).

4. Check the wheel bearing for excessive play and the disc bolts for looseness.

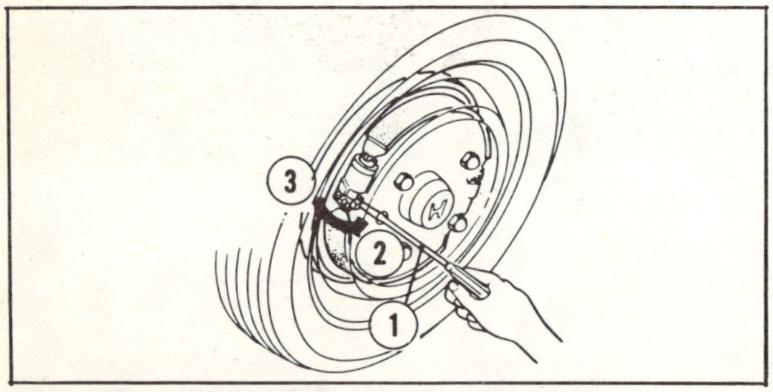

Fig. B-43
① Standard type screwdriver ② Tighten ③ Loosen

Fig. B-44
① Adjuster wrench ② Tighten ③ Loosen

h. Brake drums and shoes

h-1. Adjust the drum-to-lining clearance

Too large a clearance may increase the effective stroke of the brake pedal. In such a case, adjust the clearance between the drum and lining.

● **Procedures**

1. Jack up the car to raise a wheel off ground.
2. Depress the brake pedal two or three times.
3. Turn the adjuster clockwise until the wheel no longer turns.
4. Then back off the adjuster (**four clicks on the front wheel and two cliks on the rear**), and depress the brake pedal two or three times.
5. Turn the wheel to check for any sign of dragging. Then back off the adjuster one click more.
6. Similarly adjust the clearance on the other wheels.
7. After adjusting, drive the car and check all brakes for proper operation.

• Note that the adjustment of the service brakes may vary the effective stroke of the parking brake lever.

h-2. Inspect the brake shoes and linings

Remove the brake drum and check the linings and the sliding surfaces of the shoes for wear. Whether or not it is time to replace the linings must be determined on the basis of the most worn part.

Unit: mm (in.)

	Specification	Service limit
Lining thickness	5.0 (0.197)	Replace when it is below 2.0 (0.079).

- Check the linings for peeling or any other damage, and replace a damaged lining, if any.

Generally all the brake shoes must be replaced on four wheels at one time.

Fig. B-45
① Lining ② Checking for wear with a caliper

h-3. Inspect the brake drums

Check the brake drums for wear, cracks or any other damage and replace a damaged drum if any.

Unit: mm (in.)

Drum inside diameter	Specification	Service limit
Front	215.0 (8.46)	216.0 (8.50)
Rear	180 (7.09)	181.0 (7.13)

NOTE:

Measure the worn part of a drum as shown in Fig. B-46.

Fig. B-46
① Brake drum ② Caliper

i. Backing plates

i-1. Inspect the backing plates

1. Check the backing plates for deformation, cracks or any other damage.
2. Check the four backing plate bolts for looseness and retighten a loose bolt, if any. See the Torque Specifications.

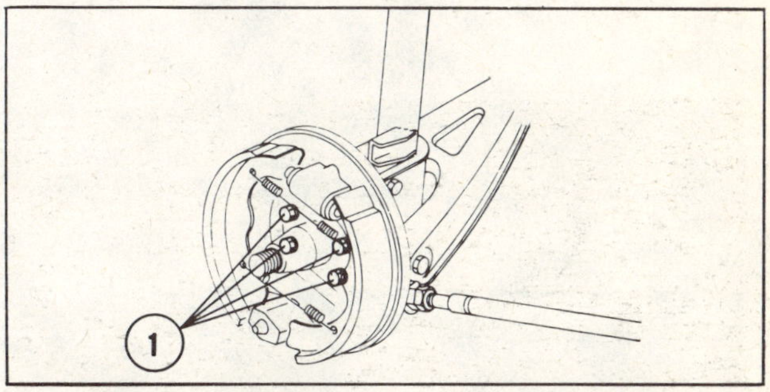

Fig. B-47
① Backing plate bolts

Wheels

a. Inspect the tire inflation

1. Measure the air pressure of a tire with a tire gauge. If necessary inflate the tire to the specified pressure.

   See the Specifications.

   - Over or under inflated tire pressure may affect the driving efficiency.
   - Check the spare tires for pressures and inflate them 0.2 to 0.3 kg/cm² (2.84 to 4.27 lb/in²) higher than specified.

Fig. B-48
① Tire gauge

b. Inspect the tires

Raise the wheel off the ground and check the tire for cracks or any other damage. Replace the tire if excessively damaged.

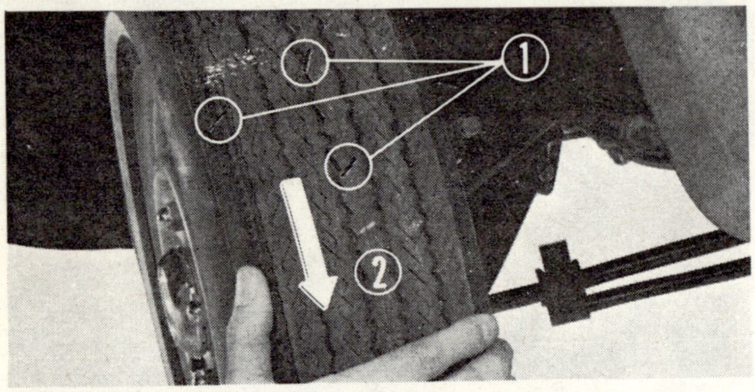

Fig. B-49
① Cracks on tire ② Rotate tire for checking

c. Inspect for excessive tire wear

d. Inspect the tire tread pattern

If the tire tread wear indicators appear as more than two solid bands, replace the tire.
Raise the wheel off the ground and check the tire for uneven, side, cornering or center wear.

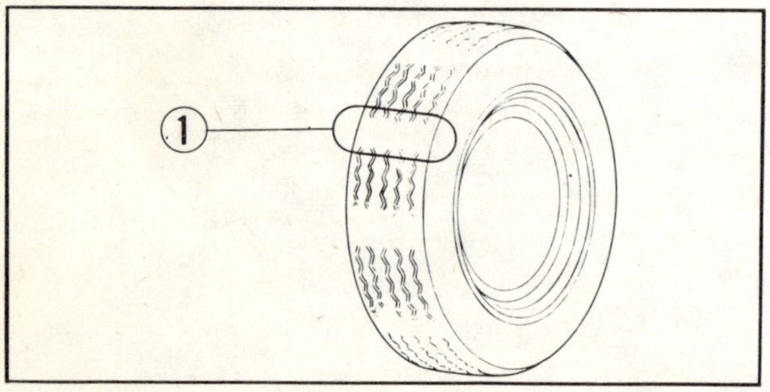

Fig. B-50
① Tire tread wear indicator

e. Inspect for foreign materials

Check to see if foreign materials such as metal particles and small stones are sticking in the tread grooves or in wall. Remove a foreign material, if any, with a screwdriver, pliers or the like.
If necessary check the tires for air leaks.

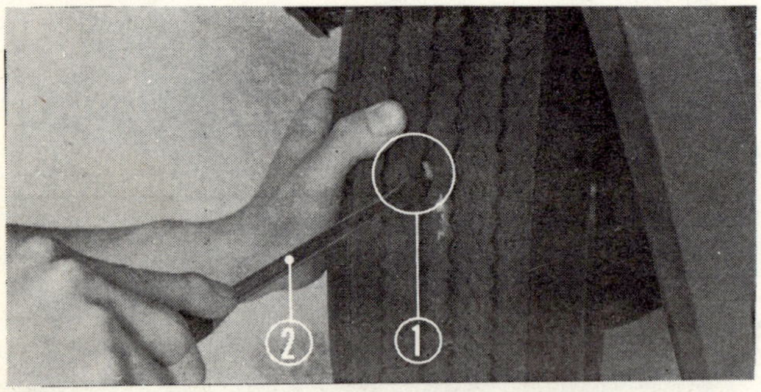

Fig. B-51
① Foreign material ② Standard type screwdriver

f. Inspect the clip nuts and hub nuts

f-1. Front wheels

1. Check the wheel clip nuts for looseness and retighten them if necessary.

   See the Torque Specifications.

Fig. B-52
① Clip nut ② Torque wrench

2. Pull out the hub nut cotter pin. Check the hub nut for looseness and retighten it if necessary.
3. With the gearshift lever placed in the neutral position, hold the drive shaft with hand and rotate it to check the shaft splines for play. If the shaft is excessively loose or if abnormal noise is heard, remove the shaft and check the splines for condition. If necessary, replace the hub.

Fig. B–53
① Spindle ② Rotate drive shaft to check for play

4. Tighten the hub nut. Tighten the nut to the specified torque with a torque wrench, and then check the cotter pin hole location. Tighten the nut until the pin holes are aligned. The torque at this time must be within specification.
See the Torque Specifications.
NOTES:
1) Hub nuts must be tightened to the specified torque. If the cotter pin holes are not aligned, change the right and feft nuts or replace them with new ones.
2) Do not turn the nuts back for aligning the pin holes.
5. Use a new cotter pin when reassembling.

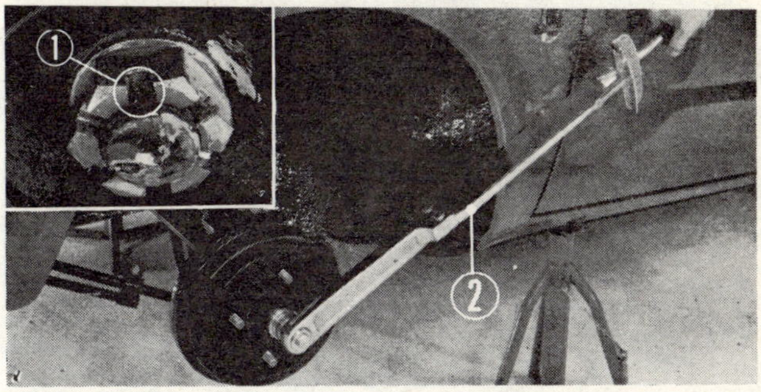

Fig. B–54
① Do not turn nut back ② Torque wrench

f–2. Rear wheels
1. Jack up the car to remove the wheel.
2. Remove the center cap. Check the hub nut for looseness with an inspection hammer. Also check the cotter pin for condition.
3. If the hub nut is loose, pull out the cotter pin, tighten the nut to the specified torque and install a new cotter pin in place. (See Par. "Front wheels" above.)

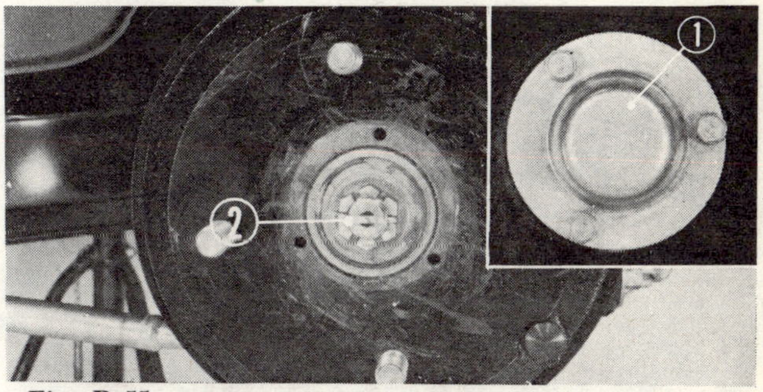

Fig. B–55
① Wheel center cap ② Cotter pin

g. Inspect the rims and wheel discs

1. Raise the wheel off the ground and check the rim for runout. The runout must not exceed **3 mm (0.117 in.)** at the outer end of the rim.
2. When removing the wheel for tire rotation, check the rim and wheel discs for corrosion, deformation, cracks or any other damage. Repair or replace a damaged wheel, if any.
3. Pay special attention to the fact that a deformed wheel rim may affect the pressures of the tubeless tires.
NOTE:
When repairing or replacing the **wheel, the** wheel balance must be checked.

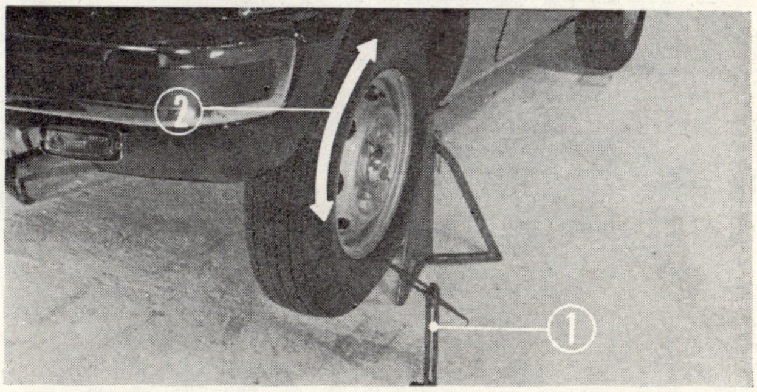

Fig. B–56
① Scribing block ② Slowly rotate

Fig. B–57
Checking wheel bearing for play

h. Inspect the front wheel bearings

See the rear wheel bearing below.

i. Inspect the rear wheel bearings

Raise the wheel off the ground and attempt to shake the wheel with hands as shown to check the wheel bearing for play. If it is excessive, check the hub nut for looseness. If the play still is excessive even after the hub nut is tightened to the specified torque, replace the wheel bearing.

When replacing the bearing, take care not to install it in wrong direction. Apply a coat of grease to the bearing.

See the Lubricatin Chart.

j. Tire rotation

In order to evenly distribute the wear of tires and help to obtain the maximum tire life, it is recommended to rotate the tires from one wheel to another periodically.

NOTE:

Spare tires must be inflated to a pressure 0.2 to 0.3 kg/cm^2 (2.84 to 4.27 lb/in^2) higher than recommended pressure.

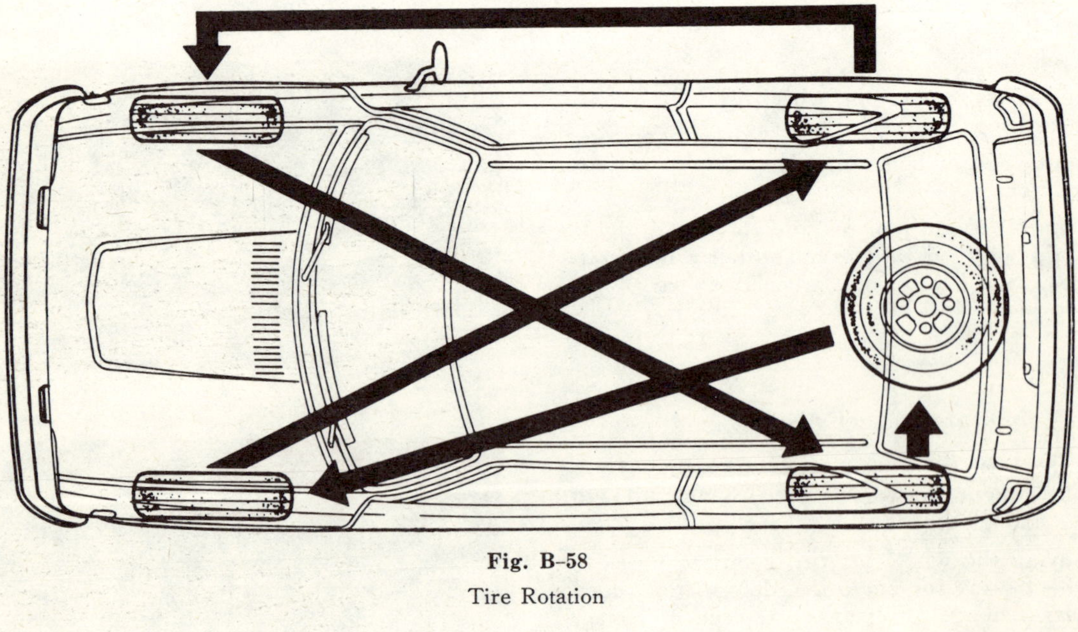

Fig. B–58
Tire Rotation

Suspension System

a. Chassis springs

a–1. Front suspension

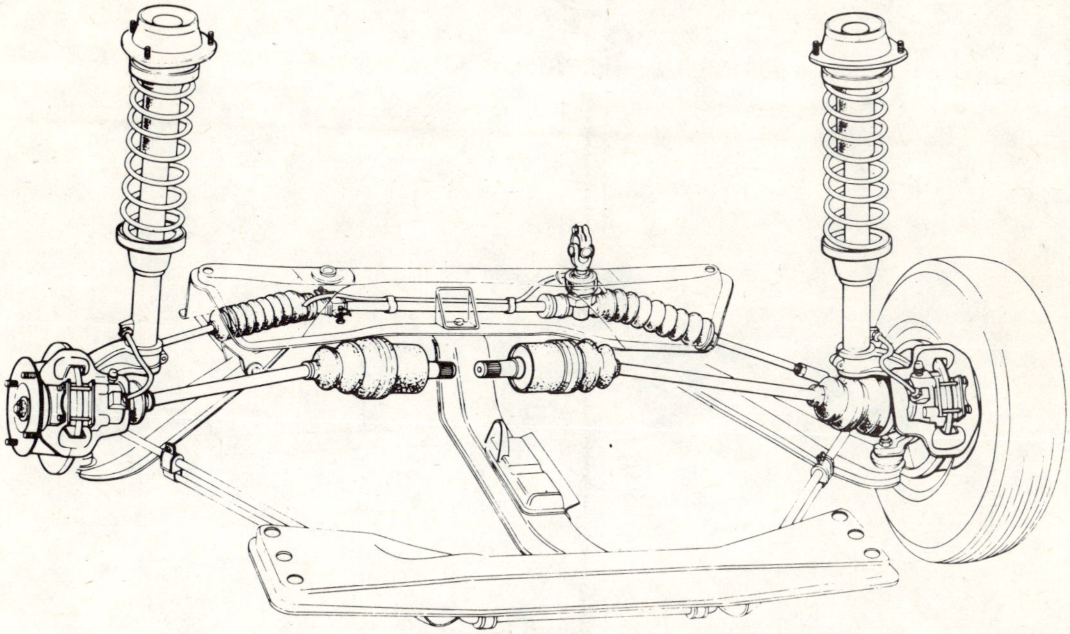

Fig. B–59

1. Inspect for damage

Check the coil springs for breakage, cracks, wear, or any other damage.

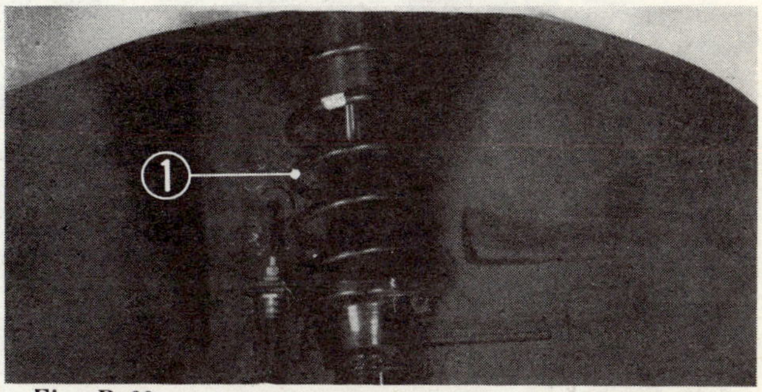

Fig. B–60
① Coil spring

2. Inspect for deflection

Place the car on level ground and check the springs for weakness judging from the inclination of the car, as viewed from the front. When checking this, make sure all tires must be inflated to the recommended pressure.

Fig. B–61
Checking springs for deflection

a-2. Rear suspension

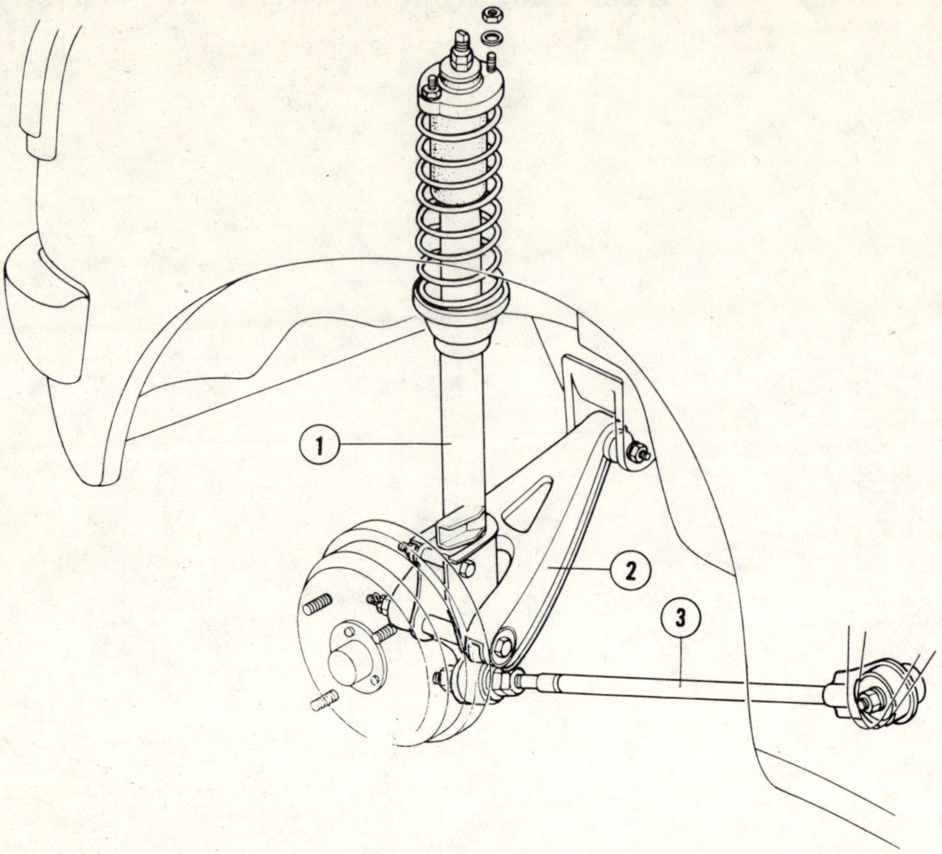

Fig. B-62

① Rear shock absorber ② Rear lower arm ③ Radius rod

Fig. B-63
① Rear lower arm ② Radius rod
③ Rear shock absorber

1. Inspect for damage

- Check the rear shock absorbers for breakage, cracks or any other damage.
- Check the radius rod and rear lower arm for breakage, cracks or wear and the attaching nuts for looseness.

See the Torque Specifications.

Fig. B-64
Checking springs for deflection

2. Inspect for deflection

Inflate all tires to the recommended pressure, place the car on level ground and check it for inclination as viewed from the rear. If the inclination is excessive, check the rear shock absorbers for proper installation. Repair it if necessary.

Check the lower arm and radius rod for deformation.

b. Connections and joints

b–1. Inspect the front shock absorbers

Check the connections of the front shock absorbers for looseness or damage, and retighten or replace them if necessary.

See the Torque Specifications.

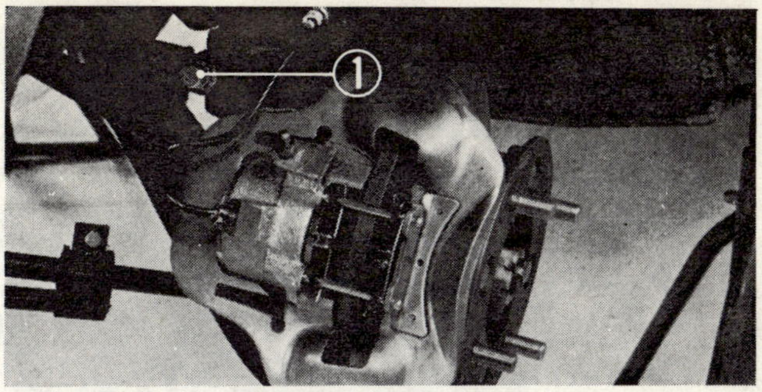

Fig. B–65
① Front shock absorber lock bolt

b–2. Inspect the joints

Check the front shock absorbers and mounting rubber nuts for looseness.

1. Check the tightening torque of the front shock absorber nuts with a torque wrench.

   See the Torque Specifications.

2. Check the rear shock absorber double nuts and its attaching nuts for looseness.

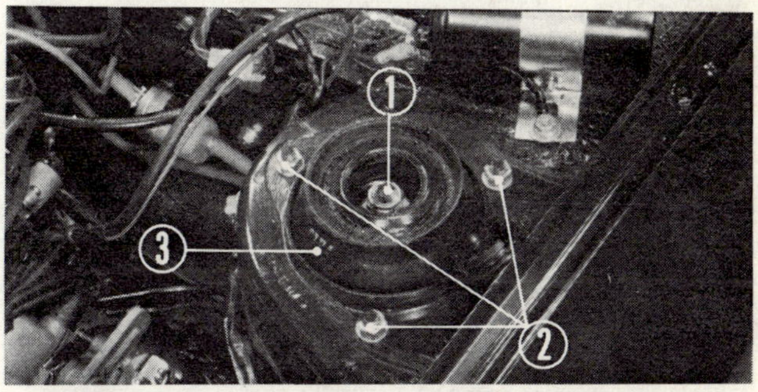

Fig. B–66
① Front shock absorber nut ② Mounting rubber nuts
③ Mounting rubber

3. Check the rear shock absorbers and rear lower arm attaching bolts for looseness, and retighten them if necessary.

   See the Torque Specifications.

b–3. Inspect the brackets

1. Check the shock absorber brackets for looseness or damage using an inspection hammer.
2. Check the lower arm bracket for looseness or damage.

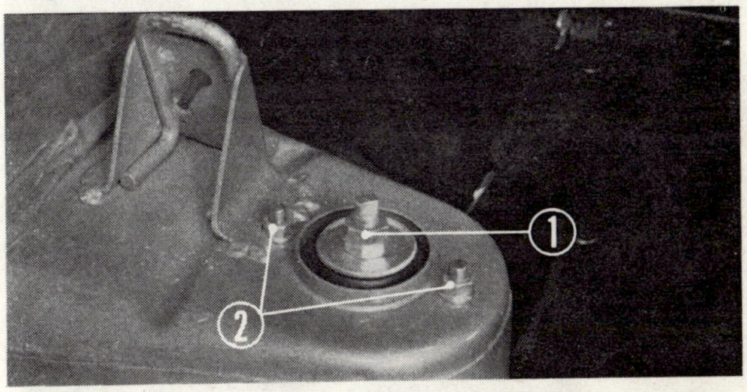

Fig. B–67
① Rear shock absorber double nut ② Attaching nuts

c. Suspension arms and knuckle support

c–1. Inspect the joints

Check the lower arm for deformation or damage and arm bushing and ball joint for play. Retighten or replace if necessary. Check the lower arm bolt for looseness.

Check the shock absorber double nuts for looseness and retighten them if necessary.

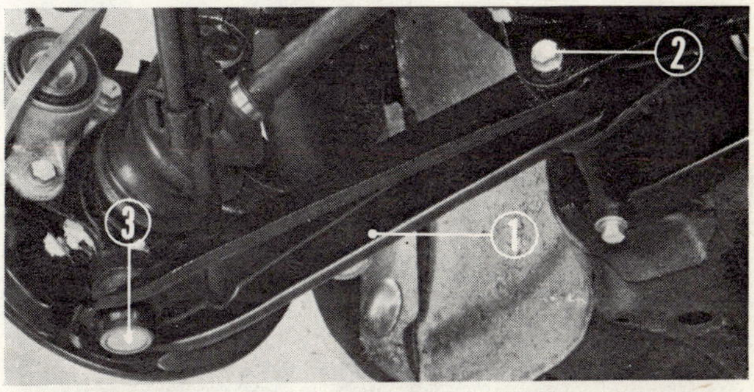

Fig. B–68
① Lower arm ② Attaching bolt ③ Ball joint

c–2. Inspect the boots

Check the lower arm ball joint boots for damage or improper installation. Replace the boots if defective.

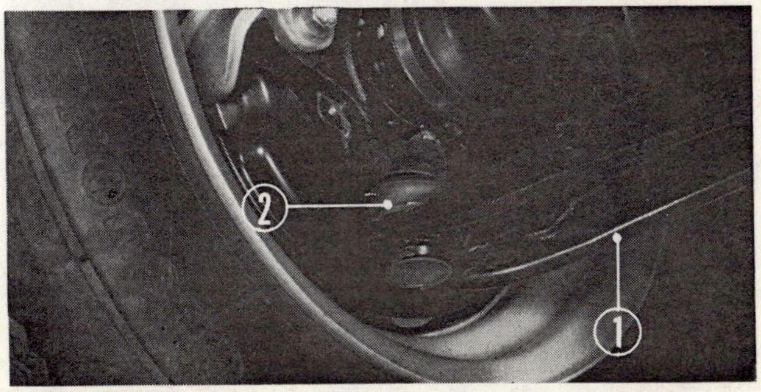

Fig. B–69
① Lower arm ② Lower arm ball joint boot

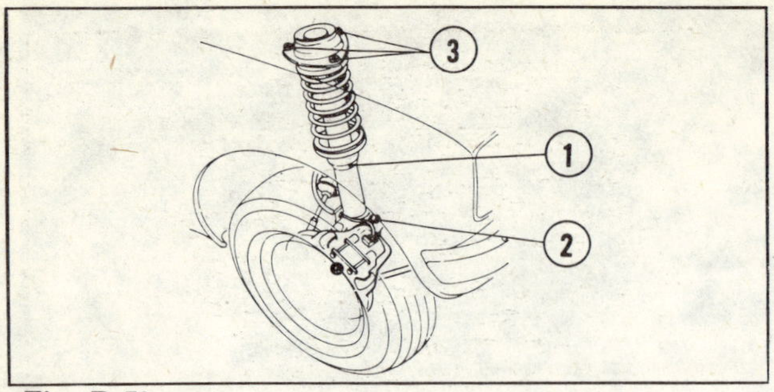

Fig. B-70
① Front shock absorber ② Lock bolt
③ Attaching nuts

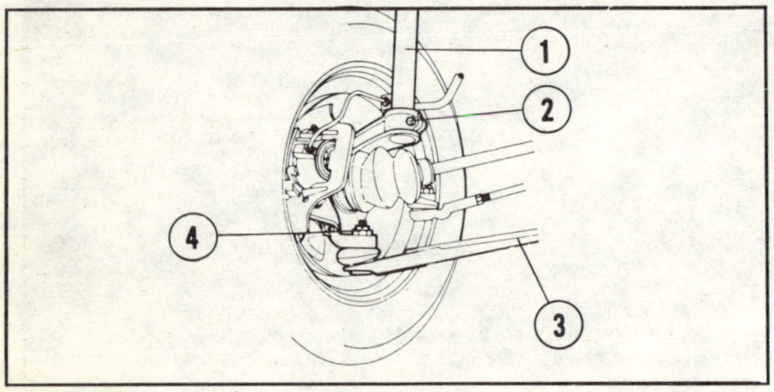

Fig. B-71
① Shock absorber ② Lock bolt ③ Lower arm
④ Ball joint nut

Fig. B-72
① Rear shock absorber

d. Shock absorbers

d-1. Inspect for oil leaks

Check the front and rear shock absorbers for oil leaks and replace a defective shock absorber, if any.

d-2. Inspect for damage

Check the shock absorbers for cracks or any other damage. Replace a damaged shock absorber, if any.

d-3. Inspect for loose connections

Check the connections of the shock absorbers for looseness and retighten them if necessary.

See the Torque Specifications.

d-4. Inspect the function

Move the car body up and down and check the shock absorbers for smooth operation and effect. Replace a defective absorber if necessary.

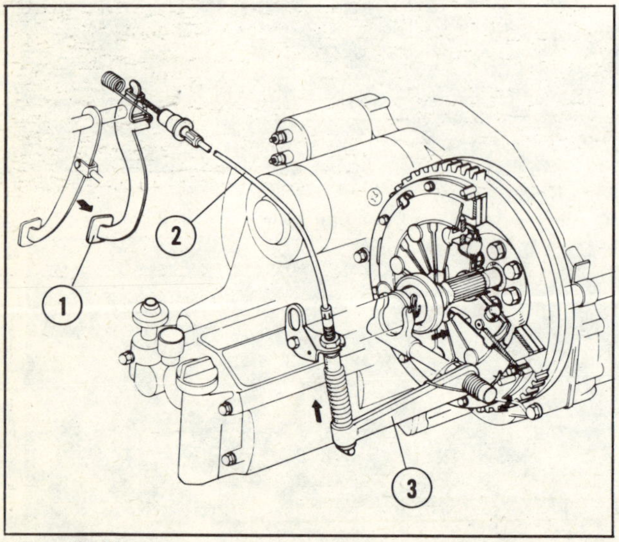

Fig. B-73
① Clutch pedal ② Clutch control cable
③ Clutch release lever

B-5. Clutch, Transmission and Drive Shafts

a. Clutch

a-1. Inspect the clutch operation

Run the engine at idle speed and check following items:

1. Depress the clutch pedal and check for abnormal noise.
2. With the clutch pedal depressed, check if the transmission gears are shifted to any desired position without any sign of binding. Also check if the car moves or if the engine stalls under the condition mentioned above.
3. Check to see if the car rolls off without any sign of slippage of the clutch, when releasing the pedal slowly and reviving up the engine.

4. Check to see if the engine stalls when depressing the brake pedal and releasing the clutch pedal slowly with the gearshift lever placed in **1ST** or **REVERSE** position.

   If not, it is an indication that the clutch is slipping. This inspection must be made at idle or a slightly higher speed of the engine.

   If any abnormal condition is found in the above inspections, check the clutch for maladjustment, the friction disc for wear, the diaphragm spring for weakness and the control cable for damage or function.

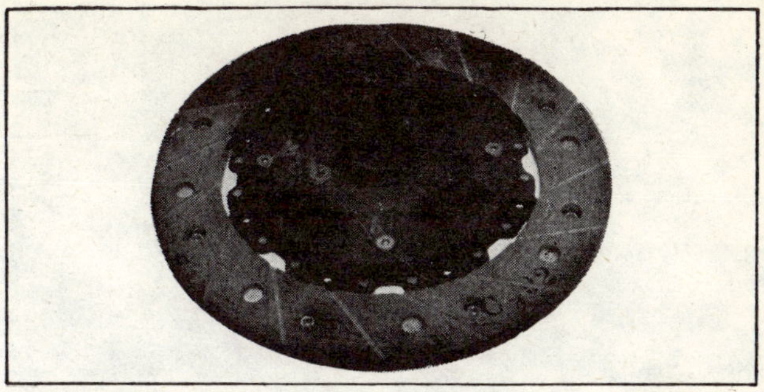

Fig. B-74
Friction disc

a-2. Inspect the clutch pedal and height

Unit: mm (in.)

Pedal play	15~20 (0.591~0.787)
Clutch release lever play	3~4 (0.118~0.157)
Pedal height above floor mat when depressed	Over 30 (1.181)

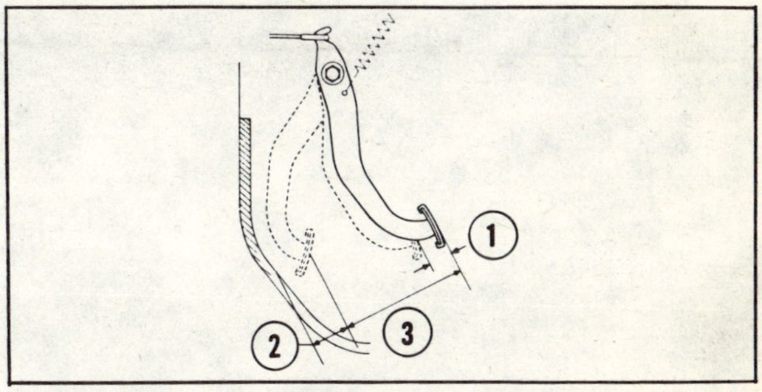

Fig. B-75
① Clutch pedal play ③ Effective stroke
② Pedal height above floor mat when depressed

Adjustment

The clutch is adjusted by loosening the adjusting nut. After adjusting, be sure to tighten the adjusting lock nut.

If the pedal height above the floor mat with the clutch disengaged is out of specification even after the release lever has been adjusted, friction disc, diaphragm spring or diaphragm surface that comes in contact with the release bearing is damaged.

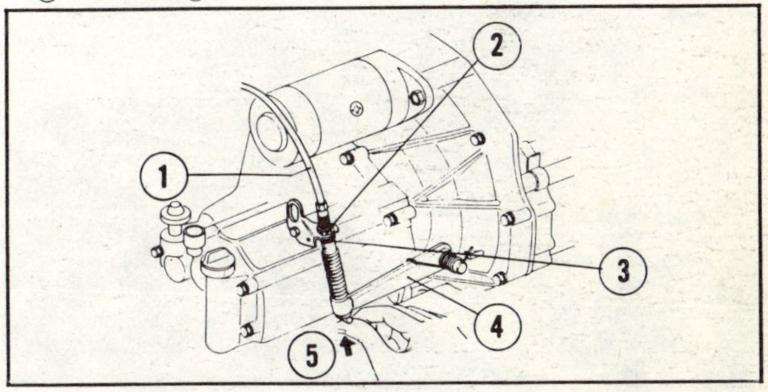

Fig. B-76
① Clutch control cable ② Adjusting nut ③ Lock nut
④ Clutch release lever ⑤ Clutch release lever play

b. Transmission

b-1. Inspect for oil leaks from transmission case

1. Jack up the front of the car and support it with safety stands.
2. Check the mating surface of the transmission case and clutch case for condition.
3. Check the mating surface of the transmission case and cover for condition.
4. Check for oil leaks from the transmission oil drain bolt.
5. Check for oil leaks from the drive shaft inboard.

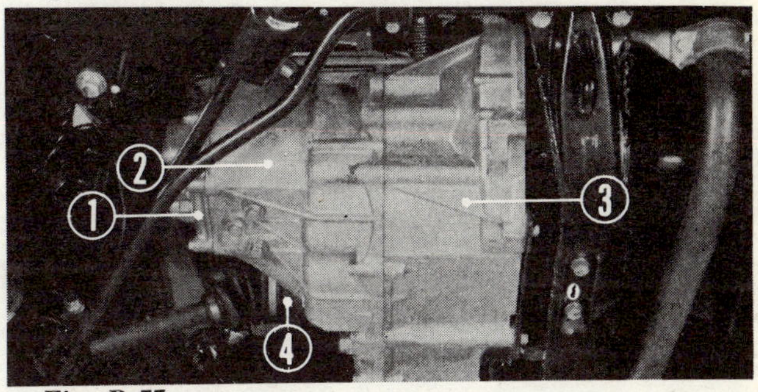

Fig. B-77
① Transmission cover ② Transmission case
③ Clutch case ④ Drive shaft

b-2. Inspect the transmission linkage

1. With the gearshift lever in the neutral position, check if the lever returns without any sign of binding when slightly tilted or if the transmission linkage is loose. Repair or replace a defective part, if any.
2. Start the engine and check if the transmission gears are shifted to any desired position without any sign of abnormal noise or binding.
3. Check the gearshift rod for bend or damage.

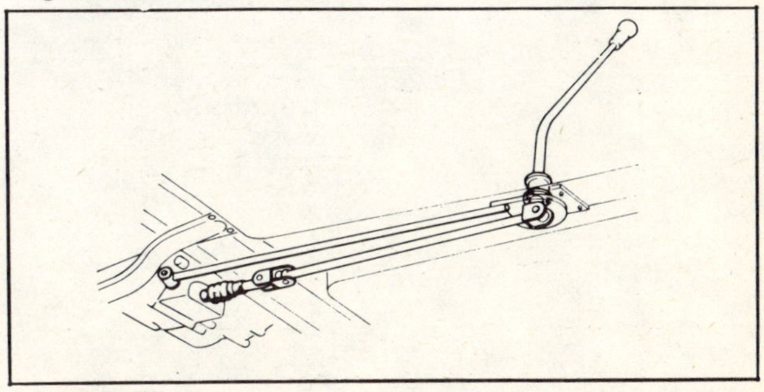

Fig. B-78
Checking for looseness of transmission linkage

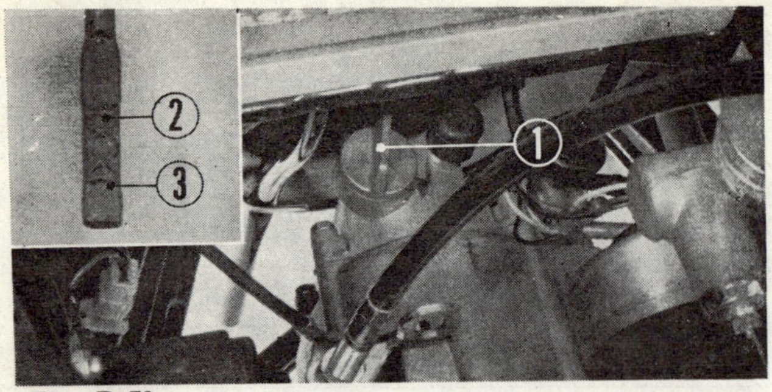

Fig. B-79
① Dipstick ② Upper level ③ Lower level

b-3. Inspect the transmission case oil level

Pull out the oil level gauge and wipe it off with a clean cloth. Then insert the gauge into position and check the oil level. If the oil is near the lower level, add oil up to the upper level. If oil is excessively dirty, change it.

Unit: ℓ (gal)

Up to upper level: 2.5 (0.66)

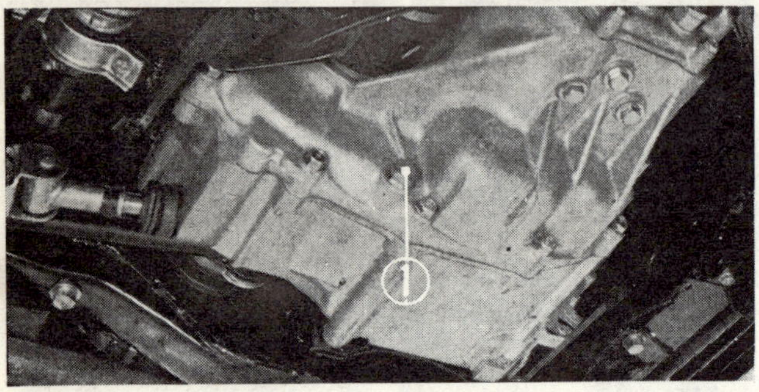

Fig. B-80
① Drain bolt

b-4. Change the transmission oil

To change, drain oil while the engine is warm.

1. Drain oil by removing the drain bolt located at the side of the transmission case.
2. After draining, tighten the drain bolt securely.
3. Pour oil through the level gauge opening up to the upper level on the gauge.

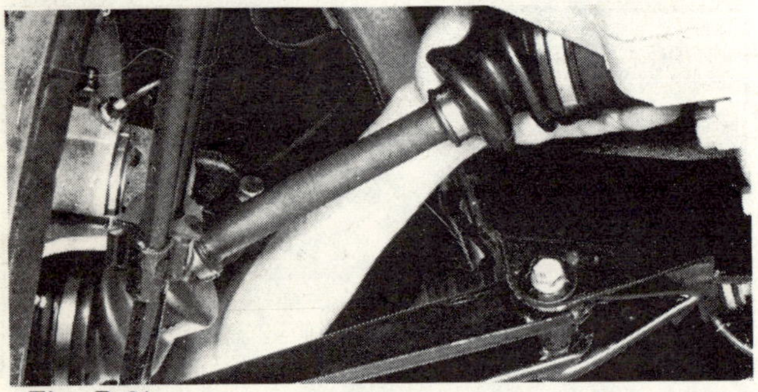

Fig. B-81
Checking looseness of drive shaft connection

c. Drive shafts

c-1. Inspect the shaft connection

Check the drive shaft connection for wear by attempting to move the shaft axially.

If the connection is loose or if the shaft comes off, the snap ring located on shaft tip may be defective. Remove the shaft and replace the snap ring.

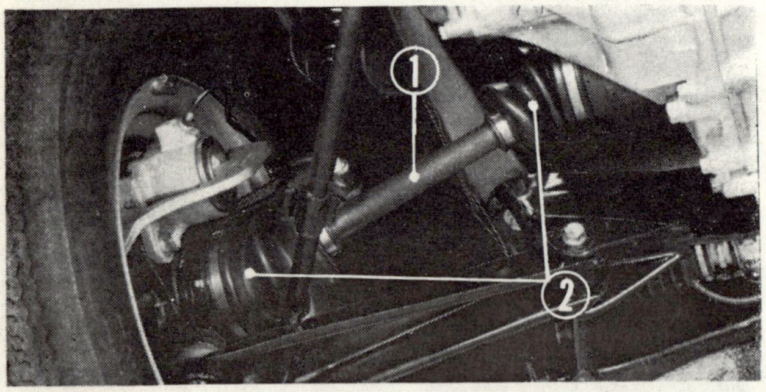

Fig. B-82
① Drive shaft ② Checking boots for damage

c-2. Inspect the boots

1. Hand-rotate the drive shaft and check the inboard and outboard joints and drive shaft splines foragainst wear or damage. Replace if necessary.

c-3. Inspect the drive shafts

Check the drive shaft for twist, bend or cracks and replace it if necessary.

a. Adjustment of idle speed

NOTES:

1) This adjustment should be made after the engine has been warmed up thoroughly. (Water temperature: Above 75°C or 167°F)
2) To provide heavy load on the AC generator, turn on the head light high beam.
3) Adjust the idle speed by means of the throttle stop screw and pilot screw. After adjusting, measure the content of carbon monoxide (CO) in exhaust gas.
4) To meet the exhaust gas regulation, the idle limiter limits the number of turns of the pilot screw within one turn. Adjust the idle speed within the permissible range.
5) Fully pull up the parking brake lever, chock up the wheels and place the selector lever in the [1] position.
6) Use a tachometer to measure the engine rpm.

1. Turn the throttle stop screw counterclockwise to obtain low speed at which the engine can continuously run smoothly.
2. Turn the pilot screw in either direction to obtain the maximum engine speed.
3. Turn the throttle stop screw to obtain the specified idle speed. Specified idle speed: 750 rpm at [1] position
4. Again turn the pilot screw to see if the engine speed rises. If so, repeat as in steps 1 thru 3 above.
5. To minimize the CO content in exhaust gas, further turn the pilot screw clockwise 1/16 turn (to the position where the idle speed just starts dropping). After adjusting, rev up the engine by depressing the accelerator pedal two or three times and check to see if the specified idle speed is maintained.
6. Measure the CO content in exhaust gas.

Fig. B-83
① Throttle stop screw ② Idle limiter

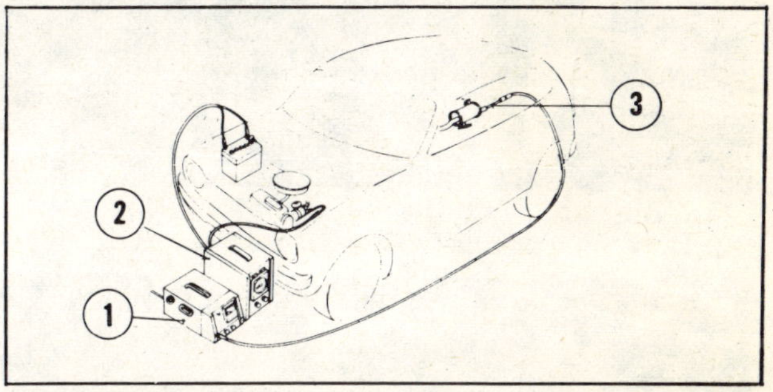

Fig. B-84
① CO meter ② Engine tachometer
③ Sampling probe

b. Inspection and adjustment of selector lever position

b-1. Inspection

1. Fully pull up the parking brake lever. Run the engine at idle speed while depressing the brake pedal.
   By moving the selector lever slowly forward and backward from the [N] position, make sure that the distances between the [N] and the points where the "D" clutch is engaged toward both [2] and [R] positions, are the same.

Range	"D" clutch engaging point
[N] → [2]	Just before slight response is felt.
[N] → [R]	Just before slight response is felt (reverse gears will issue sound).

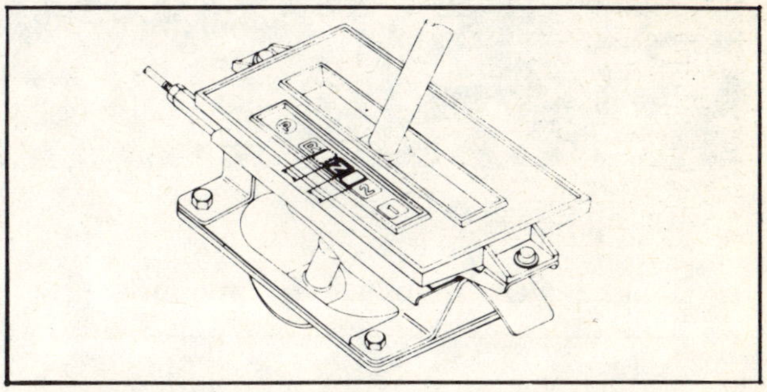
Fig. B-85

2. Be sure that the selector lever cannot be moved from [N] to [R] even if it is so moved without pushing in the button provided on the left end of the selector lever knob.

3. The engine must not start in any position other than [N] or Ⓟ.

4. The back lights must come on in [R] position.

5. Place the selector lever in Ⓟ position on down grade and check for proper braking effect.

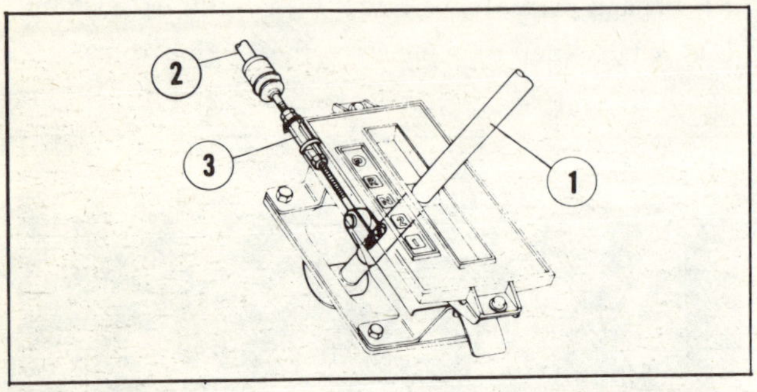

Fig. B-86
① Selectore lever ② Control cable ③ Turnbuckle

b-2 Adjustment

Adjust the length of the control cable by turning the turnbuckle located at the front bottom of the selector lever assembly. After adjusting, inspect as follows;

Cable	Selector lever position on indicator :
⇨	To [2] position
➡	To [R] position

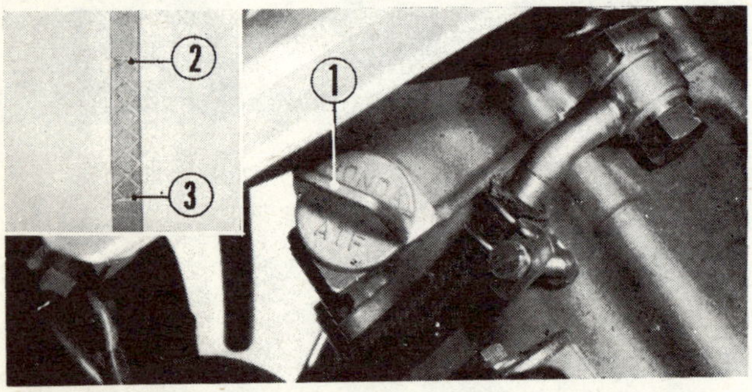

Fig. B-87
① Dipstick ② Upper level ③ Lower level

c. Inspection of automatic transmission fluid level

With the car on a level ground, loosen the automatic transmission fluid dipstick and check the level of automatic transmission fluid immediately after (within one minute) the engine was shut off.

Add the recommended fluid, G.M. Dexron, to the upper limit if necessary.

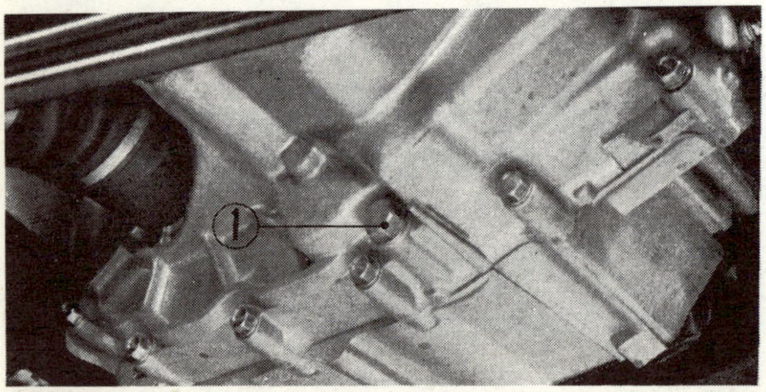

Fig. B-88
① Drain plug

d. Change of automatic transmission fluid

Remove the automatic transmission drain plug and drain the transmission system. Fill new automatic transmission fluid to the upper level through the level gauge opening.

Specified quantity : 2.5 lit. (2.2 US. qt)

NOTE :

Be sure that the required quantity of the fluid is always slightly less than the specified quantity as above due to the reminding fluid in the recess. Therefore, change the fluid while checking the level with the dipstick.

e. Measurement of hydraulic pressures

NOTES :

1) This measurement should be made after the engine has warmed up thoroughly.
(Water temperature : Above 75°C or 167°F)

2) Check the automatic transmission fluid (ATF) level before measuring.

3) Fully pull up the parking brake lever and chock up the wheels for safety.

4) Stop the engine when installing the pressure gauge attachment to the inspection hole.

5) Securely install the attachment to the inspection hole. When installing, take care not to permit dirt and dust to get inside the inspection hole.

6) When installing the plug bolt after measuring, use a new aluminum washer and tighten the bolt to the specified torque.

See the Torque Specifications.

Where the hydraulic pressures are out of specification: Repair in accordance with the Trouble Shooting **Guide**.

e-1. Hydraulic line

1. Install a tachometer.
2. Install an oil pressure gauge to the inspection hole in the hydraulic line.
3. Start the engine. Measure the oil pressure with the selector lever in [N] or Ⓟ position with the engine running at 1000 rpm.
   See the Maintenance Standard.

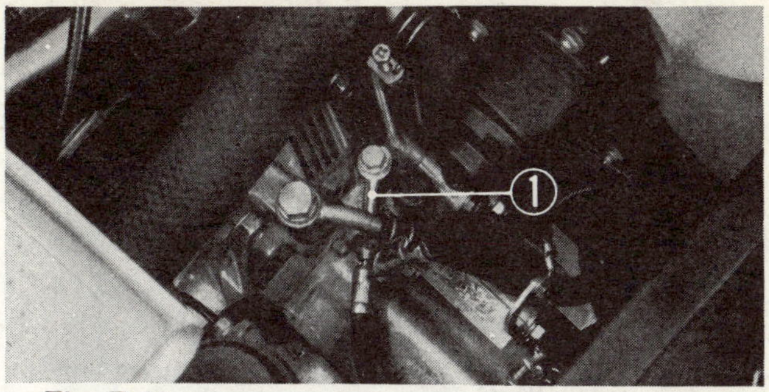

Fig. B-89
① Oil pressure joint attachment

e-2. "D" clutch

NOTE:

Jack up the front of the car and support it with safety stands.

1. Install a tachometer.
2. Install an oil pressure gauge to the inspection holes in the hydraulic line and in the "D" clutch.
3. Start the engine and move the selector lever to [2] position.
4. Measure each oil pressure while running the engine at 1000 rpm with the accelerator pedal depressed. See the Maintenance Standard.

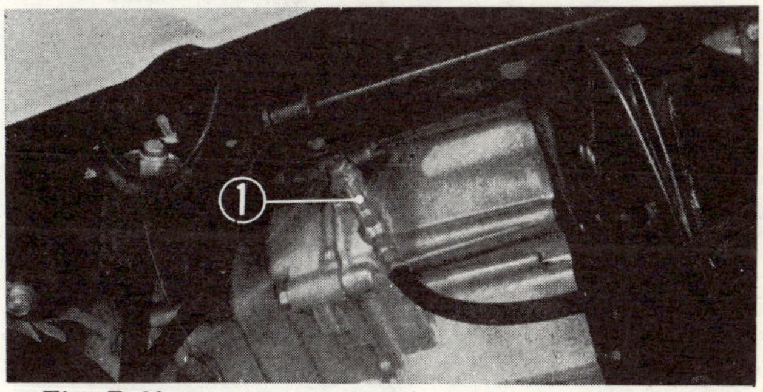

Fig. B-90
① Oil pressure joint attachment

● Test on chassis dynamometer

1. Set the car on a dynamometer and install oil pressure gauges to the inspection holes in the hydraulic line and in the "D" clutch.
2. Start the engine and move the selector lever to [2] position.
3. Measure each oil pressure while running the engine at 1000 rpm with the accelerator pedal depressed. See the Maintenance Standard.

e-3. "L" clutch

NOTE:

Jack up the front of the car and support it with safety stands.

1. Install a tachometer.
2. Install oil pressure gauges to the inspection holes in the hydraulic line and in the "L" clutch.
3. Start the engine and move the selector lever to [1] position.
4. Measure each oil pressure while running the engine at 1000 rpm with the accelerator pedal depressed. See the Maintenance Standard.

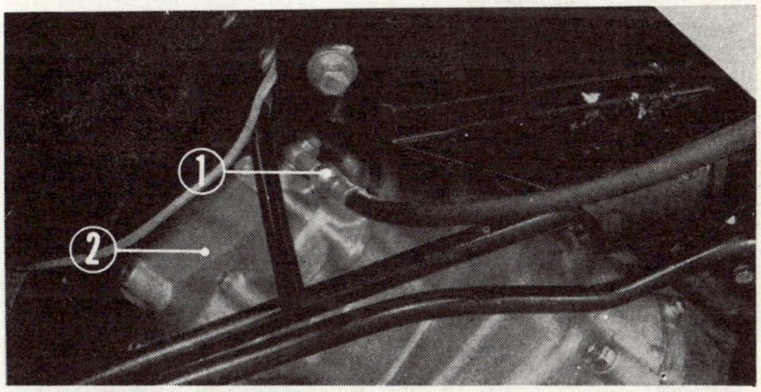

Fig. B-91
① Oil pressure joint attachment ② Transmission cover

● Test on chassis dynamometer

1. Set the car on a dynamometer and install oil pressure gauges to the inspection holes in the hydraulic line and in the "L" clutch.
2. Start the engine and move the selector lever to [1] position.
3. Measure each oil pressure while running the engine at 1000 rpm with the accelerator pedal depressed. See the Maintenance Standard.

f. Stall rpm

NOTES:

1) This inspection should not be continued **more than 10 seconds** at any one time.

2) Do not move the selector lever into any position while revving up the engine.

1. Stop the car on level ground. Fully pull up the parking brake lever and chock up the wheels.
2. Warm up the engine until the water temperature rises up to 75°C (167°F).
3. Install a tachometer in place.
4. Start the engine and move the selector lever to [2] position.

   Fully depress the brake pedal and accelerator pedal, and read the maximum engine rpm on the gauge.
5. Move the selector lever to [1] position and measure the stall rpm as in step 4 above.
6. Move the selector lever to [R] position and measure the stall rpm as in step 4 above. Check to see if the stall rpm is equal in [1], [2] and [R] positions and does not exceed the specification.

   See the Maintenance Standard.
7. If out of specification, the following causes are suspected:

Complaint	Probable causes
1. Stall rpm is high in [1], [2] and [R] positions.	• Insufficient fluid in torque converter • Air in fluid
2. Stall rpm is normal in [1] position, but high in [2] and [R] positions.	• Slippage of "D" clutch
3. Stall rpm is normal in [2] and [R] positions, but high in [1] position.	• Slippage of "L" clutch
4. Stall rpm is low in [1], [2] and [R] positions.	• Insufficient engine output (poor acceleration at low and high speeds) • Slippage of torque converter one-way clutch (poor acceleration)

• When the transmission system is disassembled, perform the following inspections.

1. Make sure the floor mat is not over the accelerator pedal stopper. Fully depress the accelerator pedal and check to be sure the carburetor throttle control lever is fully opened.
2. Release the accelerator pedal and check to be sure the carburetor throttle inner control cable has a slight play.
3. Fully pull up the parking brake lever and chock up the wheels. Move the selector lever to [2] position while depressing the brake pedal. Depress the accelerator pedal and release it suddenly. Inspect if the engine does not stall under this condition.
4. Move the selector lever in any of [1], [2] and [R] positions and check to be sure that the car moves without any sign of abnormal condition in each position.
5. On up and down grades, depress the brake pedal to stop the car. Move the selector lever to Ⓟ position and release the brake pedal to check for proper parking effect.

NOTE:

When attempting to move the selector lever from Ⓟ to any other position with the car standstill on a grade, be sure to depress the brake pedal or to pull up the parking brake lever fully.

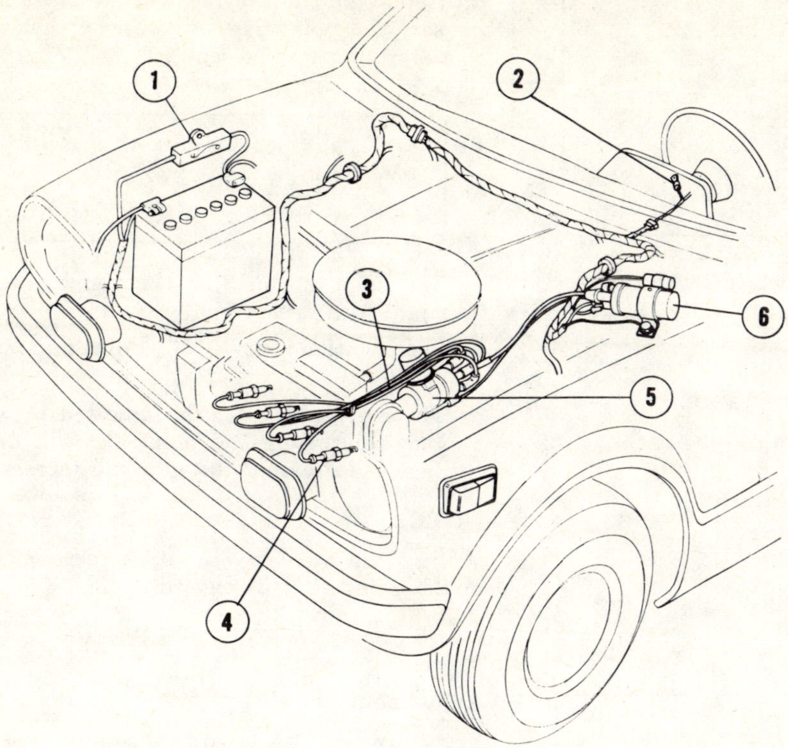

Fig. B-92

① Main fuse ② Ignition switch ③ High tension ④ Spark plug ⑤ Distributor ⑥ Ignition coil

a. Ignition system

a-1. Inspect the distributor cap

Check to see if the high-tension cords are properly connected to the distributor cap. Also check the cap for crack or contamination. Remove the cap and check the distributor terminals for burning. Repair or replace the cap if necessary.

Fig. B-93
① Distributor terminals ② Distributor cap

a-2. Inspect the contact breaker

Dirty point surfaces may be the cause of misfiring. Always keep them clean.

1. Remove the distributor cap, pull out the rotor and check the points for burning or wear.

   If the points are excessively burnt, measure the capacity of the capacitor and replace the contact points if necessary.

   NOTE:

   The condition of the point surfaces can be checked visually or by measuring the resistance with a tester.

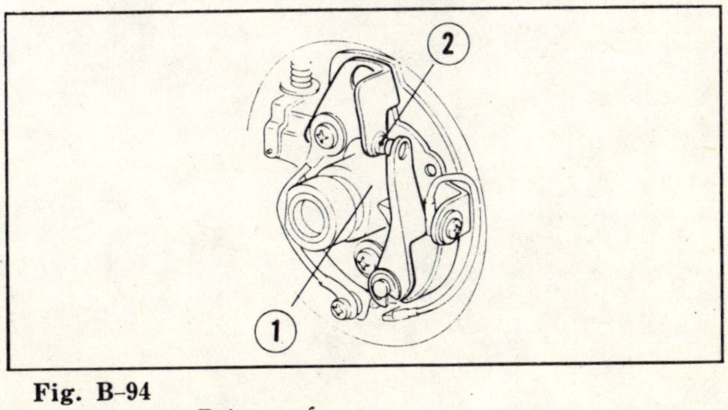

Fig. B-94
① Cam ② Point surfaces

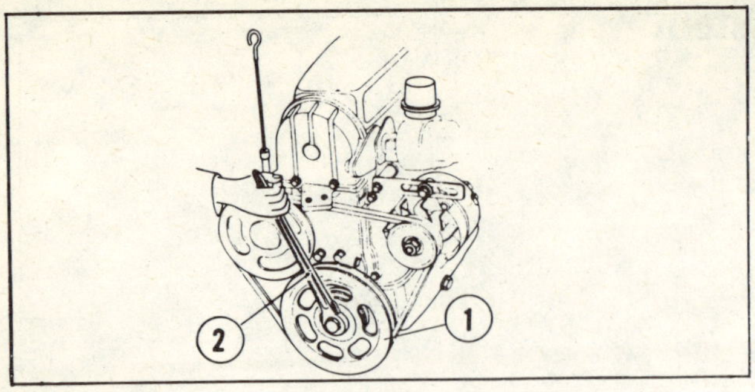

Fig. B-95
① Crankshaft pulley ② Ratchet wrench

2. Check the breaker cam sliding surface for wear. Apply a thin coat of grease to the cam surface.
3. Rotate the crankshaft pulley as shown in Fig. B-95, measure the point gap with a feeler gauge with the points fully lifted by the cam.

Unit: mm (in.)

	Specification
Point gap	0.45~0.55 (0.018~0.022)

- The point gap may be checked by measuring the dwell angle with a cam angle tester.

  Specification: 49°~55° (Point gap: 0.45~0.55 mm or 0.018~0.022 in.)

  The dwell angle can be adjusted by varying the point gap. Increasing the point gap will decrease the dwell angle; decreasing the gap will increase the angle.

NOTE:

If the points are wet with oil, decrease in current or burning of the points will be caused.

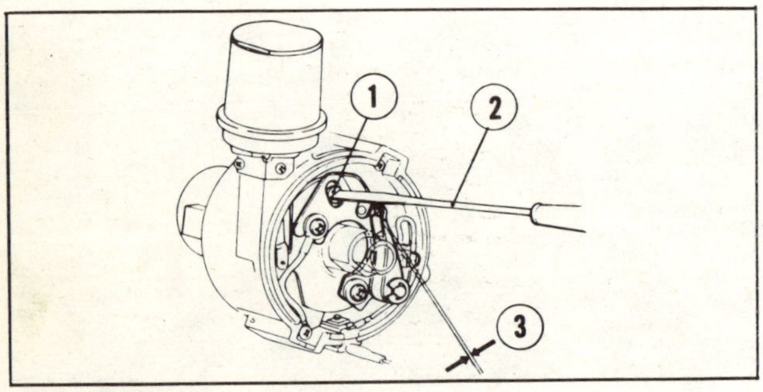

Fig. B-96
① Adjuster ② Standard type screwdriver
③ Point gap (0.45~0.55 mm)

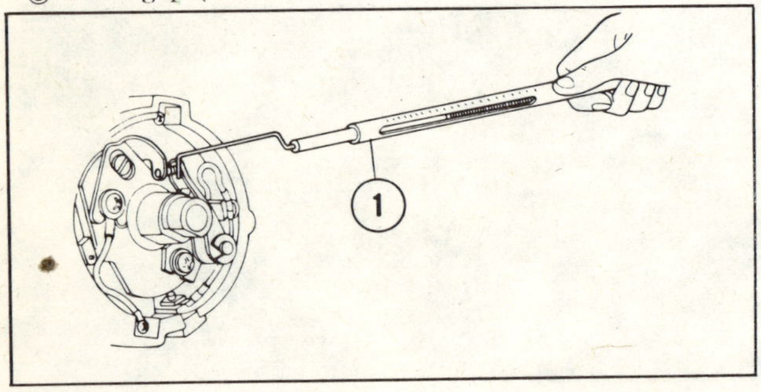

Fig. B-97
① Spring tension gauge

Adjustment

Loosen two screws. Turn the adjuster with a standard type screwdriver and adjust the point gap correctly. After adjusting, tighten the screws.

NOTE:

Every time the point gap is adjusted, the ignition timing must be adjusted.

4. Using a spring tension gauge, measure the point pressure. If it is out of specification, replace the breaker assembly.

   Specification: 500~600 g (17.64~21.16 oz)

a-3. Inspect the capacitor

Defective capacitor will result in rapid point wear. When inspecting, insert an insulator such as cardboard between the points.

1. Measurement of insulator resistance

   Measure the resistance of the insulator terminal of the capacitor and the case. If the resistance is more than 5MΩ, the insulator is normal.

2. Measurement of capacity

   Using a capacitor tester, measure the capacity of the capacitor. If it is below the specification (0.20~0.24μF), replacethe capacitor.

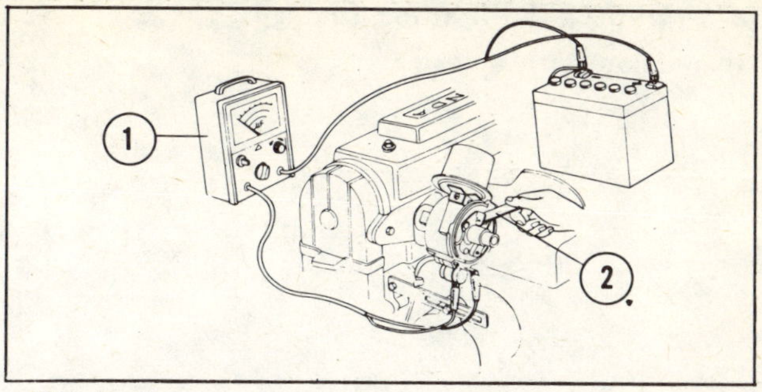

Fig. B–98
① Capacitor tester ② Card paper

a–4. Inspect the spark plugs

Remove the spark plug with a plug wrench, and check the electrodes for wear, gap or fouling and the ceramic insulator for breakage.

1. If the plug is foul, clean it with a plug cleaner or a wire brush.
2. Measure the plug gap with a feeler gauge. If it is out of specification, regap by tapping the side electrode lightly.

Unit: mm (in.)

	Specification
Spark plug gap	0.7~0.8 (0.028~0.031)

3. If the insulator or gasket is broken, replace the plug. Recommended spark plugs:

   NGK BP–5ES or BP–6ES
   BPR–6ES (France type only)
   DENSO W–20EP or W–22EP

4. When installing, attach the plug to the plug wrench and lightly screw it in, taking care not to damage the threads. Then, retighten the plug to specification.

NOTES:

1) Use the special wrench with the plug catch.
2) Be careful not to force the insulator more than necessary.

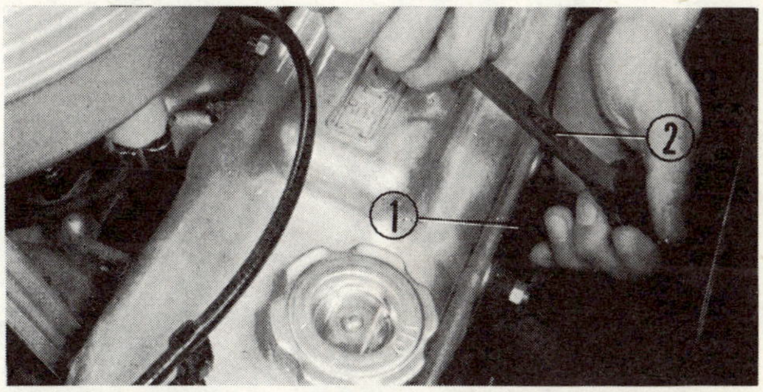

Fig. B–99
① Plug wrench ② Ratchet wrench

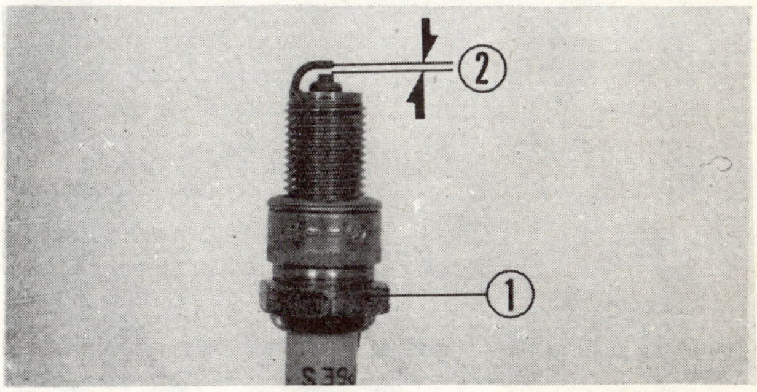

Fig. B–100
① Spark plug ② Plug gap (0.7~0.8 mm)

a-5. Inspect the ignition timing

Inspection with a tester

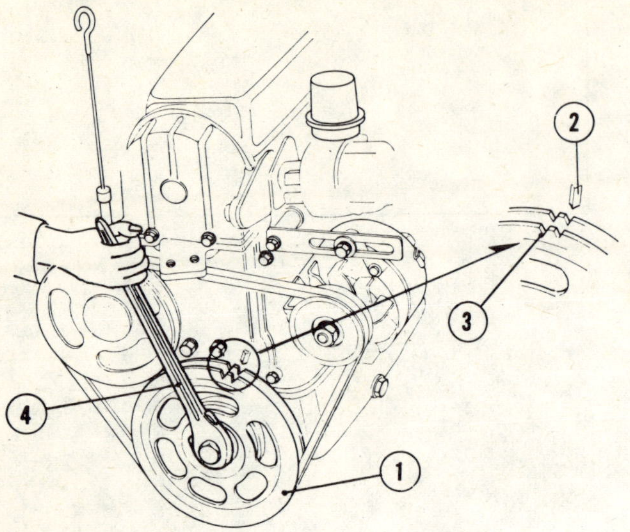

Fig. B-101
① Crankshaft pulley ③ Mark "F"
② Index mark ④ Ratchet wrench

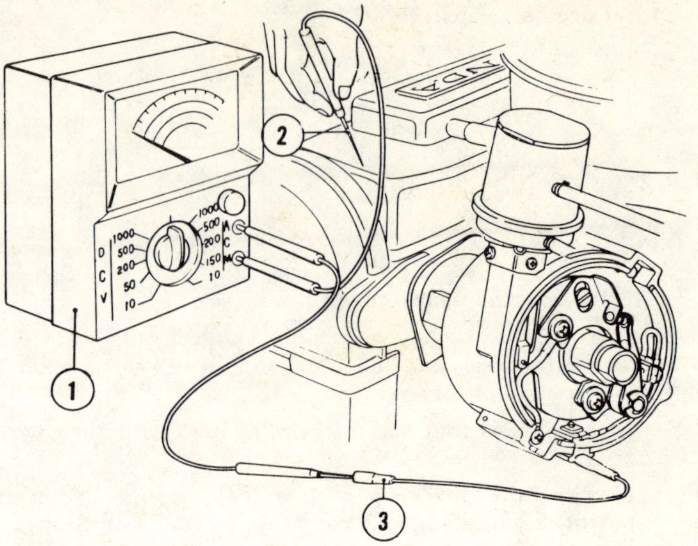

Fig. B-102
① Radio tester ② Negative ground lead
③ Primary winding

1. Disconnect the contact breaker primary winding (blue) at the ignition coil side connector, and connect the positive (+) terminal of the lead of a tester to that connector.
2. Ground the negative (−) cord to the engine body.
3. Attach a wrench to the crankshaft pulley bolt, and rotate the crankshaft in the direction of rotation of the engine.
   • It is advisable to use a ratchet wrench as shown in Fig. B-101 above to have the job done better.
4. Check to see if there is no continuity when the mark " F " on the crankshaft pulley is aligned with the index mark on the timing cover. The correct ignition timing is 5° BTOC.
   • A timing light is recommended in checking the timing, but a lamp may also be used when a timing light is not available.

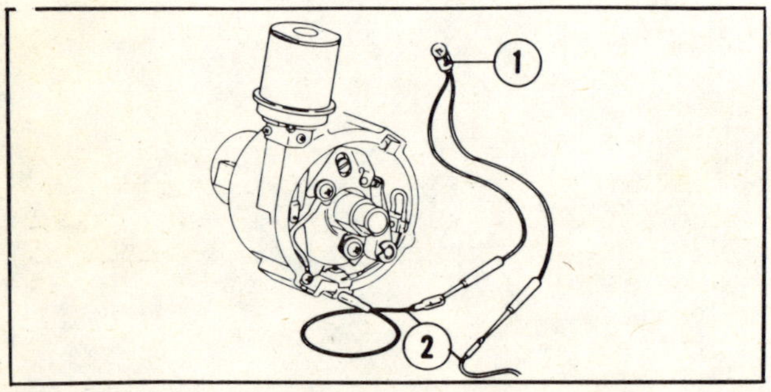

Fig. B-103
① 12V-3W lamp ② Primary winding

Inspection with a lamp

1. Disconnect the distributor primary winding (blue), connect a 12V-3W lamp as shown and turn on the ignition switch.
2. Rotate the crankshaft pulley and check to see if the lamp goes out when the index mark and the mark " F " are aligned.

Inspection with a timing light

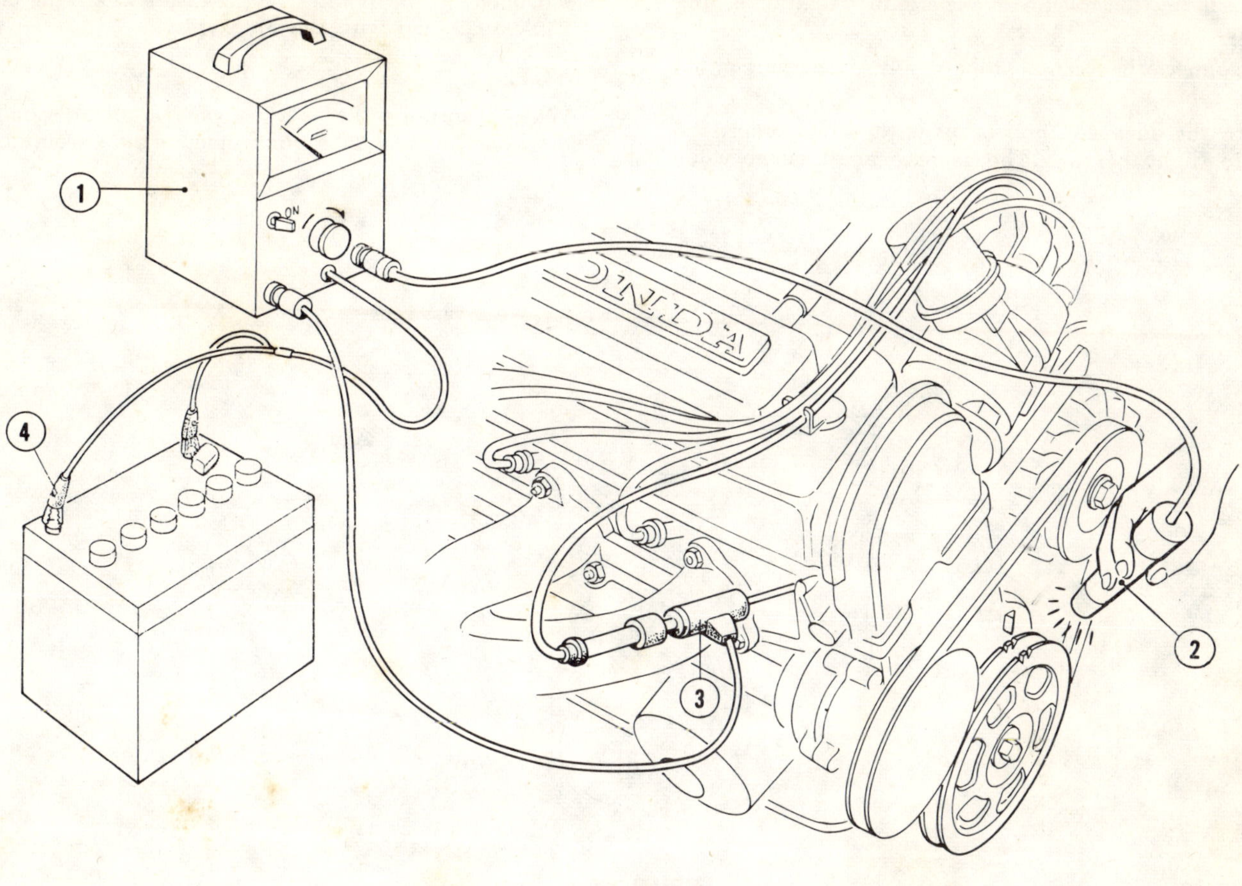

Fig. B-104

① Timing tester
② Timing light
③ Positive (+) power source
④ Negative (−) power source

1. Disconnect the vacuum advancer boost control tube.

   NOTE:

   This inspection may be affected by the vacuum advancer even at idle speeds of the engine.

2. Use the battery as a power source as shown in Fig. B-104 above.

3. Attach an attachment to the No. 1 or No. 4 high-tension cord, and connect the timing tester cord to the attachment.

4. Start the engine and aim the timing light to the mark "F" on the crankshaft pulley with the engine running at idle speed (750 rpm). Under such a condition, check to see if the mark "F" and the timing mark are aligned.

Adjustment (with a lamp)

1. Loosen the distributor mounting bolt, move the distributor body as shown in Fig. B-105 until the lamp goes out and tighten it.

2. Again rotate the crankshaft pulley one turn with a ratchet wrench or by the starting motor, and check to see if the lamp goes out when the mark "F" and the index mark are aligned.

Fig. B-105
① Bolt ② Advance ③ Retard

a-6. Inspect the spark advancer

Inspection of advance angle

1. This inspection must be made after the ignition timing adjustment.
2. Disconnect the vacuum tube and run the engine at 4000 rpm.
3. Turn the advance knob of the tester while aiming the timing light thereto. The advance angle is the value indicated by the pointer of the tester where the mark " F " is aligned with the index mark.

NOTE:

When inspecting the spark advancer only, disconnect the boost control tube connected to the carburetor.

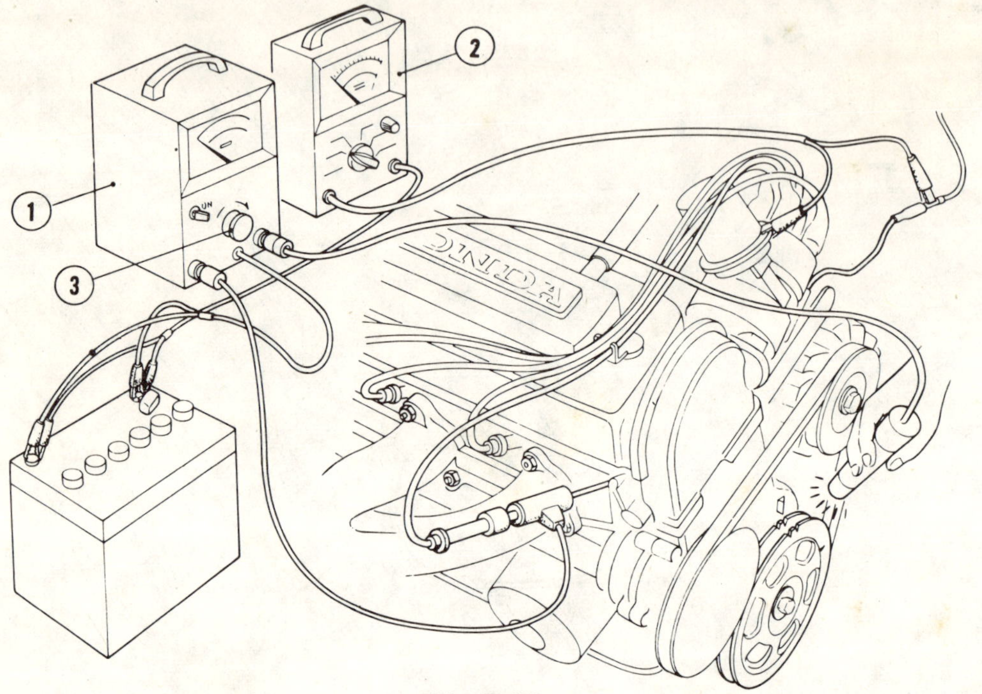

Fig. B-106

① Timing tester ② Engine tachometer ③ Advance knob

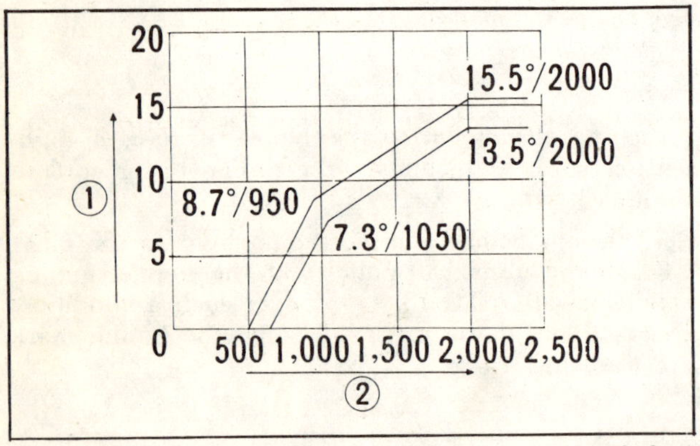

Fig. B-107 Governor advancer characteristics (camshaft)
① Advance angle ② Speed (rpm)

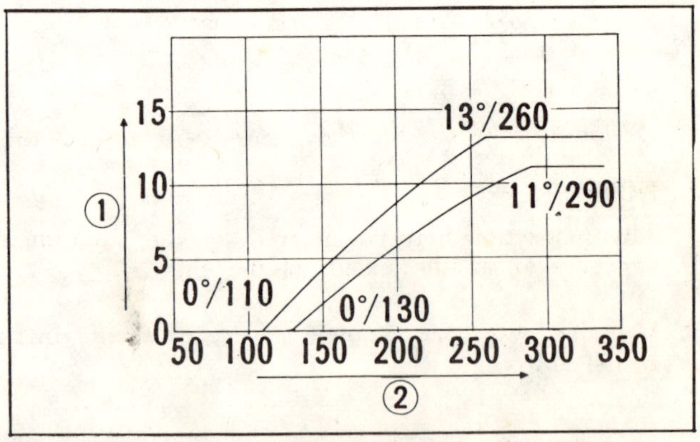

Fig. B-108 Vacuum Advancer characteristics (camshaft)
① Advance angle ② Degree of vacuum (mmHg)

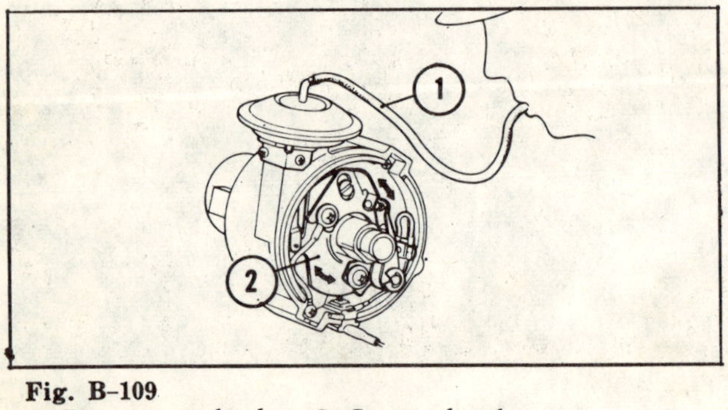

Fig. B-109
① Boost control tube ⑫ Contact breaker

Instant method for inspecting vacuum advancer

1. Suck the boost control tube as shown in Fig. B-109 and check the vacuum advancer for function by observing the movement of the contact breaker.

b. Charging system

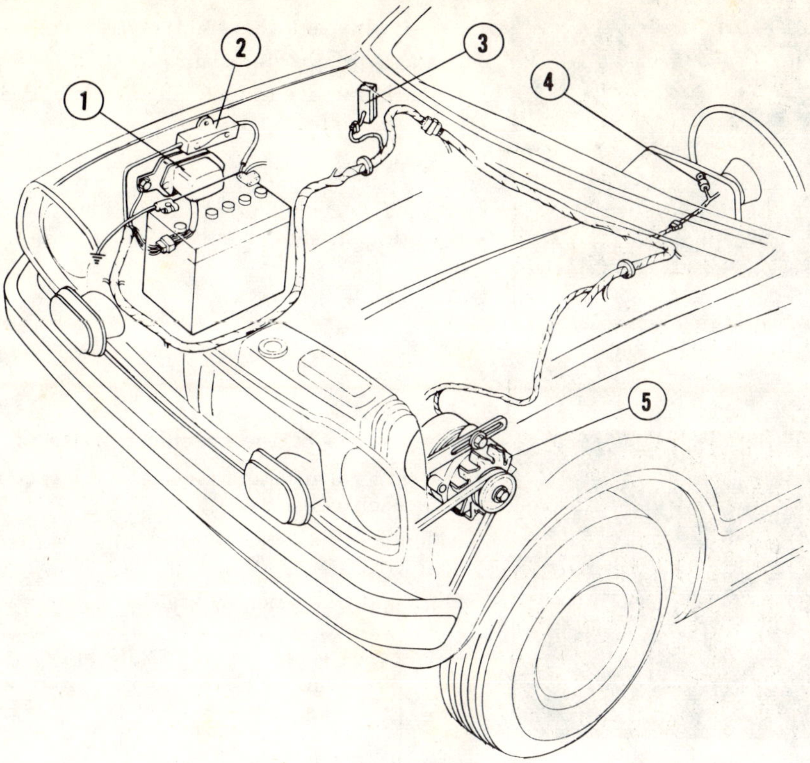

Fig. B–110

① Voltage regulator ② Main fuse ③ Fuse box ④ Ignition switch ⑤ AC generator

b–1. Charging

The Honda CIVIC uses an AC generator, which is able to charge the battery even at idle speed.

Check the pilot lamp for condition. The lamp must come on when the ignition switch is turned on and go out when the engine starts.

Also check the generator wiring coupler for proper connection, and reconnect if necessary.

- When the lamp does not go out even after the engine starts, check the generator output voltage

b–2. Inspect the drive belt

Measure the deflection of the belt at the point shown at right. If it is out of specification, readjust.

Unit: mm (in.)

	Specification
Deflection of belt	12∼17 (0.472∼0.669) *

* Depress the belt with a force of 9∼11 kg (20∼24 lb).

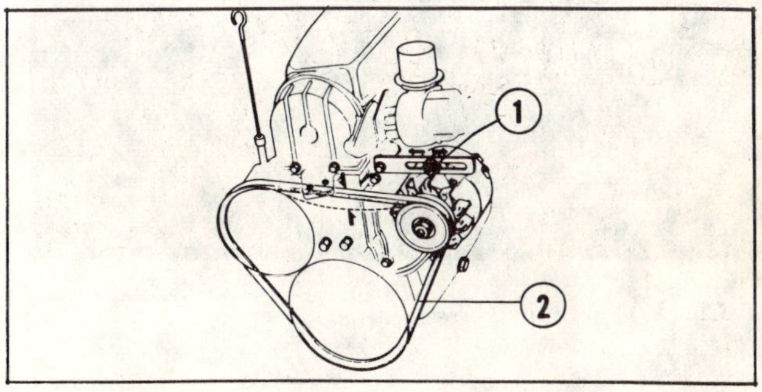

Fig. B–111

① Adjusting bolt ② Drive belt

Adjustment

Loosen the adjusting bolt and move the generator forward and backward to obtain correct belt tension.

NOTE:

If the belt sqeaks during operation, it is advisable to give a little tension to it.

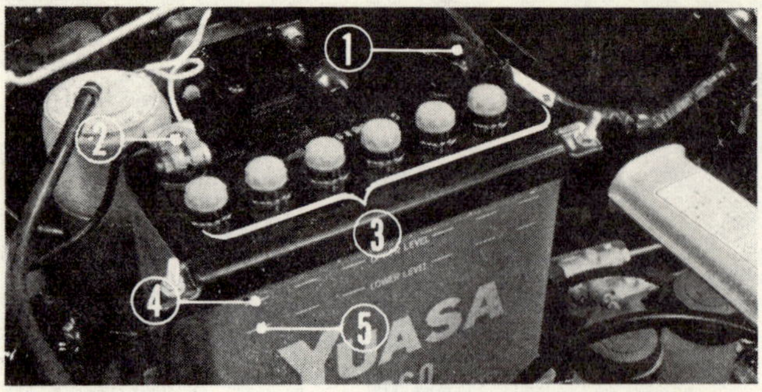

Fig. B–112
① Positive (+) terminal ② Negative (−) terminal
③ Filler caps ④ Upper mark ⑤ Lower mark

Battery

c–1. Inspect the electrolyte level

Visually check the electrolyte level. If it is near the lower mark, remove the six filler caps and add distilled water up to the upper mark.

NOTE:

Excessively filling of distilled water may cause it to spill during driving, resulting in corrosion.

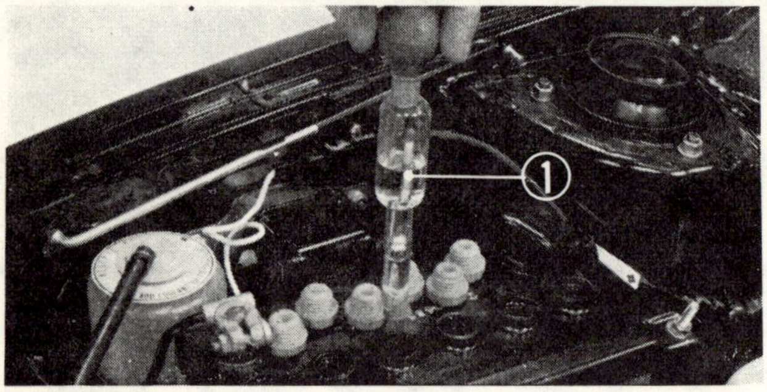

Fig. B–113
① Hydrometer

c–2. Inspect the specific gravity of electrolyte

Using a hydrometer, measure the specific gravity of electoyte in each cell.

The specific gravity of a fully charged battery is 1.25~1.27 at 20°C.

Recharge the battery if the specific gravity is below 1.20.

State of charge	Fully charged	Require charging
Electrolyte specific gravity	1.25~1.27 (20°C or 68°F)	Below 1.20

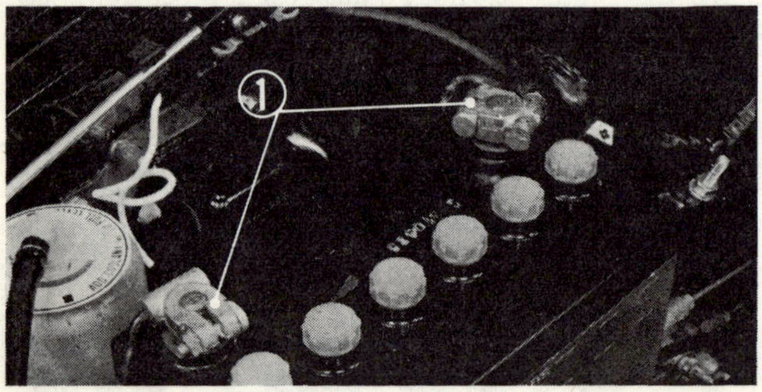

Fig. B–114
① Checking terminals for looseness

c–3. Inspect the terminals

Check the positive and negative terminals of the battery for corrosion or looseness.

Clean the terminals if corrosion is evident.

Engine

a. Engine proper

a-1. Inspect the starting and noise

Check the choke valve and throttle valve for function by operating the choke cam and throttle control lever. Pull the choke button and start the engine to check for starting and noise.

- If the engine fails to start properly, check the starting, ignition or fuel system for abnormal condition.
- If any abnormal noise is heard, check for:

1. Valve tappet clearance.
2. Leak from carburetor insulator and manifold connectign tube (for intake air control).
3. Entry of secondary air due to disconnection or damage to bypass breather pipe.

a-2. Inspect the acceleration

After warming up the engine, place the gearshift lever in the neutral position, depress the accelerator pedal and rev up the engine to check the pedal linkage for binding and the engine for stalling or knocking.

a-3. Inspect the exhaust gases

After warming up the engine, place the gearshift lever in the neutral position and check the color of the exhaust gases at idle speed or at higher speeds. When combustion process goes well, they are colorless or light blue in color. If they are black, gray or white, incomplete combustion, or burning of oil etc. may be the cause.

1. When exhaust gases are black or gray:

   Fuel-air mixture is rich or combustion is incomplete.

2. When exhaust gases are white:

   Too much oil is in the crankcase or oil is pumped up into the cylinders. Check the oil level. This failure may also be caused by excessive wear of the cylinders, piston rings or valve guides. Check the compression pressure. When the color of the gases is black at idle speed, maladjusted pilot screw or too higher a fuel level may be the cause. At the medium or higher speeds, too rich fuel mixture would be the probable cause.

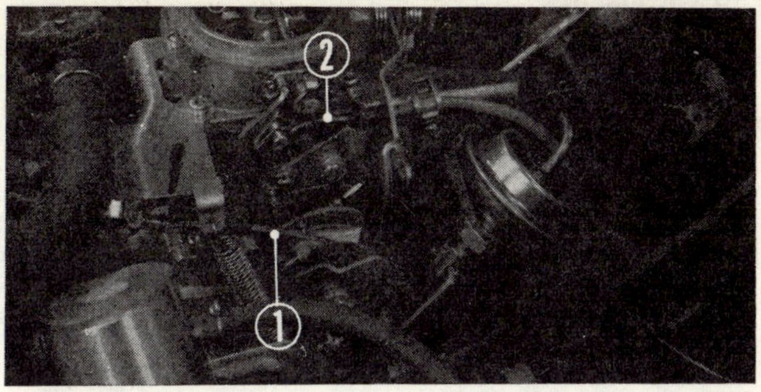

Fig. B-115
① Throttle control cable ② Choke control cable

● Inspect the content of carbon monoxide (CO)

CO meter

1. Perform a carbon monoxide test on the car, using a conventional CO meter available. See the instructions provided with the device. Fig. B-116 shows the most conventional type of CO meter.

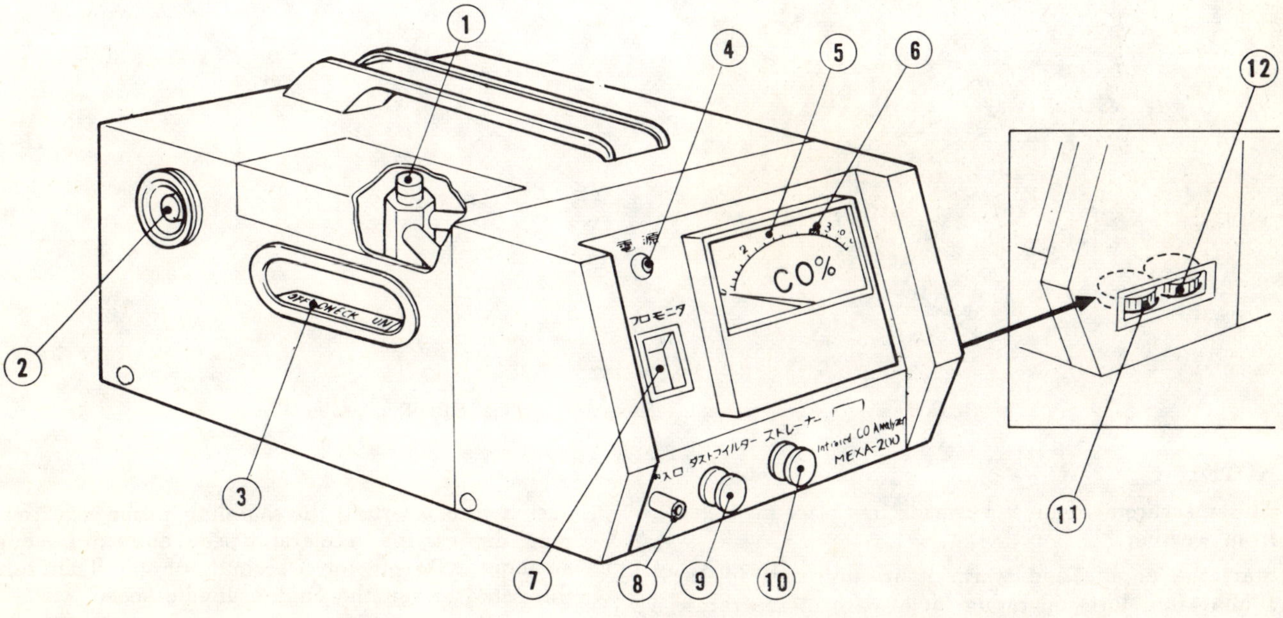

Fig. B-116

① Span gas (sample gas) inlet port
② Checker knob
③ Measuring switch OFF◀CHECK▶ON
④ Pilot lamp
⑤ Meter
⑥ Green belt
⑦ Flow monitor (for checking sample gas flow)
⑧ Sample gas
⑨ Filter
⑩ Strainer
⑪ 0 Dial
⑫ Span dial

Measurement

Measurement and adjustment of the idling-time content of CO contained in exhaust gases must be made as follows:

- Prior to measurement, the following items must be inspected and adjusted:

1. Condition of air cleaner
2. Ignition timing and function of spark advancer
3. Condition of spark plug electrodes
4. Adjustment of carburetor
5. Valve clearance
6. Condition of breather tube and drain tube

Adjustment of idle speed

1. Warm up the engine thoroughly before making the adjustment (where the water temperature is over 75°C or 167°F). In consideration of the load on the generator, turn on the head light high beam.
2. Set the engine rpm to the lowest steady speed by turning the throttle stop screw. Then, bring the engine rpm to the maximum by turning the pilot screw in either direction.
3. Turn the throttle stop screw so that the engine rpm becomes within the specified idle rpm. Finally check to see by turning the pilot screw that the adjusted engine idle speed is still maximum.
4. Then turn the pilot screw 1/16 turn clockwise.

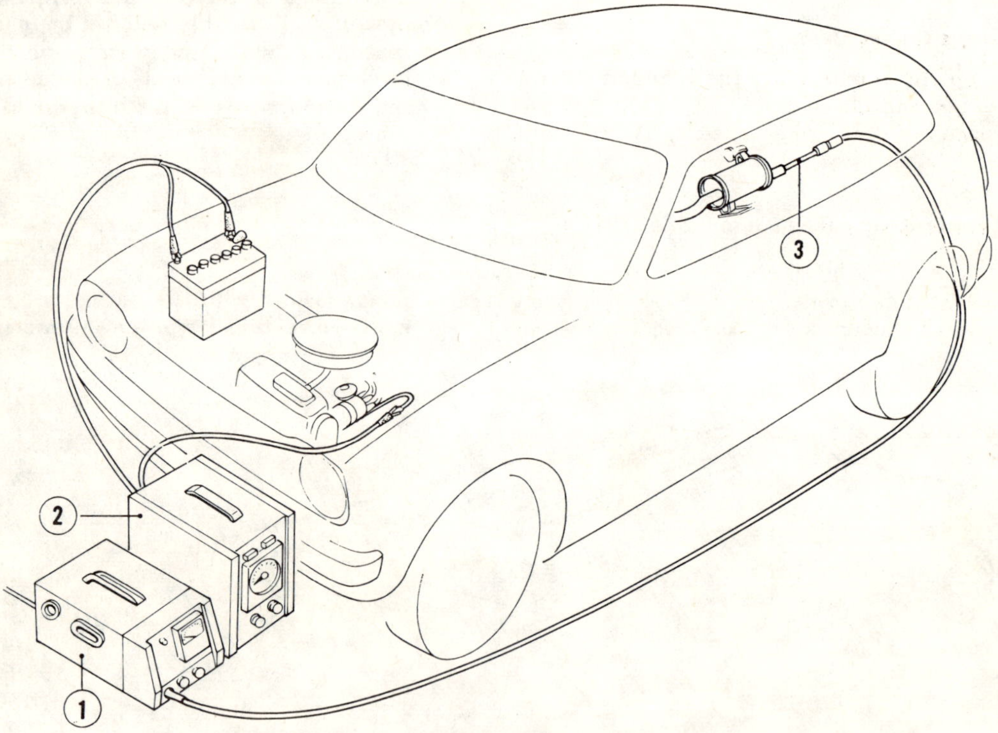

Fig. B–117

① CO meter ② Engine tachometer ③ Sampling probe

NOTE:

This measurement must be made in a place protected from weather.

1. Start the engine and warm it up until the radiator cooling fan starts operating (at a water temperature of 88.5~91.5°C or 191~197°F).

This measurement can be made as long as the engine is running, regardless of cooling fan operation.

2. After warming up the engine, install an engine tachometer as shown and adjust idle speed.
3. Just before inserting the sampling probe into the tail pipe, depress the accelerator pedal and run the engine at about 2000 rpm for 5 seconds or so. Then release the pedal to run the engine at idle speed.
4. Insert the sampling probe into the tail pipe more than 60 cm (23.62 in.).
5. Read the scale when the pointer of the CO meter stops oscillating.

Water temperature	Specified rpm	CO content
During operation of cooling fan	750	2.0~4.0%

- The measured value must be 2.0 to 4.0%. If it is below 2.0%, the test conditions above are not arranged properly. Check the conditions and measure the CO content again. If the measured value is over 4.0%, adjust CO content in accordance with the following step:

6. Turn the pilot screw 1/16 turn clockwise. Return the engine speed to the specified rpm with the throttle stop screw and read the scale of the meter. Repeat this operation until the meter registers below 4.0% content. When the meter is still registering over 4.0% content or when the CO content is below the limit, but drive feeling is bad, confirm the inspection and adjustment items or the test conditions, and measure the CO content again.

Fig. B-119
① Air cleaner element

a-5. Inspect the cylinder head and manifold parts

1. Check the nuts securing the intake manifold to the cylinder head for looseness, and retighten them if necessary.
2. Check the bolts securing the intake manifold to the carburetor for looseness, and retighten them if necessary.
3. Check the cylinder head bolts for looseness, and retighten them if necessary.

   See the Torque Specifications.

a-6. Inspect the compression pressure

1. Remove the four spark plugs.
2. Put the end of the compression gauge tube into the plug hole and hold the gauge securely with hand to prevent leak of compressed gases.
3. Place the throttle valve and choke valve in FULLY-OPEN position and turn on the starter switch. Then crank the engine until the gauge pointer registers the maximum pressure on the scale.

Unit: kg/cm^2 (lb/in^2)

Compression pressure specification (at 400 rpm)
10~14 (142~339)

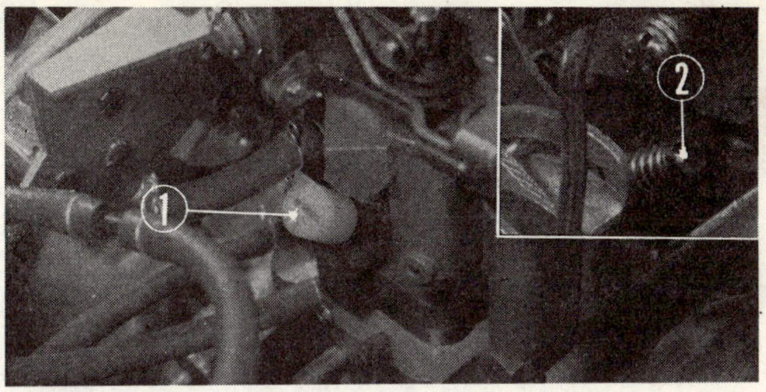

Fig. B-118
① Pilot screw ② Throttle stop screw

a-4. Inspect air cleaner element

The air cleaner element is of a filtering paper type. A clogged element may decrease the amount of air to be drawn into the engine, resulting in insufficient acceleration and in excessive fuel consumption.

1. Clean the paper element by tapping it while directing a blast of compressed air from inside to blow off dust particles accumulated in the paper. Use an air nozzle. Replace the element, if excessively dirty.
2. Clean the interior of the air cleaner case, and check the case packing and seal washer for damage.

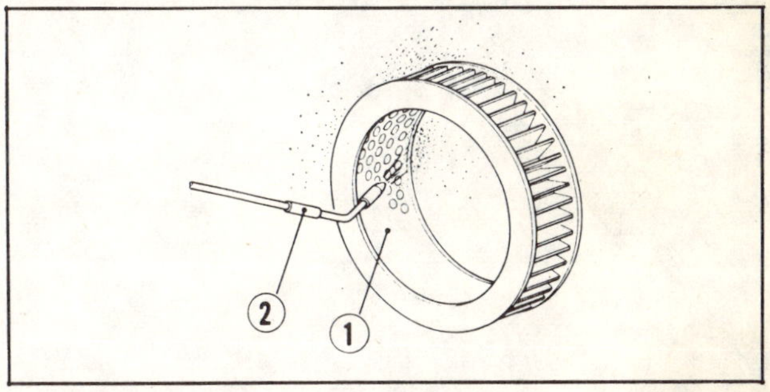

Fig. B-120
① Air cleaner element ② Air nozzle

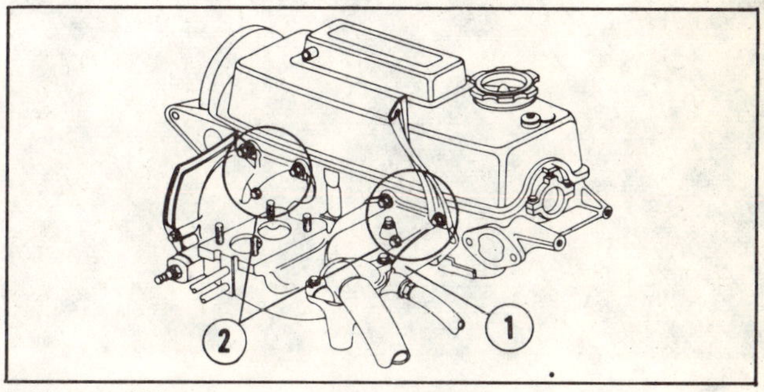

Fig. B-121
① Intake manifold ② Intake manifold nut

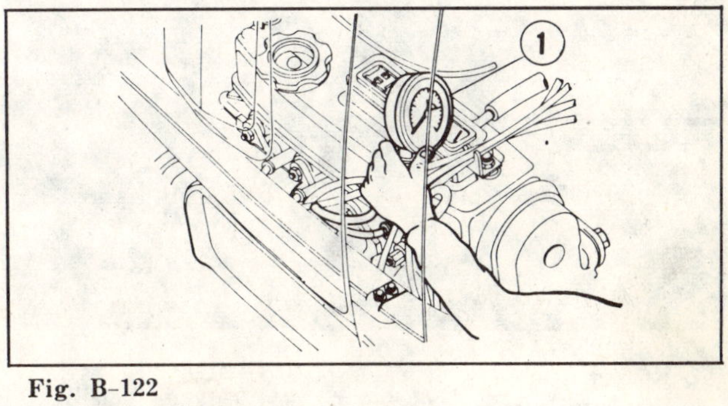

Fig. B-122
① Compression gauge

NOTES:

1) This measurement must be made when the engine is warm after the valve tappet clearance is adjusted.
2) Place the throttle valve and choke valve in FULLY-OPEN position. (Fully depress the accelerator pedal and push the choke button completely.)

If the pressure is beyond specification, it is an indication that carbon deposits are on the combustion chamber walls or on the piston heads. If the pressure is below specification, compressed gases leak from the valves, piston rings or from the cylinder head gasket.

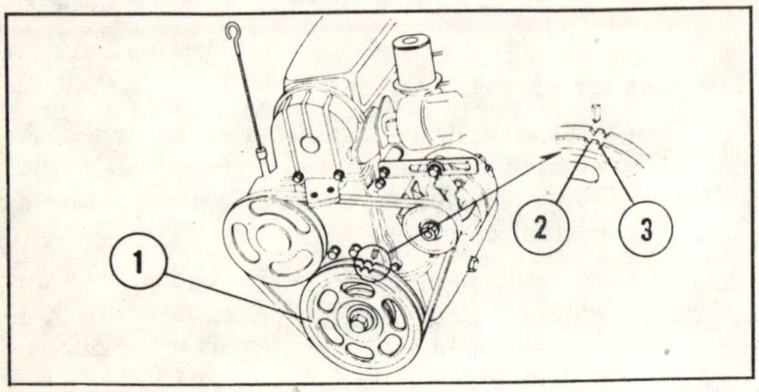

Fig. B–123
① Crankshaft pulley ② Mark "F" ③ Mark "T"

Fig. B–124
① Feeler gauge ② Adjusting screw ③ Lock nut

Fig. B–125
① Rocker arm
② Standard type screwdriver
③ Tappet lock nut wrench
④ Clearance (large)
⑤ Clearance (small)

a–7. Inspect the valve clearance

Too large a valve clearance may cause insufficient output, increased fuel consumption, noisy tappet operation or damaged valve. Too small a clearance may cause insufficient compression, erratic engine idling or backfiring.

Inspection

Rotate the crankshaft pulley as shown in Fig. B–123 until the mark "T" on the crankshaft pulley is aligned with the index mark.

When the piston is at the top dead center position on its compression stroke (when the rocker arm is not lifted at both the intake and exhaust sides).

In such a condition, measure the clearance between the valve stem and rocker arm with a feeler gauge.

- When No. 1 cylinder is at the top dead center position on its compression stroke, the valve clearance can be inspected on the following valves:

  Intake valves of Nos. 1 and 2 cylinders.

  Exhaust valves of Nos. 1 and 3 cylinders.
- When the crankshaft is further rotated one turn:

  Intake valves of Nos. 3 and 4 cylinders.

  Exhaust valves of Nos. 2 and 4 cylinders.

Unit: mm (in.)

Valve clearance	0.10~0.14 (0.004~0.006) (in cold setting)

Adjustment

1. Loosen the rocker arm lock nut and turn the adjusting screw to adjust the valve tappet clearance.

NOTES:

1) When the lock nut is tightened, the correct adjustment will be disturbed. Therefore, tighten the lock nut with the adjusting screw held with a screwdriver.
2) After tightening the lock nut, again check the clearance with a feeler gauge.
3) After inspecting and adjusting, install the cylinder head cover in place and operate the engine to check for leakage of gases from around cylinder head cover or noisy tappet operation.

a-8. Inspect the timing belt

The Honda 120 uses a timing belt instead of a cam chain. The adjustment of the belt tension is required only on the initial periodic inspection.

Too tight a belt tension may result in shortened belt life.

1. Remove the cylinder head cover.
2. Remove the timing belt upper cover.
3. Check the timing belt for damage while rotating the crankshaft pulley slowly. Especially check the belt teeth for peeling, and replace the belt if peeled.

NOTE:

About six turns of the crankshaft pulley corresponds to one round of the belt.

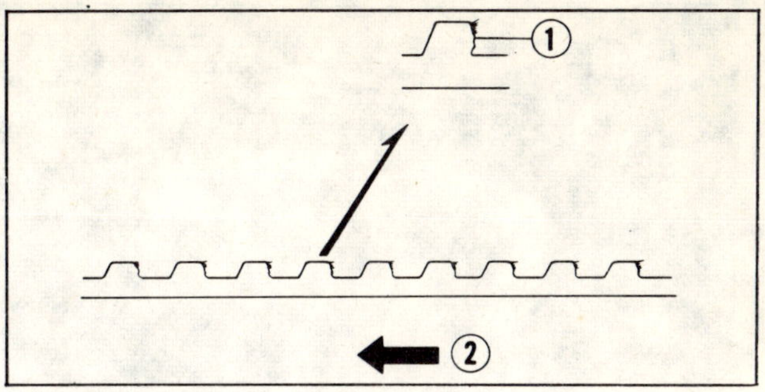

Fig. B-126
① Peeled teeth ② Direction of drive of belt

Adjustment

This adjustment can be made by loosening the tensioner lock bolt and adjusting bolt. After adjusting, tighten the bolts to the specified torque (3.5~4.0 kg-m or 25~29 lb-ft).

NOTE:

When adjusting, do not use any force other than the adjuster spring, otherwise, too tight a belt tension or improper belt pitch may be resulted.

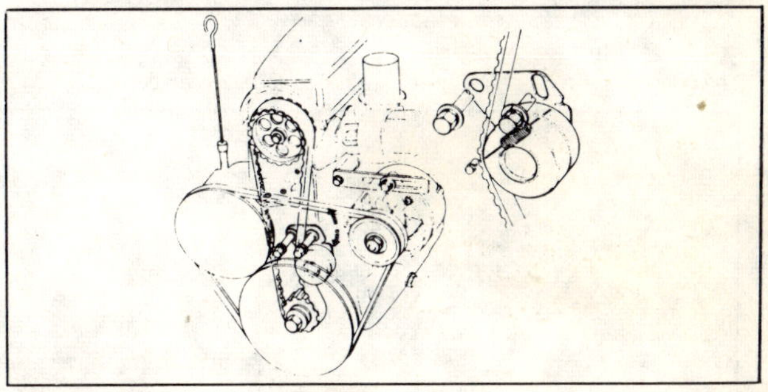

Fig. B-127

Replacement

When replacing the belt, pay attention to the following points:

1. Before removing the belt, chalk or paint the mark indicating the direction of drive on the belt to facilitate reinstallation.
2. Do not allow oil to come in contact with the belt to prevent the rubber from swelling.
3. Do not bend the belt within 13 mm (0.512 in.) or smaller radius since it is made up of glass fiber.
4. Align the matching mark on the crankshaft pulley with the mark on the crankcase and bring No. 1 and No. 4 cylinders at the top dead center position. Misalignment may allow the piston to interfere with the valve.
5. When removing or installing the belt, do not use any tool having a sharp edge such as a screwdriver to prevent damage to the belt.
6. When installing the belt on the driven pulley, rotate the crankshaft pulley to have the job done better. Do not disturb the timing.

b. Reburning of blow-by gases

b-1. Inspect the piping

Disconnect the breather tube and drain tube and check them for restriction or damage. Clean them if restricted. One end of the drain tube has a slit which serves as a valve.

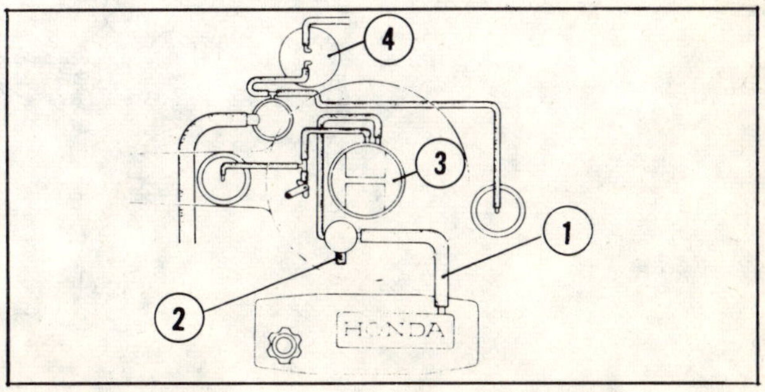

Fig. B-128
① Breather tube ③ Check valve
② Drain tube ④ Canister

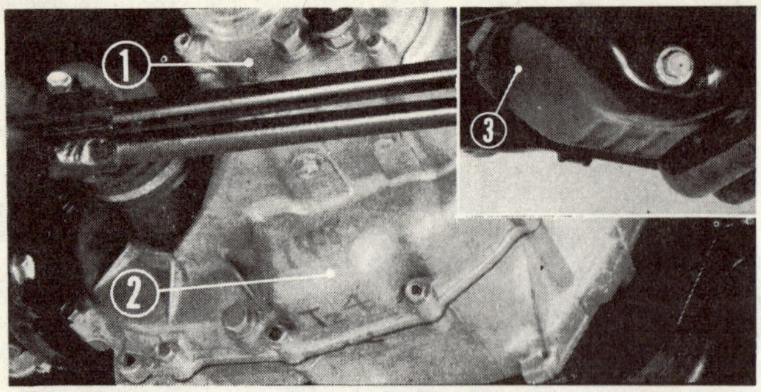

Fig. B-129
① Transmission cover ② Transmission case
③ Oil pan

Fig. B-130
① Drain plug ② Upper level ③ Lower level

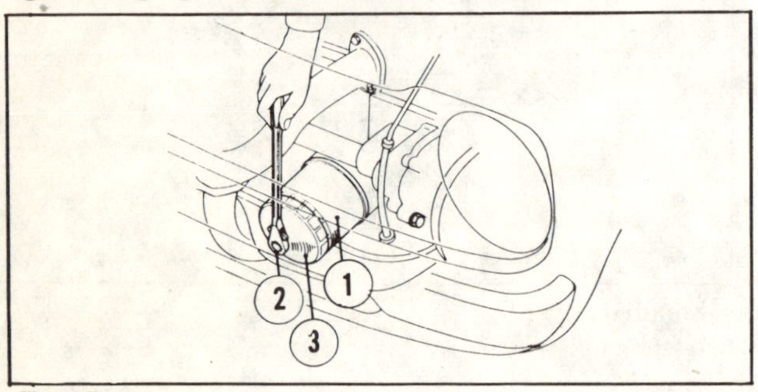

Fig. B-131
① Oil filter ② Ratchet wrench
③ Oil filter removing wrench

c. Lubricating system

c-1. Inspect for oil leaks

Check the crankcase, cylinder head, cam case, side covers, transmission case and oil pan for oil leaks.

c-2. Inspect the oil level

Remove the oil level gauge and check for dirty oil. Wipe off the level gauge with waste cloth and insert it to check the oil level. If necessary, add oil up to the upper level.

c-3. Change the oil

1. After warming up the engine, remove the drain plug and drain oil. Removal of the filler cap may facilitate draining.
2. Refill the crankcase with new oil through the filler cap.
   Capacity (up to upper level): 30 ℓ (0.792 gal)
3. When the oil is excessively dirty, replace the oil filter, too.
4. It is advisable to replace the oil filter at regular intervals.

c-4. Replace the oil filter

Replace the oil filter as shown in Fig. B-131, using an oil filter removing wrench. After replacing, operate the engine and check the oil level again.

d. Fuel system

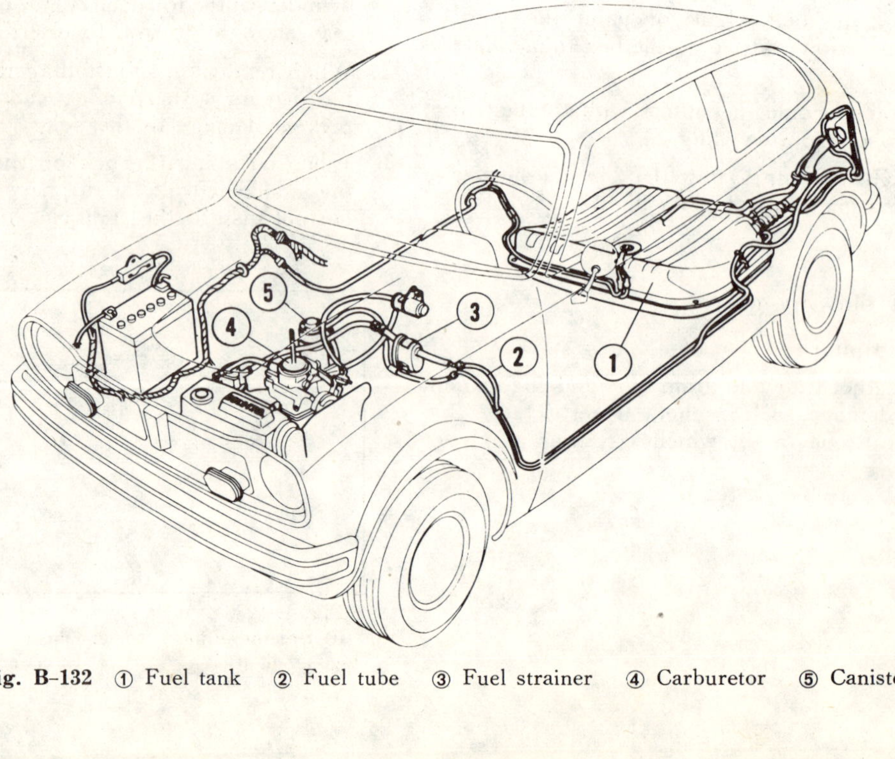

Fig. B-132 ① Fuel tank ② Fuel tube ③ Fuel strainer ④ Carburetor ⑤ Canister

d-1. Inspect for fuel leaks

1. Check the fuel tank, fuel pump, fuel strainer, carburetor, canister, fuel pipe, breather tube and check valve for fuel leaks.
2. Check the ports and joints on the carburetor and joints on the fuel tubes for cracks.
3. Check for leaks from the fuel filler cap when the tank is topped up. Also check the filler cap rubber for cracks, and replace the cap if necessary.

d-2. Inspect the carburetor

Check to see if the inside of the carburetor is dirty, and disassemble and clean the carburetor if necessary.

d-3. Inspect the carburetor linkage

1. Check the throttle valve, choke valve and fast idling mechanism for excessive play, binding or damage.
2. Check the throttle control cable for wear, binding or damage.
3. Check the carburetor linkage for abnormal condition due to freezing. Also check the throttle control cable boot for breakage and replace it if necessary.

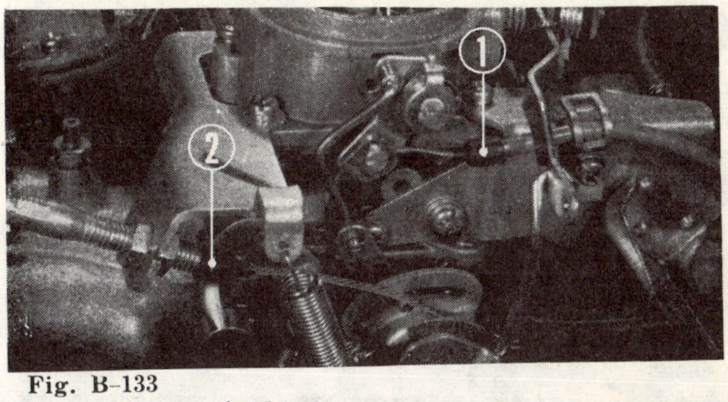

Fig. B-133
① Choke control cable boot
② Throttle control cable boot

4. Suck the vacuum tube as shown in Fig. B-134 and check the air cleaner intake air heating control valve for proper operation. (To facilitate sucking, remove the check valve.)

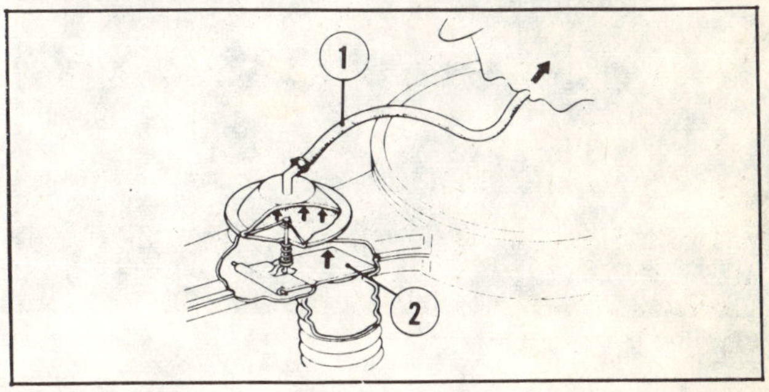

Fig. B-134
① Vaccum tube ② Control valve

d-4. Inspect the throttle valve and choke valve

1. Depress the accelerator pedal and check if the throttle valve opens fully.
2. Fully pull the choke button and operate the relief control lever by hand to check if the relief valve opens fully.
3. Move the throttle valve and choke valve shafts to check for excessive play.
4. Throttle valve must open fully when the accelerator pedal is fully depressed and the valve must be fully closed when the pedal is released.
5. Check to see if the secondary throttle valve opens fully when the primary valve opens fully. (The adjustment can be made by bending the connecting rod.)

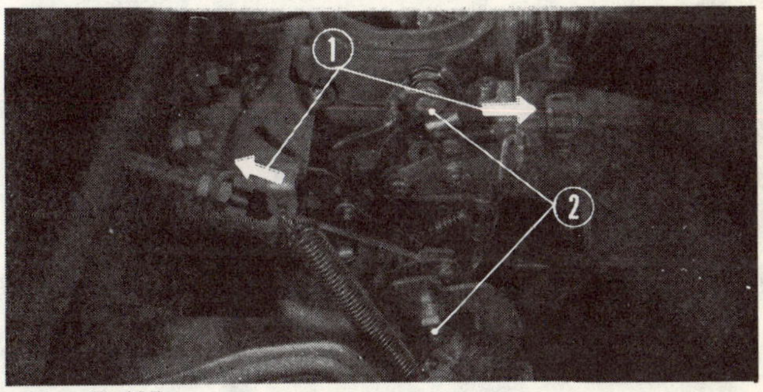

Fig. B-135
① Fully pull ② Checking for excessive play

d-5. Inspect the float level

1. If overflow, excessive fuel consumption or black exhaust gases are noted, check the float level.
2. The float level must be at the position marked (●) on the level mark glass when the engine is running.
3. To adjust the float level, bend the float seat (See page 161). After adjusting, again check the float level while the engine is running.

Fig. B-136
① Float level line

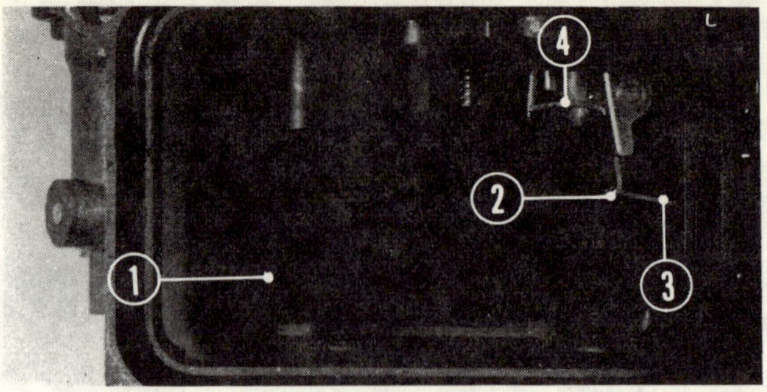

Fig. B-137
① Float ③ Stopper
② Float arm ④ Float valve seat

d-6. Adjust the carburetor idle speed

Adjustment of idle speed. (See page B-38).

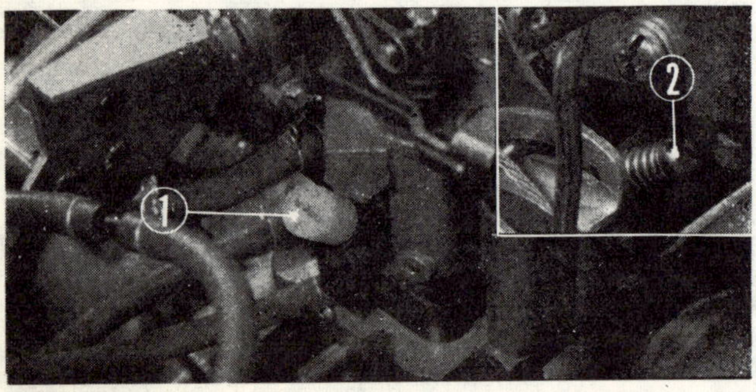

Fig. B-138
① Idle limiter ② Throttle stop screw

After adjusting, rev up the engine by depressing the accelerator pedal two or three times and check to see if the specified idle speed is maintained.

After adjusting, measure the C0 content of exhaust gases.

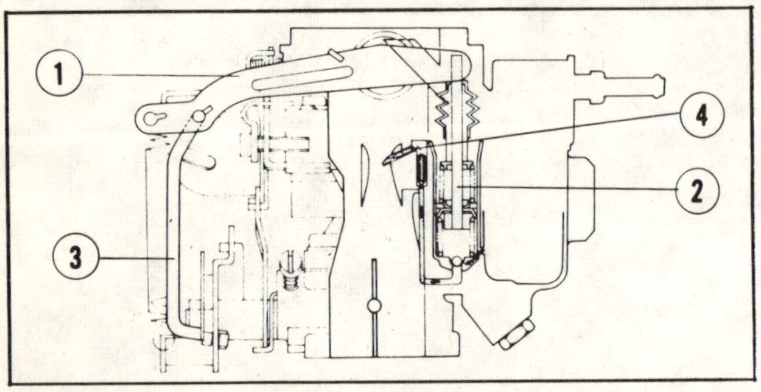

Fig. B-139
① Pump lever ③ Pump rod
② Accelerator pump piston ④ Pump ejector

● Accelerator pump displacement

1. Inspection of the pump displacement must be made with the air cleaner removed and the engine stopped.
2. Check the pump for smooth operation. See if fuel spurts out of the pump discharge nozzle when the throttle valve is moved from FULLY-CLOSED position to FULLY-OPEN position quickly.

   When the pump is operated slowly, fuel must be ocntinuously forced out until the pump comes to the end of its stroke.

3. If the pump is found to be defective, disassemble it and check for restrictions. On cars with carburetors, adjust the displacement by changing the position of the pump connecting rod if acceleration is improper or if starting is poor in cold weather.

In hot weather: Outside position (decrease)

In cold weather: Inside position (increase)

● Fast idle speed

(See page 161)

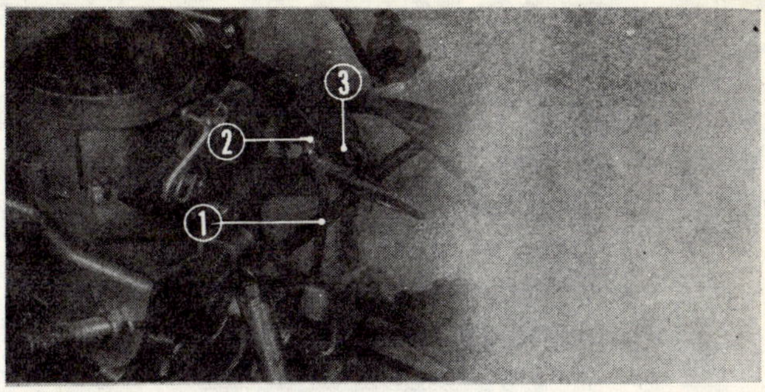

Fig. B-140
① Pump connecting rod ② Increase ③ Decrease

d-7. Inspect the fuel strainer

1. Pull out the fuel tube and remove the fuel strainer.
2. Replace the strainer if excessively dirty.
3. If water is found in the strainer, blow it down by directing a blast of compressed air.

NOTE:

When removing the strainer, prevent fuel from spilling out by using waste cloth.

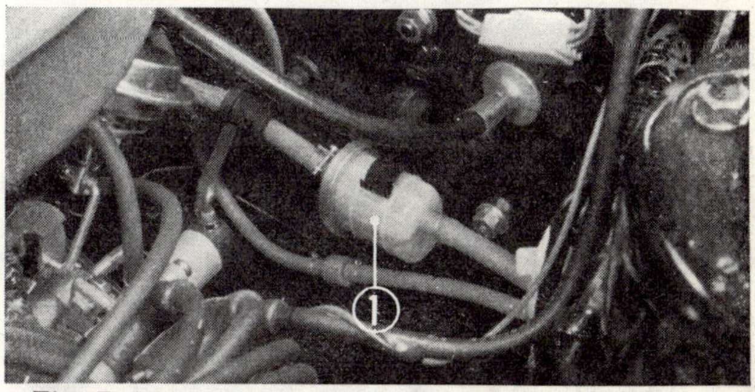

Fig. B-141
① Fuel strainer

d-8. Inspect the fuel tube

Check the fuel tube for cracks or any other damage. Especially check the tube at the carburetor inlet port for obstruction.

Fig. B-142
① Fuel tube joint

e. Inspect the cooling system

The radiator of the Honda CIVIC is provided with a spare Factory-filled radiator coolant contained is not needed to change for about two years.

e-1. Operation of thermostat

1. If the thermostat valve starts opening at room temperature, replace the thermostat.

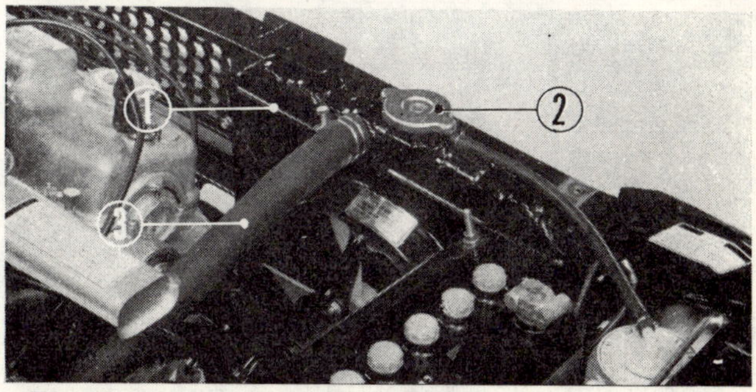

Fig. B-143
① Radiator ② Radiator filler cap ③ Water hose

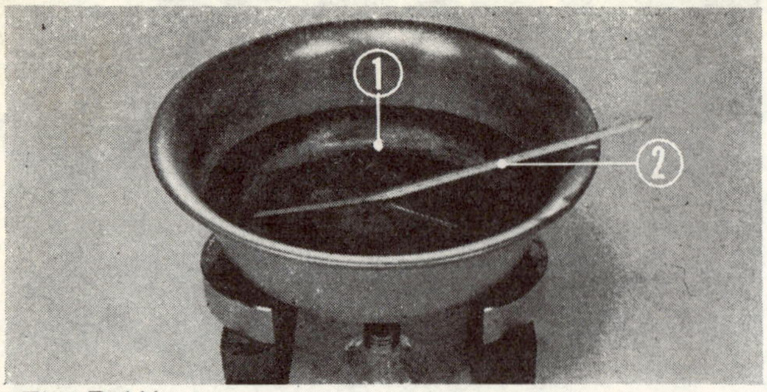

Fig. B-144
① Thermostat ② Thermometer

2. Immerse the thermostat in water and test it for operation by heating the water. The sensing part of the thermostat is small in area.

   Immerse the thermostat in water heated to about 97°C (207°F) for about five minutes, and measure the valve stroke.

   See the Maintenance Standard.

e–2. Radiator and hoses

1. Check the weldments of the upper or lower radiator tank and core for leaks. Repair a damaged part, if any, by soldering.
2. If more than 20% of heat dissipating area of the radiator is clogged, replace the radiator.
3. Check the upper and lower water hoses, especially around their bands, for cracks or deterioration. It is advisable to replace a hose earlier than at regular service time.

Fig. B-145

e–3. Repair bend of the radiator fins

Repair a bent fin with a pliers or a standard type screwdriver, if any.

NOTE:

If many fins are bent, the radiator efficiency will be reduced, resulting in engine overheating.

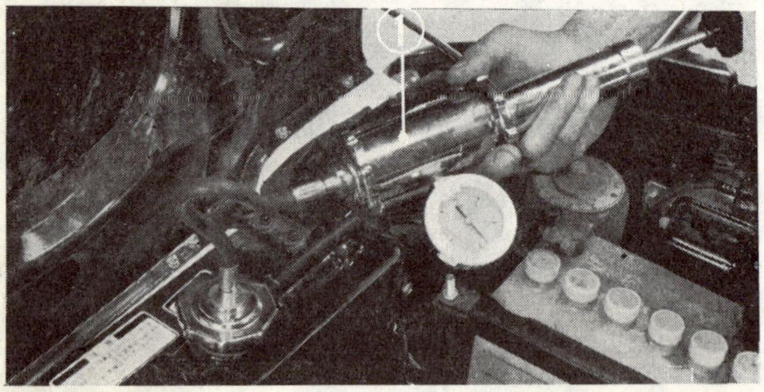

Fig. B-146
① Cap tester

e–4. Inspect the radiator for leaks

The leaks can be checked with the radiator mounted on the car by using a radiator filler cap tester.

1. Warm up the engine and stop it. Then remove the radiator filler cap.
2. Refill the radiator with cooling water up to the filler neck.
3. Attach the cap tester to the filler cap.
4. Pump up the cooling water to the specified pressure with the tester and check for leaks.

   NOTES:

   1) If the pressure drops in a short period of time, improper installation of the tester or leaks may be the cause.
   2) When checking for leaks with the radiator off the car, first plug ports in the radiator and put the radiator in a water tub. Then send air into the radiator through the reserve tank tube and check to see if air bubbles are noted. The leaks can also be checked by using the pressure of city water.

e-5. Inspect the radiator filler cap for condition

1. Attach the filler cap to the radiator filler cap tester and pump it up. If the pressure gauge does not indicate any air leaks from the cap for about six seconds, the cap is normal.

NOTE:

Before attaching the cap to the tester, wet the seal with water.

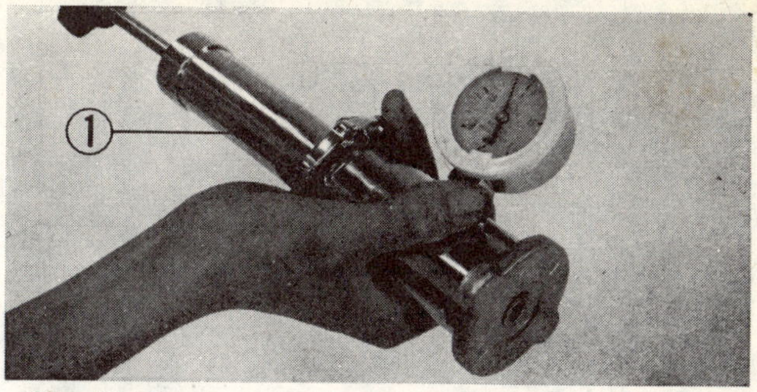

Fig. B-147
① Cap tester

e-6. Change of cooling water

NOTES:

1) When the temperature of the cooling water is deemed to have exceeded 100°C (212°F) (for example, after driving in summer), do not remove the radiator filler cap; otherwise the pressure drops suddenly, causing hot water to spout out.
2) Do dot allow radiator fluid to come in contact with painted surfaces. Wash or wipe the surface with a waste cloth upon noting that the fluid comes in contact with the surface.

1. Drain the radiator by loosening the drain cock. Also drain the spare tank.
2. Refill the radiator with city water (soft or treated water) containing up to the filler neck. At this time, loosen the intake manifold breathing water bolt to bleed the cooling system. When air bubbles are no longer noted in the cooling water coming out of the air vent hole, tighten the water bolt to the proper specifications.
3. Place the heater TEMP lever in FULLY-OPEN (HI) position.
4. Let the engine idle at 700 to 800 rpm for about 10 minutes and check to see if the thermostat valve opens. Place the hand on the upper hose to check temperature. If you feel hot, the thermostat valve is found to be open.
5. Stop the engine and remove the radiator filler cap to check the water level.

   The level will lower when the cooling system is bled. Again refill up to the filler neck and also top up the spare tank.
6. Again run the engine and check to be sure the water level does not lower. If the water level further lowers, it is an indication that the cooling system is not bled completely. Continue to bleed the system until the water level will not lower.
7. When the level stops lowering, install the radiator filler cap securely.

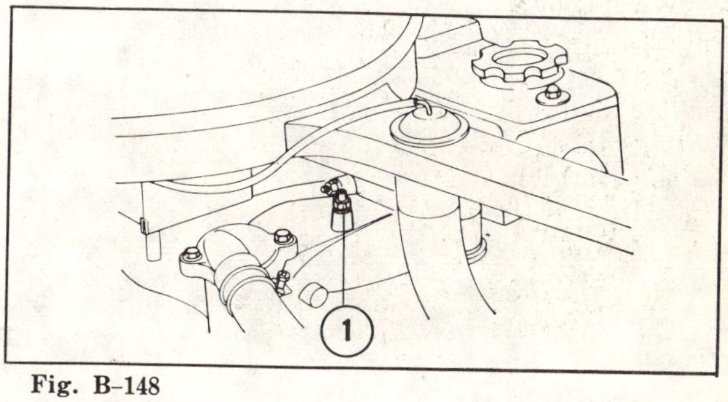

Fig. B-148
① Bleeder screw

Precautions for handling the radiator fluid

1. Do not mix radiator fluids of different brands.
2. Never drink the radiator fluid since it is poisonous.
3. Use the mixing ratio for the minimum operating temperature plus −5°C (−9°F).
4. Any particular brand of the coolant is not recommended, but high quality, anti-freeze coolant, which has been specially developed for aluminum engine, should be used.

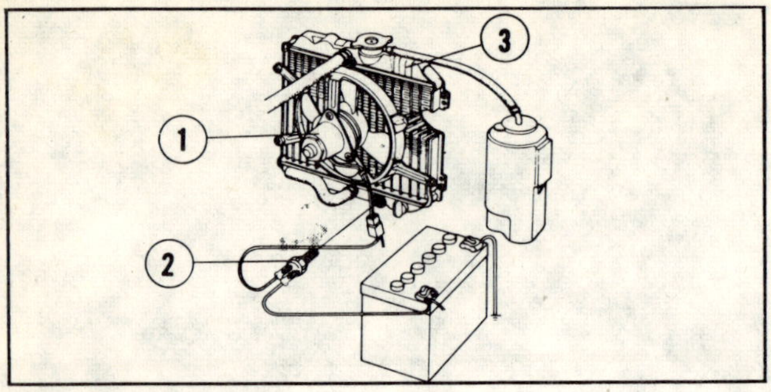

Fig. B-149
① Cooling fan motor ③ Radiator
② Cooling fan switch

e-7. Inspect the cooling fan

The cooling fan is provided with a switch which turns on or off according to the temperature of cooling water. It starts to operate when the water temperature reaches 88.5~91.5°C (191~197°F). If the fan does not operate and cooling water overflows when the engine is running at idle speeds and the water temperature gauge pointer stays in the red zone on the scale, the switch is suspected to be defective.

Remove the switch and test it by heating to a temperature of 88.5~91.5°C. Check for continuity with a radio tester. The switch is in good condition if there is any continuity at this temperature or if there is no continuity at a temperature of 83.5~86.5°C (182~188°F).

a. Lighting system

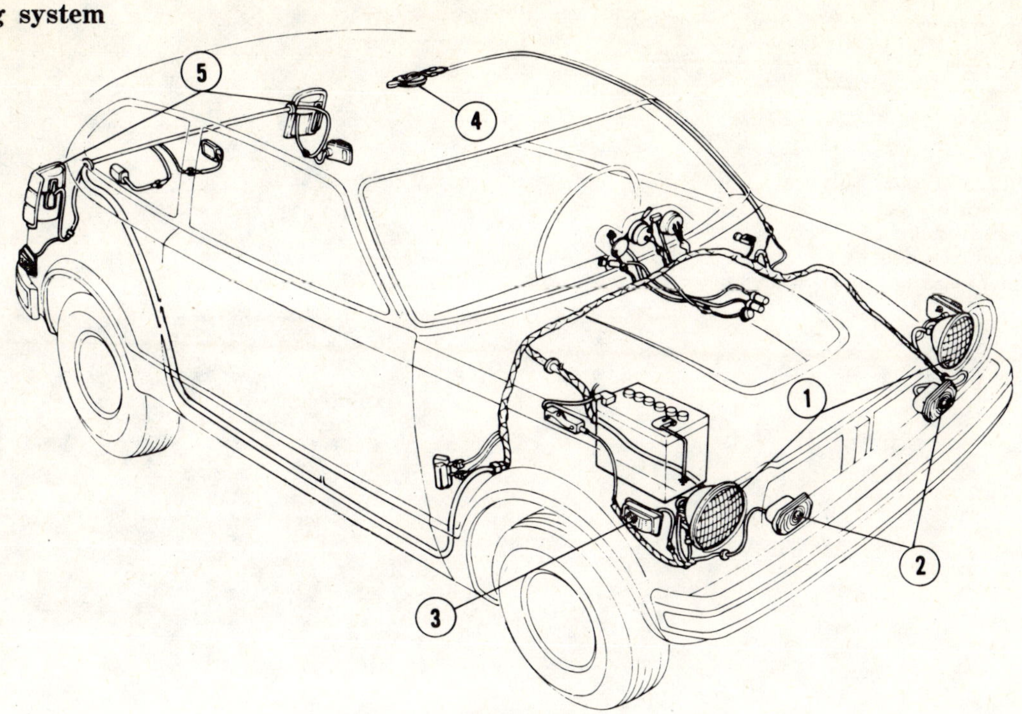

Fig. B-150

① Head light
② Front combination light
③ Side marker light
④ Room light
⑤ Rear combination light

a-1. Inspect the operation

1. Check the head lights, front and rear combination lights, backup lights, side turn signal lights, room light and meter lights for proper operation.
2. Check the head lights and accessory lights for proper installation. Also check the head light beams for proper aiming and adjust them if necessary.

a-2. Inspect the lenses

Check each light lens for breakage, dirt, blur or any other damage, and clean or replace it if necessary.

a-3. Head light aiming

Head light alignment may vary due to local regulations. Vehicles exported to U.S.A. are adjusted as follows: For specifications of the other type, refer to local regulations. According to local regulation, head light aiming adjustment will be more liberal, however, it is desirable to follow the procedure given above in consideration of safety driving.

1. Park the car in front of the wall screen with 7.62 m (25 ft) distance from the screen. Then level the car by bouncing both the front and rear. Note the head light aiming is conducted with literes fuel (less than a half-fuel) in the tank, with a driver seated in the car, the car unloaded except for the spare tire and tools. Ensure that the tires have the correct inflation pressure.
2. Aim the head light as shown in the picture. Each head-light can be adjusted by means of adjusting screws.

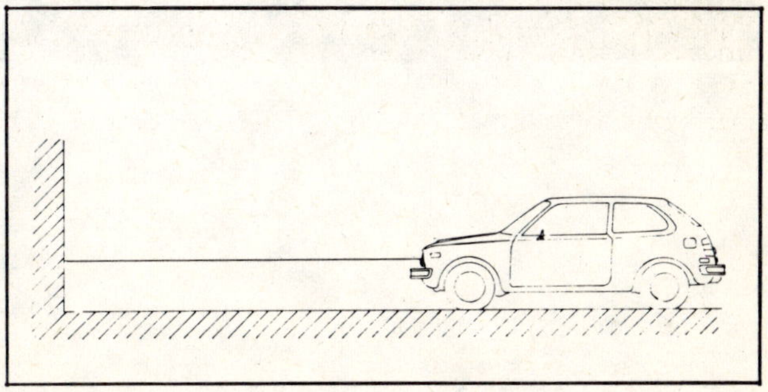

Fig. B-151

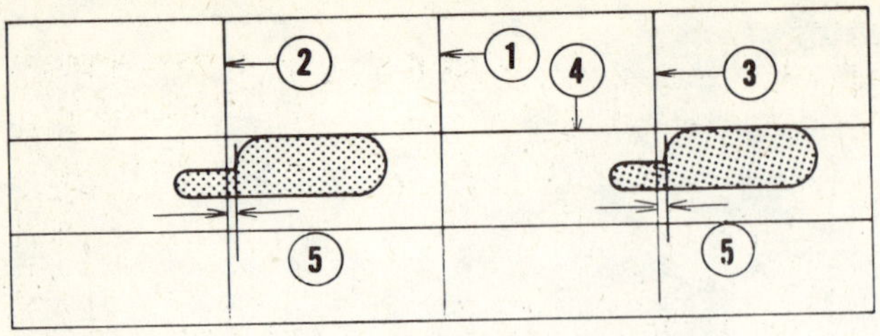

Fig. B–152 Headlight low beam diagram
① Center line of vehicle
② Center line of left headlight
③ Center line of right headlight
④ Line of adjustment and horizontal center line of headlight ⑤ 2 in.

Fig. B–153
① Retaining screw

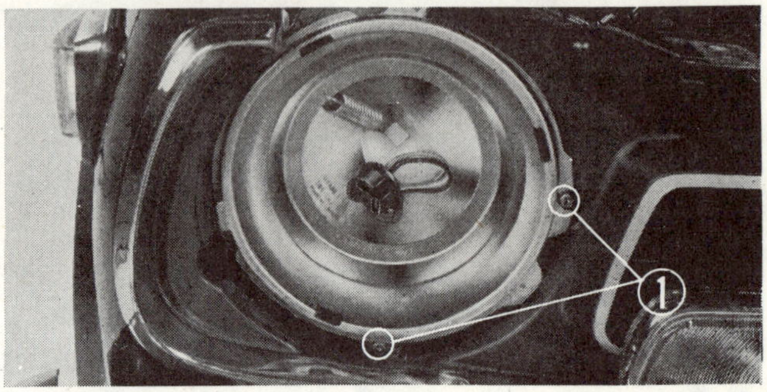

Fig. B–154
① Adjusting screw

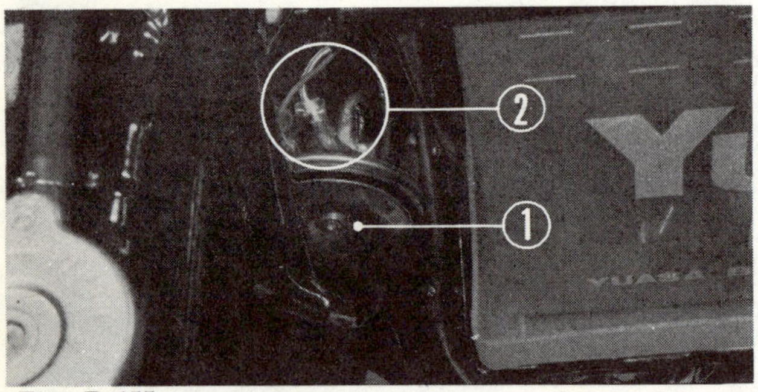

Fig. B–155
① Horn ② Horn cord connection

Fig. B–156
① Front windshield wiper arms

b. Horn, turn signal lights and windshield wipers

b–1. Inspect the horn

1. Push the horn pad to check the horn for proper operation.
2. If the horn fails to sound, remove the horn pad and check contact between the retainer and horn seat for condition. Connect the horn cord securely and replace it if defective.

b–2. Inspect the turn signal lights

Operate the turn signal control lever and check the right and left turn signal lights for proper flashing. If the number of flashes is below 50 per minute, check the turn signal relay for condition. If the number of flashes is over 120, also check the relay and bulbs for condition. The specified number of flashes per minute is 85.

b–3. Inspect the windshield wipers

1. Check to see if the windshield wiper blades are installed 10~20 mm (0.4~0.8 in) just above the lower part of the weather strip at the stopped position. If the arms are installed in wrong positions, adjust them by changing the engagement of the wiper arm serration.
2. Check the wipers for operation after spraying water on the window glasses. Do not operate the windshield wipers without water on the glasses since it may scratch the glasses or wiper blades or damage the motor.

c. Inspect the rear view mirror and door mirrors

1. Check the mounting of the rear view mirror for looseness. Also check to see if those mirrors are installed at correct angles as viewed from the driver in seat.
2. Check the mirrors for dirt or rust.

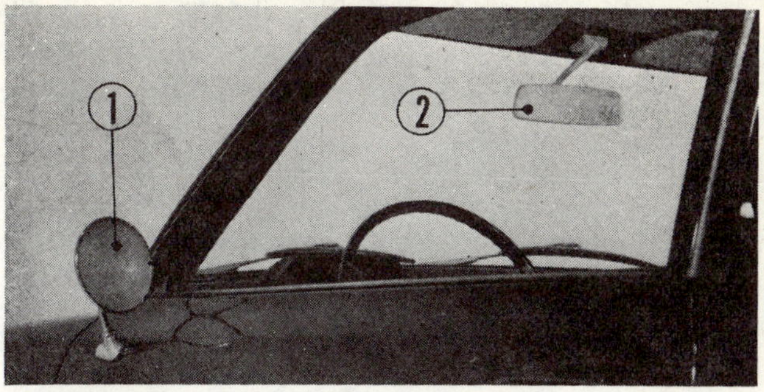

Fig. B-157
① Door mirror ② Rear view mirror

d. Reflectors and license number plate

d-1. Inspect for dirt or damage

1. Check the reflectors and license number plate for dirt or damage and replace them if necessary.

Fig. B-158 ① License number plate ② Reflector

e. Meters

e-1. Inspect for operation

1. Turn the ignition switch on and check if the battery charge light and parking light (with parking brake lever pulled up) come on or if the fuel meter pointer oscillates.
   - If the lights fail to come on, check the bulbs for damage, the wiring for connection, or the fuses for blowing.
2. Operate the engine and check if the battery charge light goes out. Drive the car and check the speedometer for operation.
   - Reading of the speedometer should be; 38~42 km/h at a car speed of 40 km/h or 39~40 mph at a car speed of 40 mph.

Fig. B-159
① Speedometer ③ Water temperature gauge
② Fuel meter

f. Exhaust pipes and muffler (silencer)

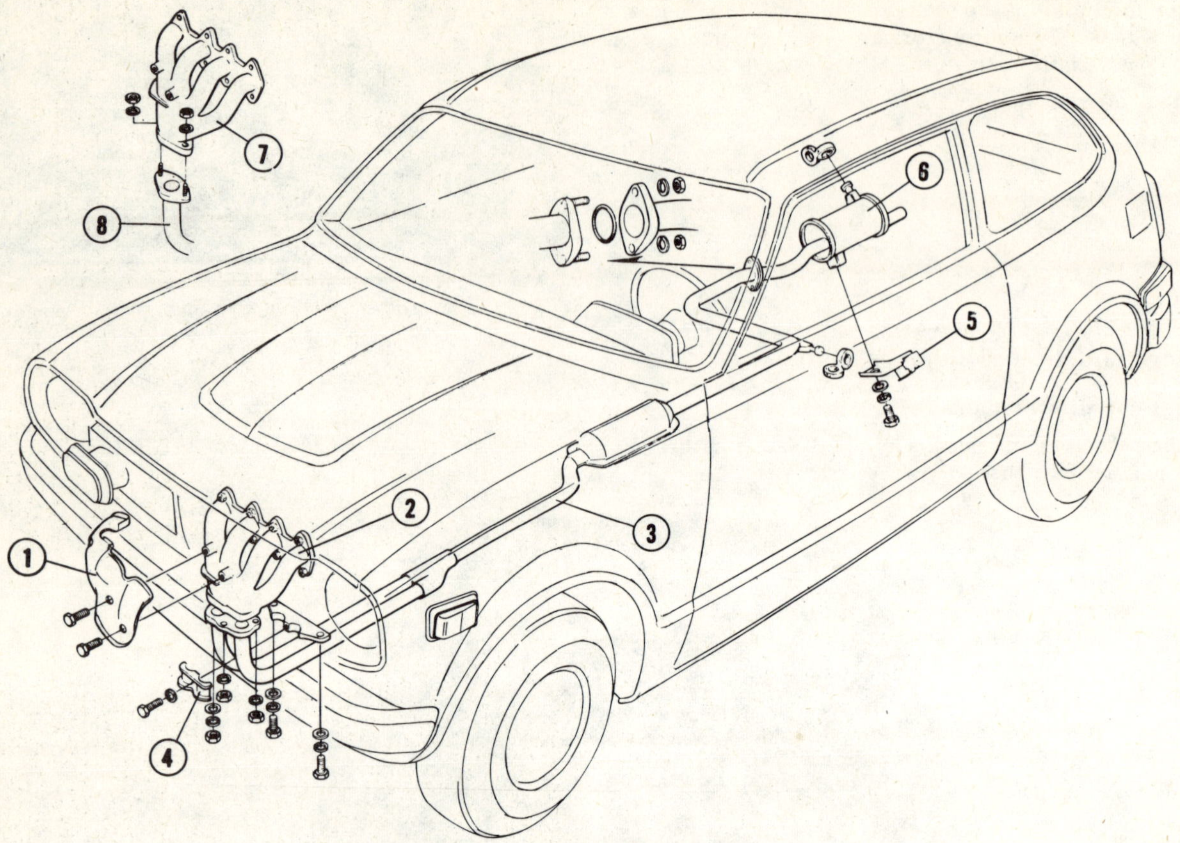

Fig. B-160

① Hot air cover
② Exhaust manifold
③ Exhaust pipe
④ Pipe clamp
⑤ Silencer mount
⑥ Exhaust silencer
⑦ Exhaust manifold
⑧ Exhaust pipe

f-1 Inspect for looseness or damage

1. Check the exhaust manifold for loose connection, cracks or breakage.
2. Check the exhaust pipe and silencer connection for looseness or breakage.
3. Check the exhaust silencer mount rubber for peeling or cracks and the attaching bolts for looseness.
4. Check the exhaust pipe and exhaust silencer welded parts for cracks, corrosion or peeling.

f-2. Inspect the exhaust silencer

Using a sound-level meter, check the exhaust silencer for function. Replace the silencer if excessively noisy or corroded.

Specifications of noise:

1. Drive the car at a speed of 35 km/h (22 mph) on level pavement and detect noise with the detector of a sound-level meter placed at a point 7 m (23 ft) apart from the center of the car, on the left side as viewed in the direction of drive. When detecting, keep the detector face at right angles to the car center. The noise must be below 70 phons on A scale.
2. When the engine is running at a speed of 3300 rpm (60% of the rpm at the maximum engine output) under no load, detect the noise with the detector placed at a point apart 20 m (66 ft) from the rear end of the tail pipe. The noise must be below 70 phons on A scale.

g. Inspect the locks

g-1. Door locks

Inspection

1. Check to see if the door lock smoothly engages the striker.
2. Lock the door and check for rattle.
   If the door is improperly locked, adjust it with shims or by means of hinge bolts. Then move the striker in either direction for adjustment of engagement.

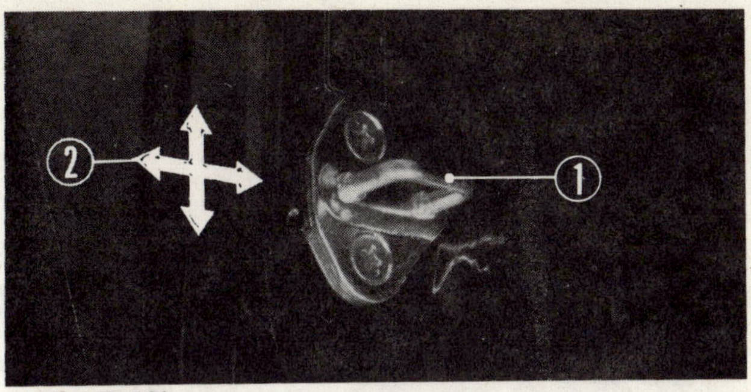

Fig. B-161
① Door striker ② Directions of adjustment

g-2. Engine hood lock

Inspection

Lock and open the engine hood to check for condition and adjust it if necessary.

● Adjustment of strikers

If the engine hood is not opened or closed properly, loosen the striker bolts and adjust the striker to the correct position, by sliding the striker to right or left, or the lock up or down. After adjusting, apply a coat of grease to the striker.

Fig. B-162
① Engine hood lock

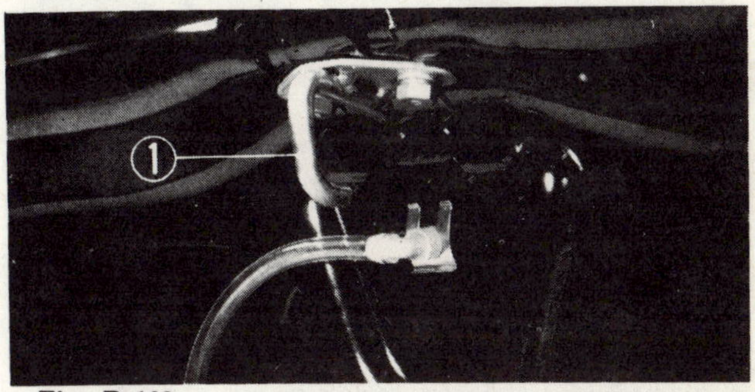

Fig. B-163
① Striker

g-3. Inspect the trunk lid lock

1. Lightly push the trunk lid to check for proper locking. The lid lock can be moved up and down for adjustment by loosening two lock striker bolts.

   NOTE:

   The tail gate lock can be moved forward and backward for adjustment by loosening two lock striker bolts.

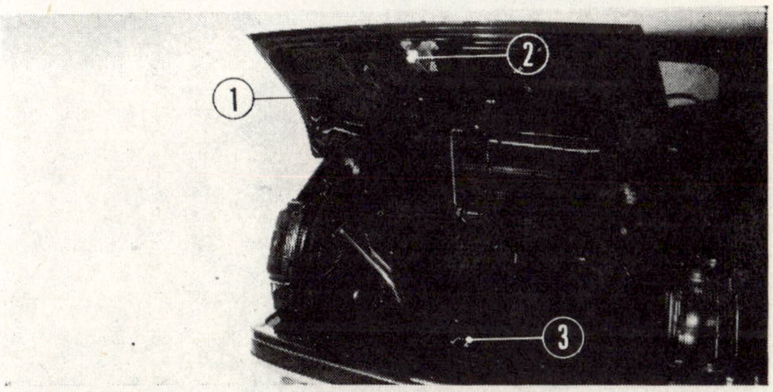

Fig. B-164
① Trunk lid ② Lid lock ③ Striker

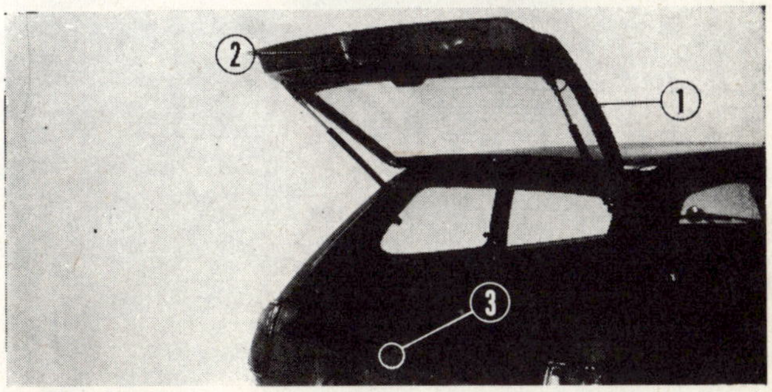

Fig. B-165
① Tail gate ② Gate lock ③ Striker

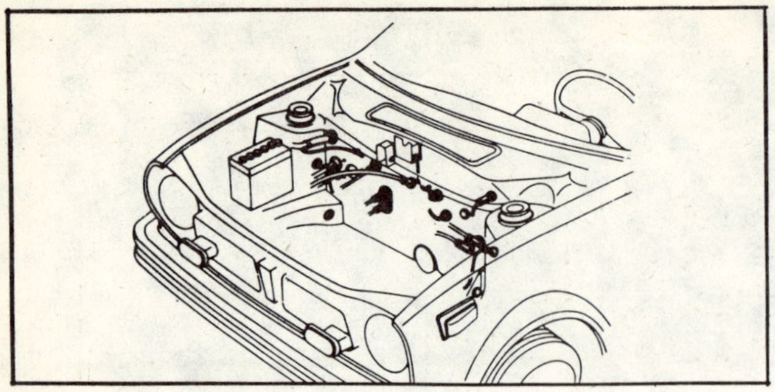

Fig. B-166 Dash board grommets

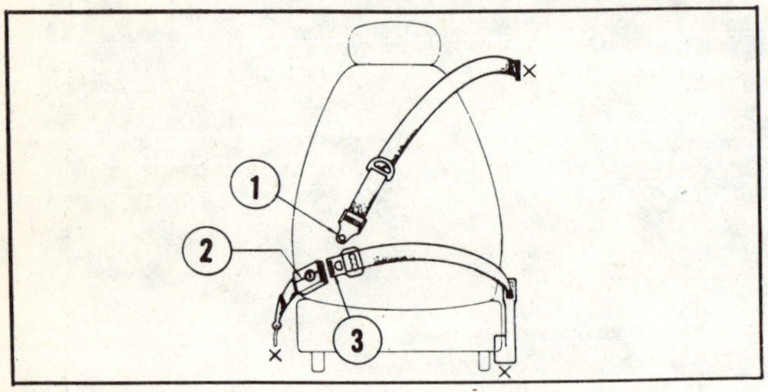

Fig. B-167
① Eye piece ② Buckle ③ Tongue plate

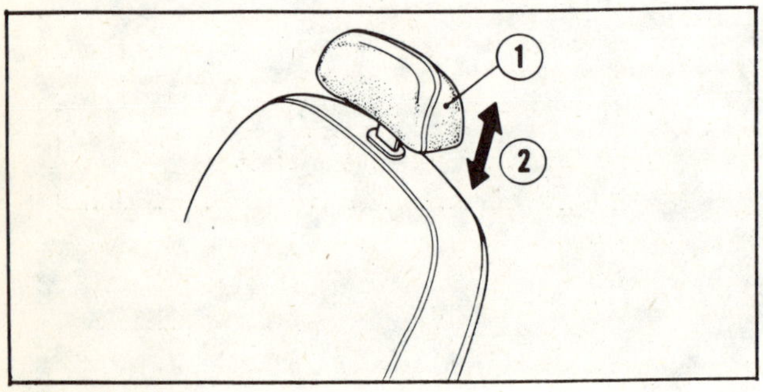

Fig. B-168
① Head rest ② Adjusting

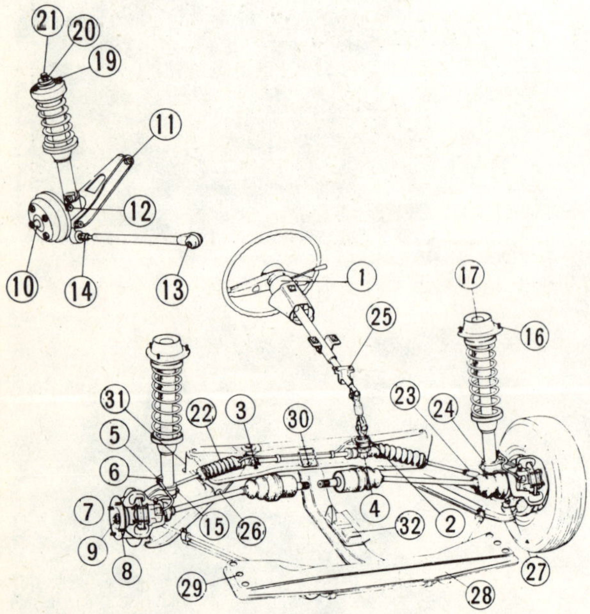

Fig. B-169

h. Dash board grommets

h-1. Inspect for proper installation

Check the grommets for proper installation. Improper installation may result in water leakage or entry of dirt and dust. Replace a grommet if its rubber is cracked. If the grommets are loosely fitted in the dash board holes, repair it with a sealant.

i. Inspect seat belts

1. Check the tongue plate and buckle for proper engagement and the belt for damage.
2. Check the belt anchor bolts for looseness and retighten them if necessary.

j. Inspect the head restraints

1. Check the head restraint for installation or height adjustment.
2. Shake the head rest fore and aft to check for rattle and the garnishes for proper installation.
3. The head restraint height should be adjusted to the ear-level of a seat occupant.

k. Car body

Inspect for looseness or damage

1. Check the front beam, center beam and engine support beam for cracks or any other damage.
2. Check the engine mount, front beam, center beam and engine support beam bolts for looseness.
3. Check the doors, trunk lid (tail gate) and engine hood for damage or loose connection.

See torque specifications at the end of this manual.

Ref. No.	Part	Ref. No.	Part
1	Steering shaft nut	17	Front shock absorber center nut
2	Steering joint bolt	18	Master cylinder
3	Steering gear box	19	Rear shock absorber bolt
4	Rack guide lock nut	20	Rear shock absorber double nut (upper)
5	Brake hose joint	21	Rear shock absorber double nut (lower)
6	Brake pipe flare nut	22	Rack end (in boot)
7	Backing plate mounting bolts	23	Tie rod end lock nut
8	Wheel clip nut	24	Tie rod ball joint
9	Front sprindle nut	25	Steering column
10	Rear axle nut	26	Lower arm
11	Rear lower arm	27	Lower arm ball joint
12	Rear shock absorber and lower arm	28	Lower arm radius rod
13	Rear radius rod and body	29	Front beam
14	Rear radius rod and hub carrier	30	Center beam
15	Front shock absorber lock bolt	31	Engine mount beam
16	Front shock absorber bolt	32	Center beam and engine mount

l. Lubrication of chassis parts

Operate the chassis parts to check for condition. If any part is noisy or inoperative, apply a coat of oil or grease to its sliding surface.

Lubricate the front door hinges. The steering gear box is provided with a grease fitting.

Lubricating method

Lubricate until clean grease comes out.

- The steering gear box can be lubricated by loosening its bellows band and removing the dust seal.

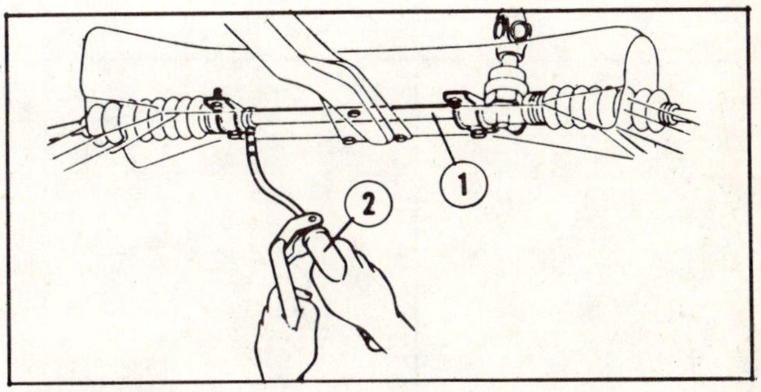

Fig. B–170
① Steering gear box ② Grease pump

m. Optional parts

- Check the electrical cords for proper connection.
- Check the wiring for condition.
- Check to see if the optional parts are properly installed without any sign of abnormal condition.

n. Evaporative emission control system

n-1. Idle cut valve operation

1. Remove the purge tube from point Ⓐ.
3. Attach a tachometer.
3. Start the engine and set to run at 1200 rpm.
4. Cover the point Ⓐ with your finger and check for suction.
   - If there is no suction, check whether the fixed orifice is clogged or not.
     1) Remove the purge tube from point Ⓒ.
     2) Start the engine and let it idle.
     3) Bring your finger close to Ⓒ and check for suction.
   - If there is suction at Ⓒ, replace the idle cut valve.
   - If there is no suction at Ⓒ, clean the fixed orifice with a piece of wire or a blast of air.

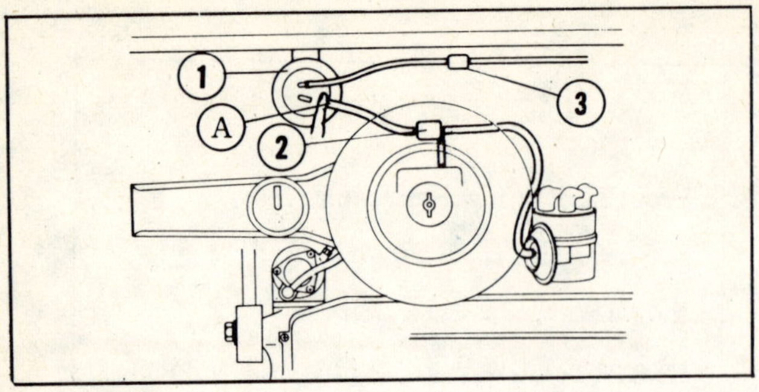

Fig. B-171
① Canister ② Idle cut valve ③ One way valve

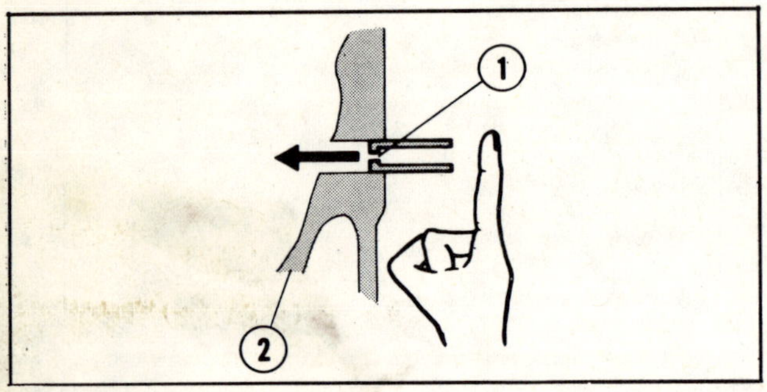

Fig. B-172
① Fixed orifice ② Intake manifold

n-2. Canister inspection

1. Remove the canister purge air guide tube from Ⓑ and purge tubes from Ⓐ and tubes from Ⓓ.
2. Cover the points Ⓐ with your fingers and blow air from the tube Ⓓ. Check that air comes out from Ⓑ.
   - If not, replace the canister.

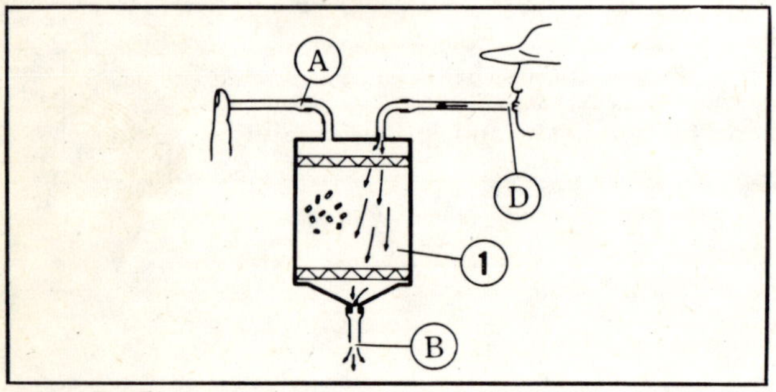

Fig. B-173
① Canister

n-3. One way valve inspection

1. Remove the one way valve from the dash board lower.
2. Blow air lightly from the Ⓣ mark side and check that the air passes through.
3. Check that no air blows through when you blow in from the opposite side.
4. If defective, replace the one way valve.

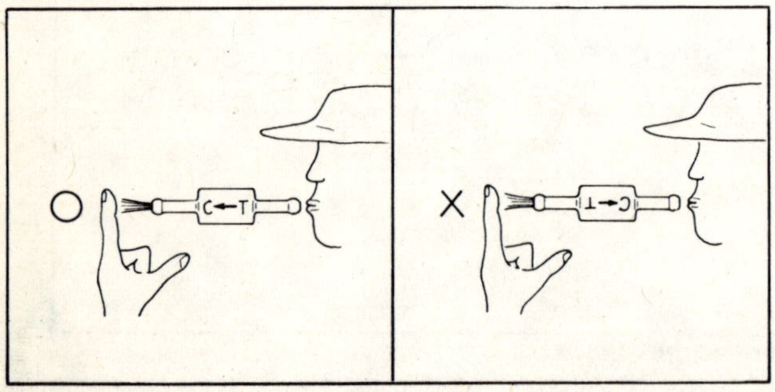

Fig. B-174

n–4. Fuel filler cap inspection

● Inlet relief valve

1. Take the fuel filler cap off the car and install in the cap adapter ⑦.
2. As the cap installed, screw the adapter into the fuel filler cap tester ⑤.
3. Blow the compressed into the accumulator ③ through the inlet ② until the relief valve ④ operates.
4. Connect the part Ⓐ and thc joint ⑨ with a tube. Then install another tube on the part Ⓑ and let the other end into a water bath.
5. Gradually open the control valve ②, and read the pressure on the gauge ① when the bubbling starts in the bath.

   The pressure should be within the following range.

Operation pressure	0.43~1.0 psi (0.03~0.07 kg/cm²)

● Outlet relief valve

1. Take the tubes off.
2. Connect the joint ⑨ and the port Ⓑ with a tube. Then install another tube on the port Ⓐ and let the other end into a water bath.
3. Gradually open the control valve ②, and read the pressure on the gauge ① when the bubbling starts in the bath.

   The pressure should be within the following range.

Operation pressure	1.85~3.27 psi (0.13~0.23 kg/cm²)

If either relief valve operation pressure is not within the described ranges, replace the fuel filler cap.

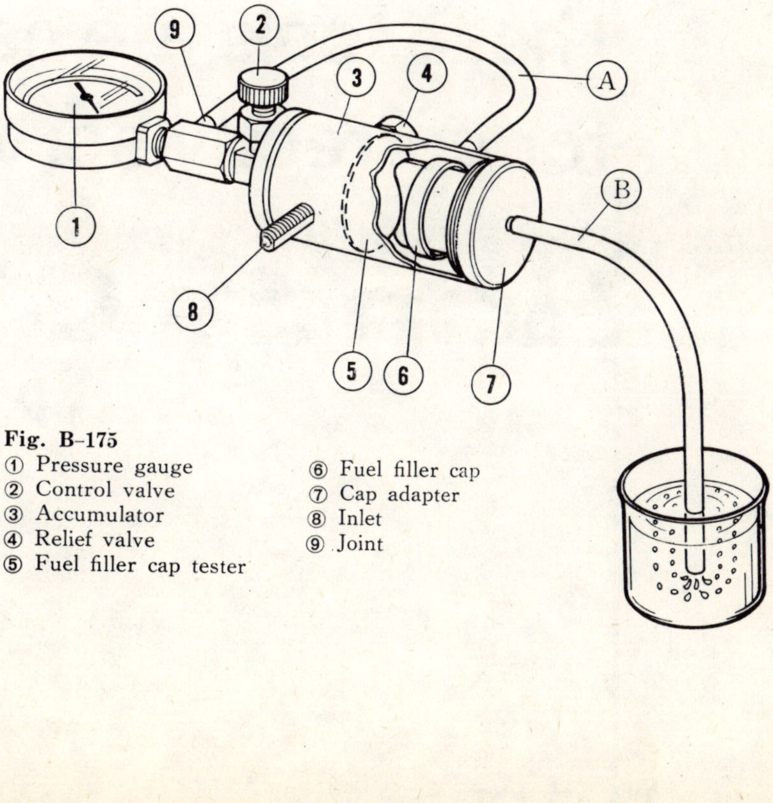

Fig. B–175
① Pressure gauge
② Control valve
③ Accumulator
④ Relief valve
⑤ Fuel filler cap tester
⑥ Fuel filler cap
⑦ Cap adapter
⑧ Inlet
⑨ Joint

SPECIAL TOOLS

Refer to page 266
for Special Tool List

These tools are essential
for correctly servicing the

Honda Civic

ENGINE REMOVAL AND INSTALLATION

Fig. C–1

① Positive cord
② Speedometer drive cable
③ Clutch control cable
④ Throttle control cable
⑤ Ignition cord assembly
⑥ Ground cable

NOTES:

1) When removing or installing the engine, use a garage jack and safety stands in correct position.

   When lifting the engine with a chain block and sling, attach the hook and sling firmly.

2) Use the fender cover to protect the painted surfaces of the car body.

C–1. Removal of Engine

1. Drain the lower radiator tank by loosening the drain cock.

   After draining, tighten the cock.

   • Since the radiator of the Honda CIVIC is of a semi-sealed type, drained water must be kept in a clean container for reuse. To facilitate draining, remove the radiator filler cap.

2. Drain the transmission case by removing the drain bolt; otherwise the oil will flow out when the drive shafts are pulled out and the engine is lifted.

Fig. C–2

3. Remove the front grille.
4. Remove the engine hood by loosening its hinge bolts.

   NOTE:

   After removing the engine hood, store it, exercising care not to damage its painted surface.

5. Disconnect the cord at the battery positive terminal and the ground cable at the transmission case.

Fig. C-3
① Fuel tube

6. Disconnect the tube running between the canister and the carburetor at the canister.
7. Disconnect the fuel tube at the pump inlet port.

NOTE:

When disconnecting the fuel tube, the fuel will flow out. Wipe it off with a clean waste cloth.

8. Disconnect the lower water hose at the connecting tube and the upper hose at the thermostat cover by loosening the hose bands.
9. Disconnect the following control cables and cords from the engine unit:
   - Throttle and choke control cables
   - Clutch control cable
   - Ignition cord assembly
   - Starting motor control cable
   - Back light switch cord
   - Speedometer drive cable
   - Tachometer drive cable
   - AC generator wire and wire harness
   - Thermounit cord
   - Cooling fan motor cord
10. Disconnect the bypass hoses at the H-type joints.
11. Disconnect the engine torque rod.
12. Leave the right and left wheel clip nuts loosened.
13. Fully pull up the parking brake lever. Jack up the front of the car and support it with safety stands.

Fig. C-4
① Engine sling

14. Remove the front wheels and place them under both side sills in back of the safety stands. Should the safety stands accidentally get out of place, the wheels will take load of the car, preventing damage to it.
15. Attach the engine sling to the clutch wire stay and to the cylinder head and attempt to lift the engine just far enough for the chain to take the load.
16. Remove the right engine rubber B.

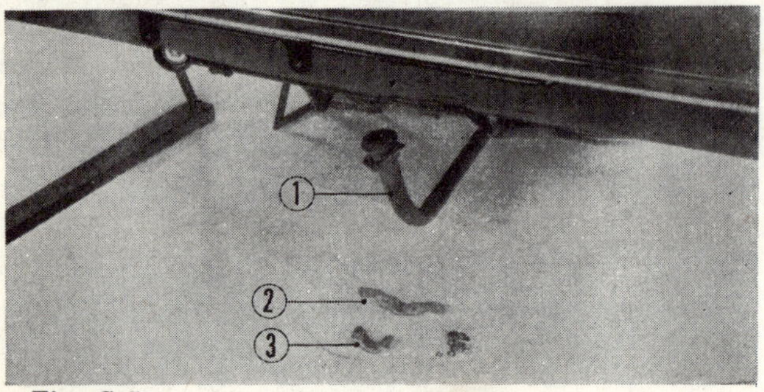

Fig. C-5
① Exhaust pipe ② Exhaust pipe clamp A
③ Exhaust pipe clamp B

17. Remove the exhaust pipe clamps A and B.
18. Remove the exhaust pipe and manifold nuts and leave the pipes lowered.

19. Remove the lower arm and tie rod end ball joints.

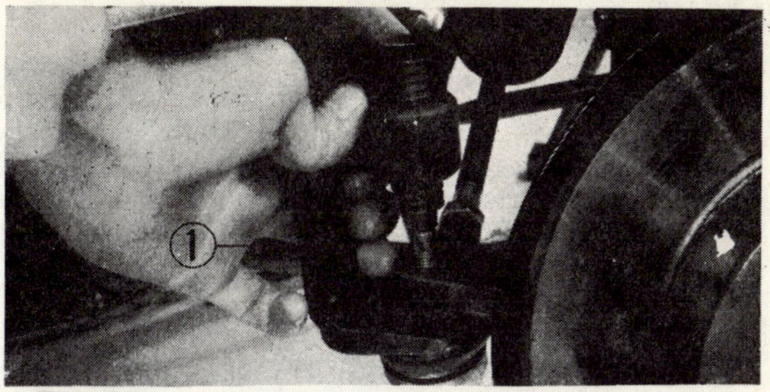

Fig. C-6
① Ball joint puller

20. Hold the disc or backing plate as shown and pull the right and left drive shafts out of the differential case.
   - The right drive shaft can easily be pulled out by inclining the engine.

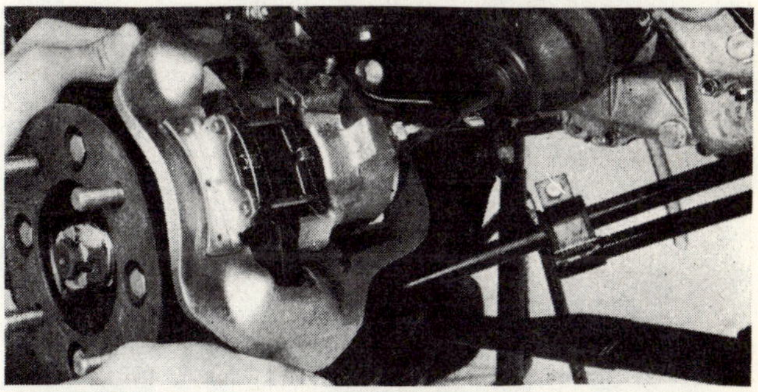

Fig. C-7

21. Pull out the gearshift rod 8 mm (0.315 in.) spring pin with a spring pin driver and disconnect the rod at the transmission case.
22. Disconnect the gearshift extension at the engine.

NOTE:

Do not disconnect the frame-side ends of the gearshift rod and extension.

Fig. C-8
① Gearshift rod ② Spring pin driver

23. Making sure the engine is lifted safely, remove two center bracket nuts.

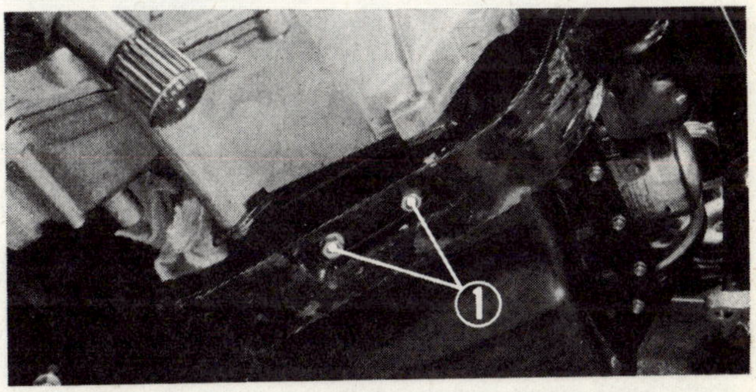

Fig. C-9
① Nuts

24. Stuff the differential case pinion shaft hole with clean waste cloth to keep the oil in and foreign materials out. Then lift the engine slowly, taking care not to allow any engine part to interfere with other parts of the car.

NOTES:

1) When lifting, the engine must be inclined so that the transmission case does not interfere with the radiator, fan motor or battery hanger.
2) When disassembling the engine, apply a thin coat of oil or grease to the machined surfaces of the drive shafts and wrap them with vinyl to protect from corrosion and dirt.

Fig. C-10

Engine Running Test

Before installing the engine to the car body, it must be run-in at about **750~2000** rpm for **20~30** minutes on a engine running test stand. During the test, check for any sign of water or oil leaks, abnormal noise or any other abnormal condition.

NOTE:

When running the engine, plug up the drive shaft holes and place the transmission gears in the neutral position.

Engine Installation

To install the engine, reverse the removal procedures. Pay attention to the following points:

Tighten mount bolts and nuts to the specified torques below.

1. Temporarily tighten the center bracket mount bolts first. Then insert the right and left drive shafts into position and tighten the right mount and torque rod bolts in this order to the specified torques.
2. Use a new spring pin.
3. Make sure all bolts and nuts are tightened to specification. Push the shock absorbers and check to be sure the drive shaft end clips snugly fit in the slots in the differential case.
4. Check to be sure the control wires, cables and cords are connected properly.
5. When connecting the bypass hose, be careful of its connection, IN or OUT side.
6. Check the engine and transmission oil levels. This check must be made in one or two minutes after the engine is stopped.
7. Fill the radiator with cooling water.

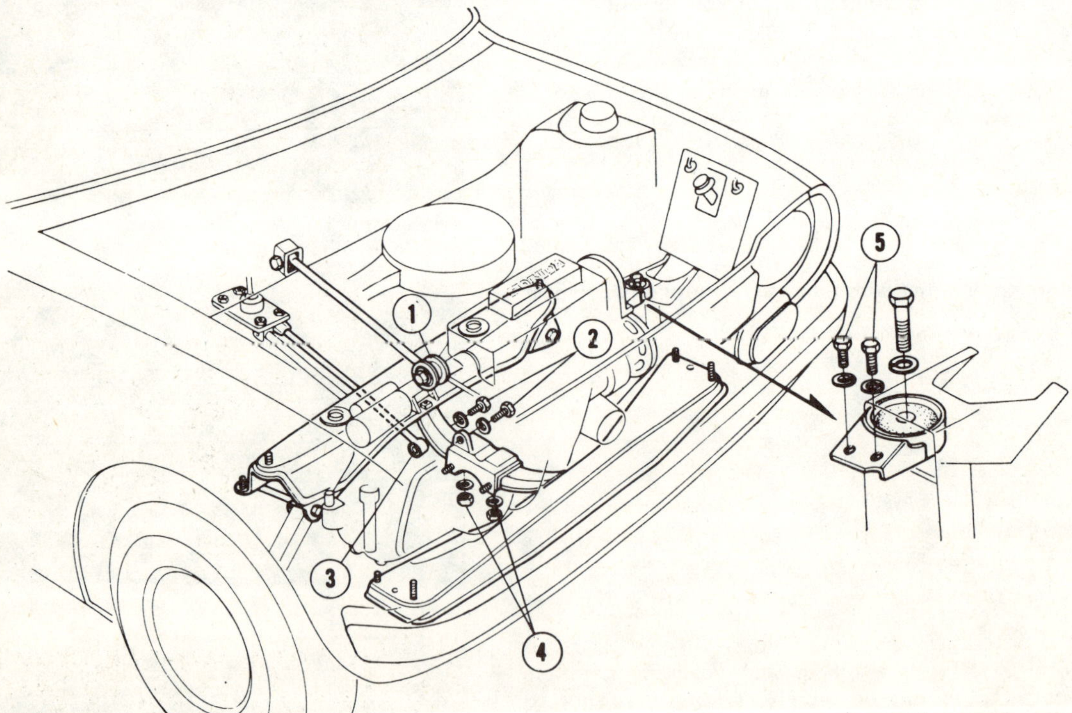

Fig. C-11

① 3.5~4.3 kg-m (25.3~31.1 lbs-ft)

② 1.9~2.5 kg-m (13.7~18.1 lbs-ft)

③ 1.8~2.5 kg-m (13.0~18.1 lbs-ft)

④ 1.0~1.6 kg-m (7.2~11.6 lbs-ft)

⑤ 1.9~2.5 kg-m (13.7~18.1 lbs-ft)

Cylinder Head

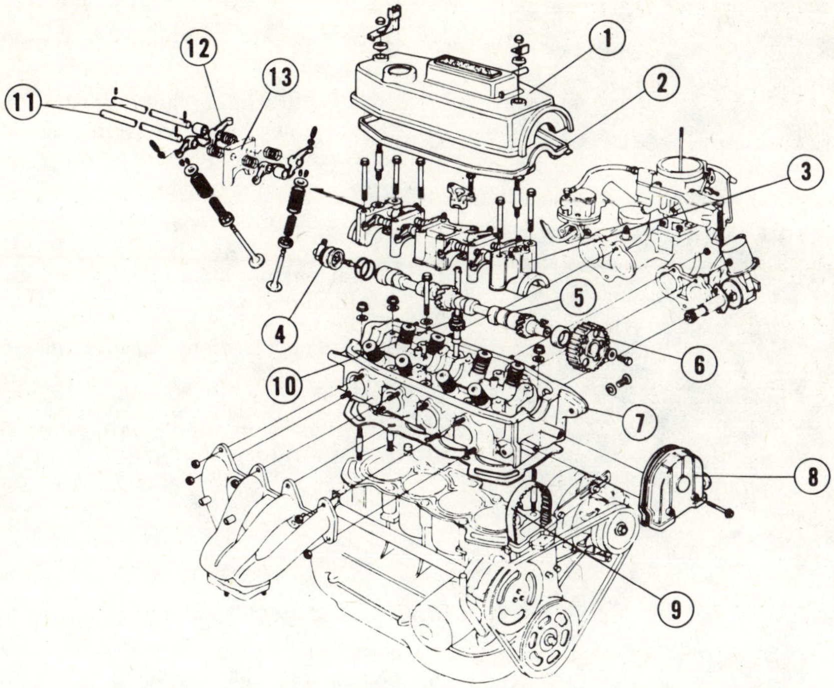

Fig. D-1

① Head cover
② Head cover packing
③ Camshaft holder
④ Tachometer drive body
⑤ Camshaft
⑥ Timing belt driven pulley
⑦ Cylinder head
⑧ Timing belt upper cover
⑨ Timing belt
⑩ Oil pump driver gear
⑪ Rocker arm shafts
⑫ Rocker arm
⑬ Rocker arm shaft holder

NOTE:

1) The cylinder head and block must be removed when the engine is cold, to avoid warpage due to heat.
2) Before washing the cylinder head and block, check them for water or oil leaks due to blow-holes in casting.
3) Remove oil, scale or carbon deposits accumulated from each part before washing. When decarboning, take care not to score or scratch the sliding surface.
4) After washing the oil holes or orifices in each part, check to be sure they are not restricted by directing a blast of compressed air.
5) The parts must be marked or set aside separately in groups, each tagged for its location, so that the same combination or set as before can be reproduced at the time of assembling. Pay special attention to keep dirt and dust off the parts.
6) When storing the parts for a long period of time after washing, apply a coat of oil or grease to their machined surfaces to protect from rust.

a. Removal (on car)

1. Remove the engine hood.
2. Drain the radiator.
3. Disconnect the upper water hose at the thermostat cover.
4. Disconnect the thermounit cord, throttle and choke control cables.
5. Disconnect the hot air tube and remove the air cleaner case.
6. Disconnect the tube running between the canister and carburetor at the canister.

Fig. D-2

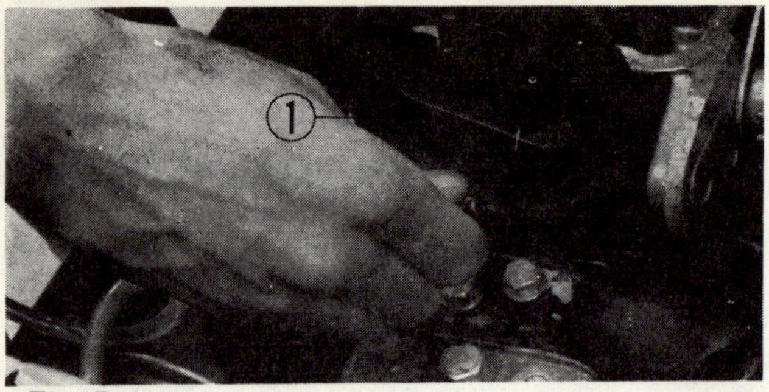

Fig. D-3
① Timing pulley puller

Fig. D-4

7. Disconnect the bypass hose at the intake manifold.
8. Disconnect the fuel tube at the pump inlet port.
9. Disconnect the exhaust pipe at the exhaust manifold.
10. Disconnect the engine torque rod.
11. Remove the cylinder head cover.
12. Bring the No. 1 cylinder piston to the top dead center position.
13. Loosen the timing belt adjusting bolt, remove the timing driven pulley bolts, and remove the belt and driven pulley with a driven pulley remover.

NOTE:

Be careful not to allow the woodruff key to fall inside the timing cover.

14. Remove the fuel pump and distributor.

NOTE:

When removing the pump, wipe off the fuel oil flowing from the pump outlet port with a clean waste cloth.

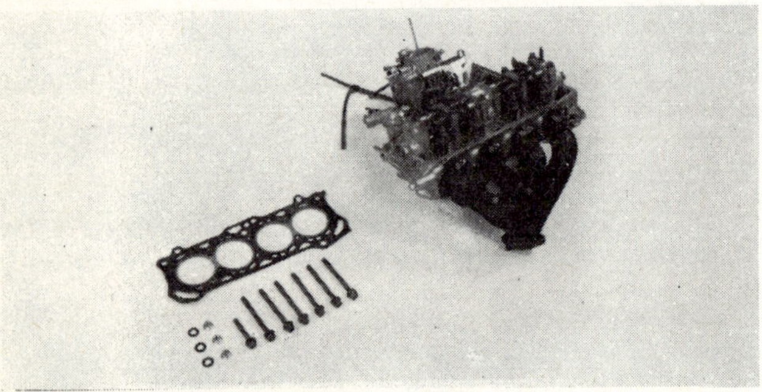

Fig. D-5

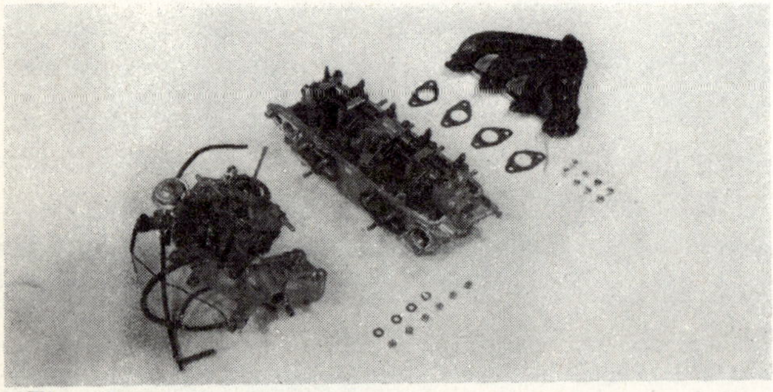

Fig. D-6

15. Remove the oil pump gear holder and remove the pump gear and shaft.
16. Loosen the cylinder head bolts and nuts and remove the head together with the intake and exhaust manifold.

NOTE:

Loosen the cylinder head bolts and nuts in a criss-cross pattern, beginning on outside bolt.

17. The cylinder head can be removed as shown.
18. Separate the intake and exhaust manifolds from the cylinder head.

NOTES:

1) After removing the cylinder head, cover it up with a clean waste cloth to prevent foreign materials from getting into the cylinders.
2) Keep the intake manifold clean.

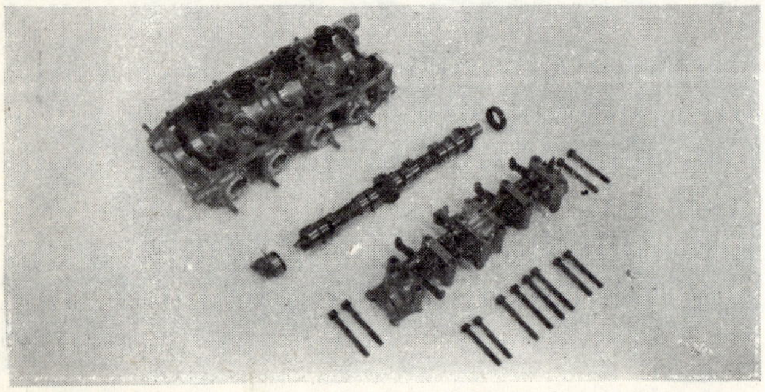

Fig. D-7

b. Disassembly

b-1. Cylinder head

1. Loosen the rocker arm shaft holders and can holder and remove the rocker arms together with the shafts.

NOTE:

Loosen the shaft holders in a criss-cross pattern, beginning on outside holder.

2. Remove the camshaft and tachometer body (if equipped) or the right head seal.

3. Using a valve lifter, compress the valve springs and remove the valve cotters, retainers, springs and seats in this order. Then remove the valves.

NOTE:

The parts must be marked or set aside separately in cylinders, each tagged for its location, so that the same combination or set as before can be reproduced at the time of assembling.

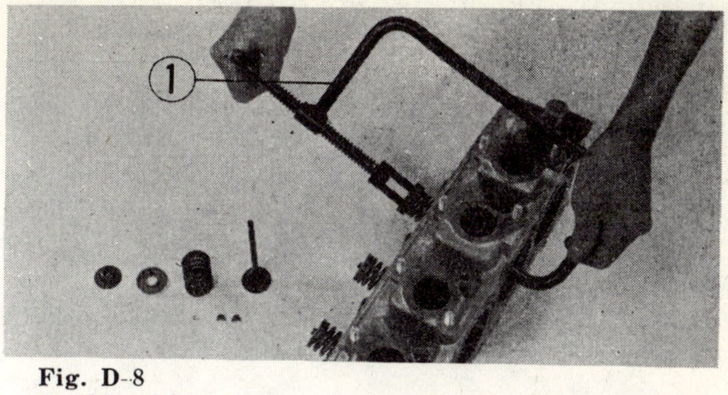

Fig. D-8
① Valve lifter

b-2. Rocker arms and shafts

To disassemble the cam holders, begin on No. 4 cylinder. Pull out 4 mm (0.157 in.) pins and remove the holders, rocker collars, rocker arms and springs in this order.

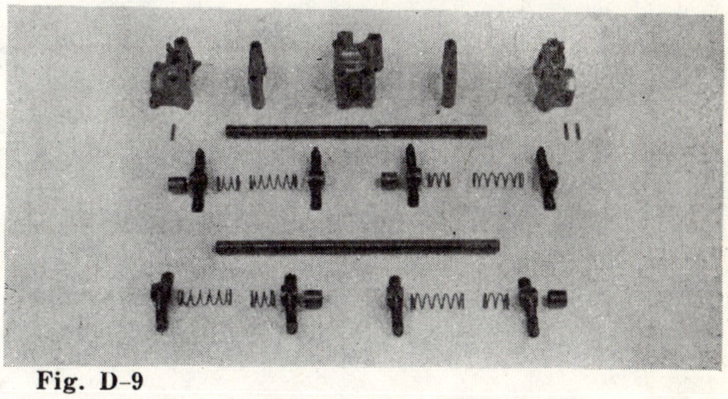

Fig. D-9

c. Inspection

NOTE:

Before inspecting the cylinder head parts, thoroughly clean them with a suitable solvent, removing dust and dirt, oil, carbon deposits and scale. Be careful not to damage the combustion chambers and valves.

The parts must be marked or set aside separately in groups, each tagged for its location, so that the same combination or set as before can be reproduced at the time of assembling.

See Maintenance Standard.

Inspection item	Remedy
Cracks in parts	Repair or replace.
Excessive warpage of cylinder head gasketed surface	Repair.
Excessive valve guide-to-stem clearance	Replace.
Worn valve seat contact surface	Repair or replace.
Worn valve	Replace.
Stepped wear of valve stem end face	Repair or replace.
Out of squareness, free length and tension of valve spring	Replace.
Runout of camshaft	Replace.
Worn cam lobe	Replace.
Excessive rocker arm shaft-to-cam holder clearance	Replace.
Worn inside diameter of rocker arm	Replace.

d. Reconditioning

d-1. Reconditioning of warped cylinder head

1. Apply a coat of Prussian blue or similar paste to a steel straightedge to check the cylinder head for warpage.
2. To recondition the warped head, refinish it by moving a fine oilstone in the pattern of the figure 8 or on a lapping lathe.

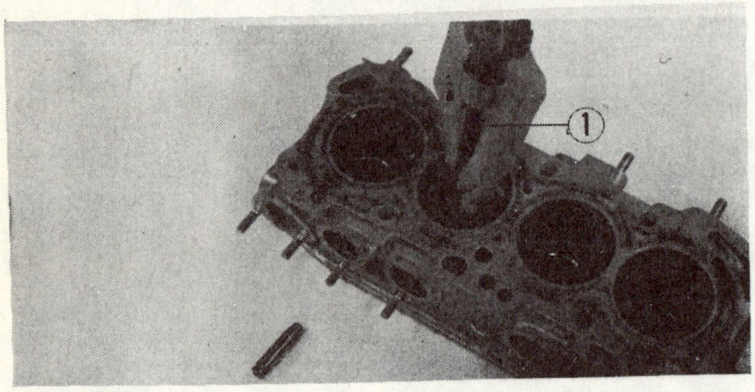

Fig. D-10
① Valve guide driver

d-2. Replacement of valve guides

1. Where driving in or out the valve guide, use a valve guide driver.
2. After driving the valve guide in, be sure to refinish it by reaming since a new valve guide has a reaming allowance. Be sure to push in and pull out the reamer while rotating it.

   For the finished dimension of the valve guide, see the Maintenance Standard at the end of this manual.

   NOTE:

   Do not forget to install the intake valve guide seals in place.

Fig. D-11
① Valve guide reamer

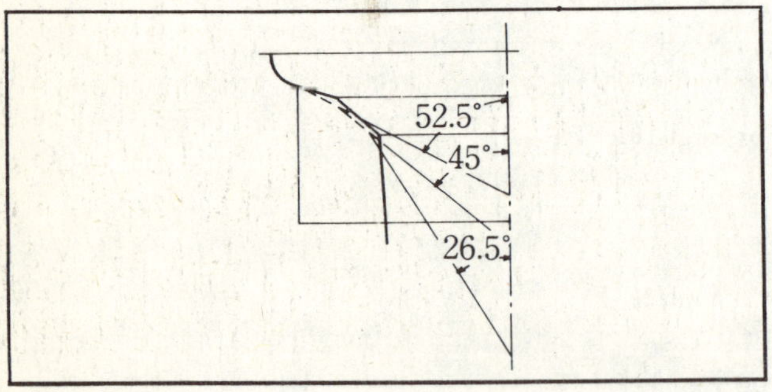

Fig. D-12

d-3. Reconditioning of valve seats

1. Where the valve seat contact width is uneven or exceeds the repair limit, recondition the valve seat with a valve seat grinder.
2. After reconditioning, apply a uniform coat of a proper grade of lapping compound to the valve face and lap the valve on the seat using a valve lapping tool.

   To lap, place the valve into position with a valve lapping tool attached as shown in Fig. D-13. Using the tool, move the valve up and down while turning it back and forth with both hands so that never turn the valve forcing it against the seat, the valve surface contacts the valve seat when it bottoms.

   NOTE:

   Take care not to allow the compound to enter into between the valve stem and guide. After lapping, wash out the compound completely and apply a coat of engine oil to the valve face and seat.

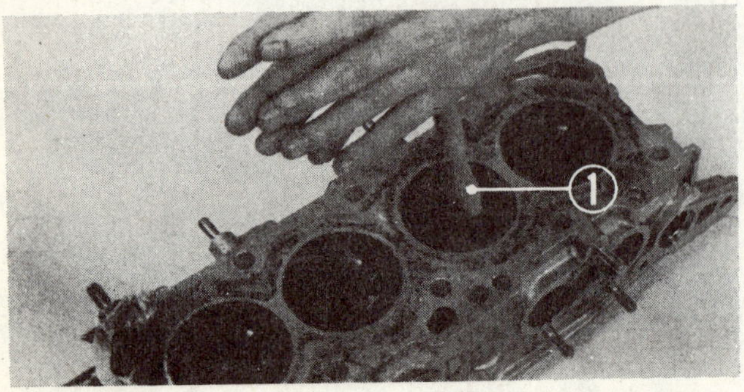

Fig. D-13
① Valve lapper

e. Assembly

To assemble the cylinder head, reverse the disassembly procedures. Pay attention to the following points:

e-1. Valves

1. When installing the valve, apply a coat of oil to the stem and insert it in the valve guide, and check to be sure the valve smoothly moves up and down. Then, install the valve spring with a valve lifter.
2. After installing the valve, give a light blows of mallet to the end face of the valve stem two or three times. This will fit the valve and cotters properly.

   NOTE:

   It is advisable to replace all valve springs at one time.

Fig. D-14

e-2. Rocker arms and shafts

1. When installing the rocker arms, be careful of the installation order and direction. The rocker arm shafts must be so installed that one end having a notch fach faces the intake valve.
2. After installing, move the rocker arms to the checked for smooth movement.

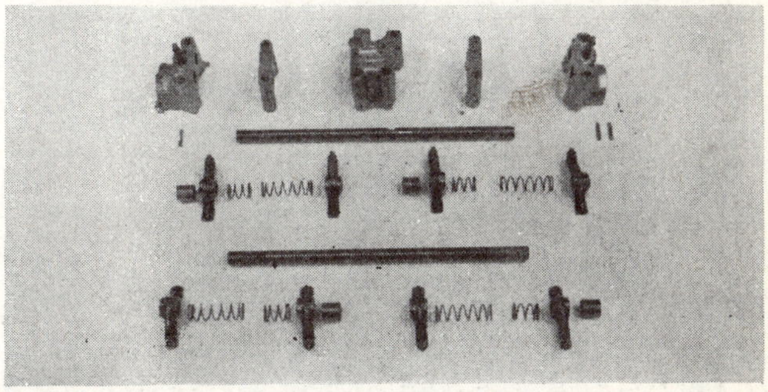

Fig. D-15

e-3. Camshaft

1. Install the camshaft, oil seal and tachometer body to the cylinder head.

   NOTE:

   Before installing, apply a coat of oil to the camshaft journals.
2. Apply a coat of sealant to the mating surfaces between the right and left cam holders and the cylinder head. When tightening the cam holders, leave the rocker arm adjusting bolt fully loosened.

Fig. D-16

e-4. Cylinder head

1. Install the intake and exhaust manifolds to the cylinder head.
2. Rotate the crankshaft pulley and align the "top dead center position" mark on the pulley with the index mark on the timing belt cover. (No. 1 and 4 cylinders are now at the top dead center position.)
3. Put the cylinder head gasket in place and install the cylinder head.

   NOTE:

   When installing the cylinder head, wipe the mating surfaces of the head, cylinders and gasket off oil or dust with a clean waste cloth.

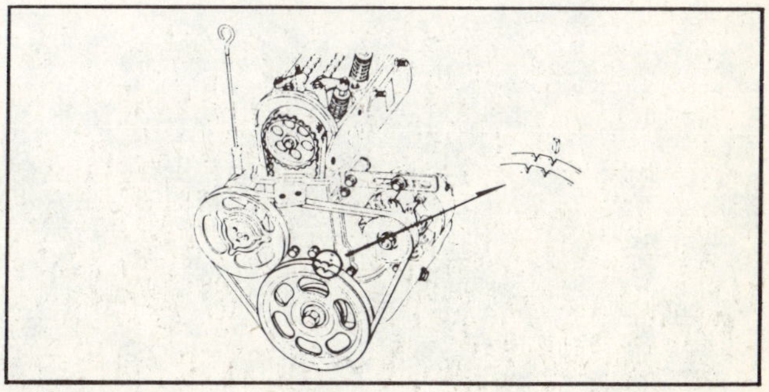

Fig. D-17

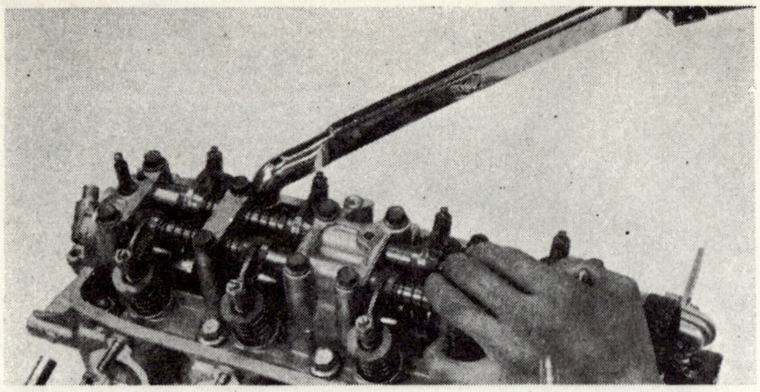

Fig. D-18

Fig. D-19

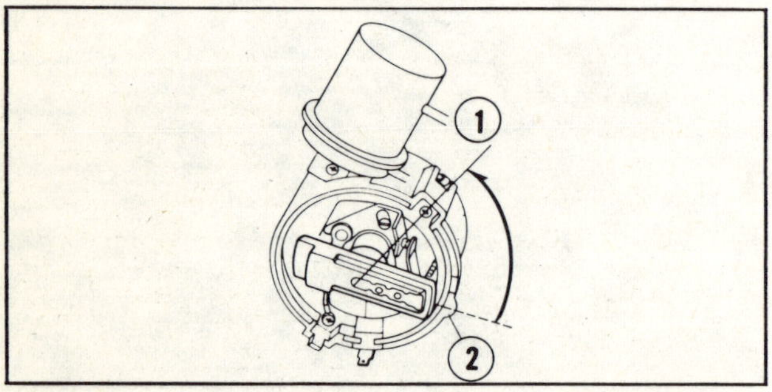

Fig. D-20
① No. 1 cylinder side ② Lug

Fig. D-21
① Tappet adjusting wrench

Fig. D-22
① Feeler gauge

4. Tighten the cylinder head bolts in a criss-cross pattern, beginning on inside bolt, to the specified torque. Use a torque wrench.
5. Tighten the timing belt driven pulley bolt together with the woodruff key.

   Install the timing belt around its pulley so that the marks on the pulley is incline with the cylinder head top surface.

   NOTES:

   1) When installing the belt, take care not to disturb the timing position already set.
   2) Do not apply oil to the belt or bend it.

e-5. Distributor

1. Install the distributor with No. 1 cylinder kept at the top dead center position on the compression stroke (the mark on the crankshaft pulley aligned with that on the belt cover).
2. Line up the metal end of the rotor head with the protrusion on the distribution housing.
3. Carefully insert the pinion of the distributor in the cylinder head opening with the fixing plate aligned with the distributor mounting base.

   Then secure the plate at the center of the slot.

   At this time, the rotor head must face No. 1 cylinder.

   NOTE:

   The pinion gear has helical teeth. The rotor will be moved when the gear is inserted.

   After installing the cylinder head, inspect and adjust the point gap and ignition timing.

f. Adjustment

f-1. Valve clearance

1. Rotate the crankshaft pulley until the "top dead center position" mark on the pulley is aligned with the index mark on the timing belt cover. This adjustment must be made on a cylinder which is at the top dead center position (a cylinder whose intake and exhaust valve rocker arms are not being lifted by the cam.)

- When No. 1 cylinder reaches the top dead center position on its compression stroke, the valve clearance can be inspected on the following valves:

  Intake valves of No. 1 and No. 2 cylinders

  Exhaust valves of No. 1 and No. 3 cylinders

- When the crankshaft pulley is further rotated one turn:

  Intake valves of No. 3 and No. 4 cylinders

  Exhaust valves of No. 2 and No. 4 cylinders

NOTE:

When rotating the pulley with the starting motor, remove the distributor cap or disconnect the center cord to prevent high voltage to be applied to the spark plugs.

2. **Check the clearance between the rocker arm adjusting bolt and valve stem with a feeler gauge.** To adjust **the clearance, loosen the adjusting bolt lock nut and turn the adjusting bolt in either direction until the correct clearance is obtained.**

After adjusting, tighten the lock nut with adjusting bolt held with a screwdriver to prevent the correct adjustment from being disturbed. After tightening the lock nut, again check the clearance with a feeler gauge.

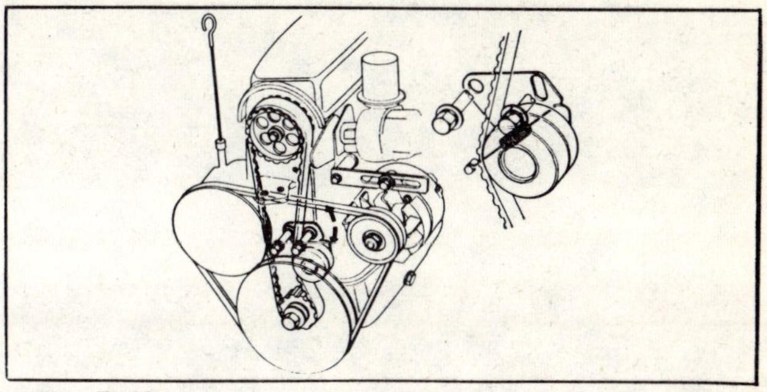

Fig. D-23

f-2. Timing belt tension

1. Loosen the tensioner bolt and adjusting bolt to adjust the belt tension. After adjusting, tighten the bolts. The belt tension will be automatically adjusted by the adjuster spring, when the bolts are loosened.

NOTES:

1) When adjusting, do not use any force other than the adjuster spring.

Too tight a belt will cause improper belt pitch to result in the shortened belt life.

2) The adjusting bolt is only loosened, but not removed. Otherwise, the removed washer will drop in to the belt and pulley to damage them.

After making the above adjustments, the following items must be inspected:

- Compression pressure.
- Ignition timing with a timing light.
- Water and oil leak, abnormal noise, etc.

D-2. Pistons and Connecting Rods

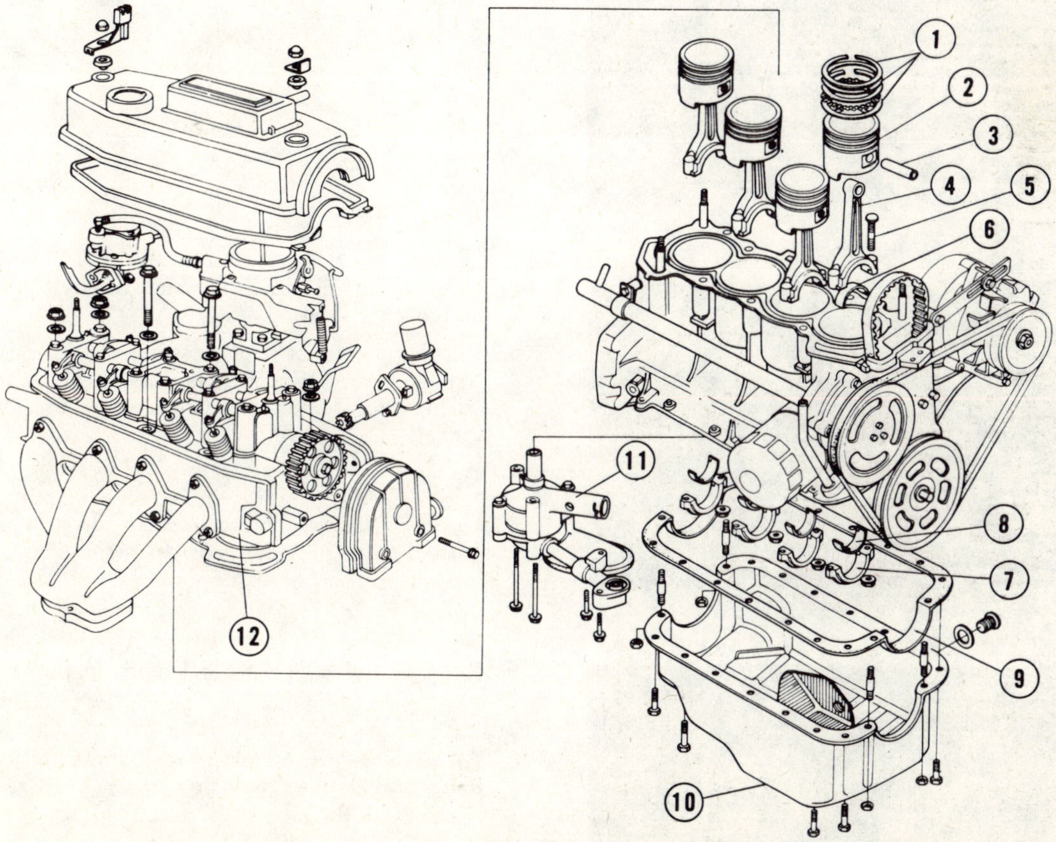

Fig. D-24

① Piston rings
② Piston
③ Piston pin
④ Connecting rod
⑤ Connecting rod bolt
⑥ Connecting rod bearing
⑦ Connecting rod bearing cap
⑧ Connecting rod bearing
⑨ Oil pan gasket
⑩ Oil pan
⑪ Oil pump assembly
⑫ Cylinder head assembly

Fig. D-25

a. Removal (on car)

1. Remove the engine hood.
2. Drain the radiator.
3. Drain the engine by removing the drain bolt.
4. Remove the cylinder head.
5. Jack up the front of the car and support it with safety stands.

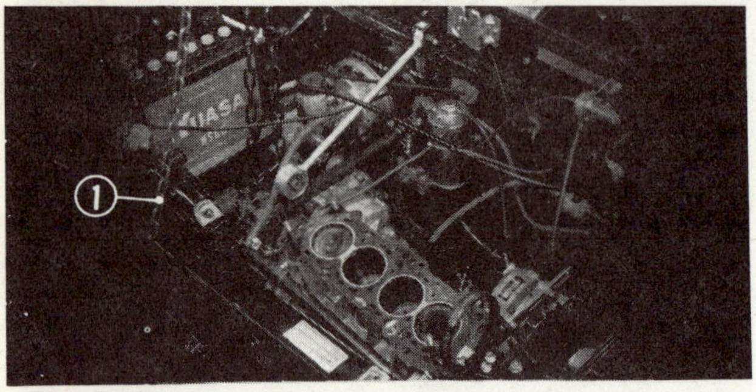

Fig. D-26
① Chain

6. Attach a chain to the clutch sling and attempt to lift the transmission case just far enough for the chain block to take load.

NOTE:

Do not remove the engine right mount.

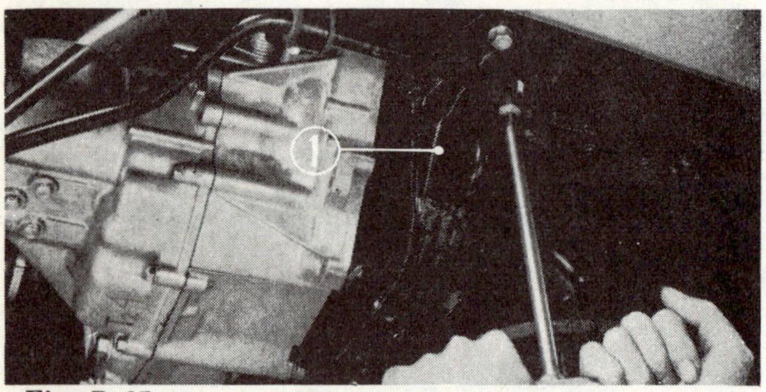

Fig. D-27
① Center beam

7. Remove the center beam and engine mount rubber A.

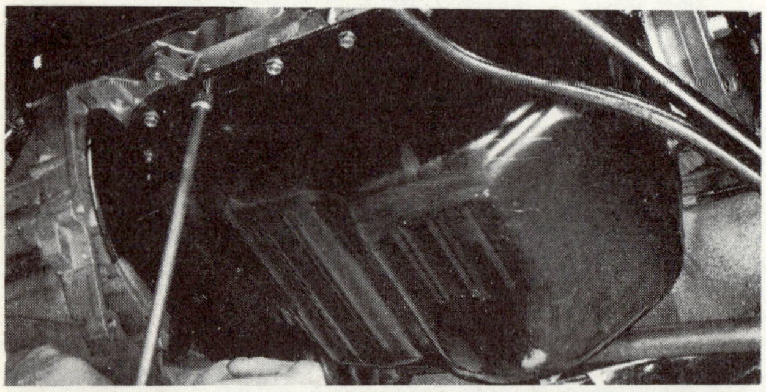

Fig. D-28

8. Loosen the oil pan bolts and remove the oil pan and clutch case cover.
   - Loosen the bolts in a criss-cross pattern, beginning on outside bolt. Where it is difficult to remove the oil pan gasket, lightly tap the corners of the oil pan with a mallet.

NOTE:

Do not pry off the gasket or do not tap the oil pan. with the tip of a screwdriver strongly.

Fig. D-29
① Oil pass block

9. Remove the oil pass block and oil pump assembly.

NOTE:

As soon as the oil pass block bolts are loosened, the oil in the oil gallery may flow out. Prepare the oil pan or a waste cloth.

10. Remove the connecting rod bearing caps. Using the handle of a hammer, pull up the pistons and connecting rods from the cylinder.

NOTES:

1) When removing the bearing caps, do not move the installation positions of the bearings. Each bearing and cap are selective-fitted so that the specified oil clearance can be obtained.
2) Do not remove the bearing caps without valid reason.
3) Before pulling up the connecting rods, check the top surfade of the cylinder for condition. Reface the cylinder if abnormally worn or decarbon it if necessary.

Fig. D-30

11. When lifting off the pistons, put the marks indicating the cylinder numbers on the piston crowns so that the pistons can be reinstalled into the same position as before.
12. Remove the piston rings from the piston.

NOTES:

1) The piston rings are made up of cast iron and, therefore, they may be broken when excessive force is applied to them. Be careful in handling them.
2) The piston rings removed must be set aside separately in pistons so that the same combination or set as before can be reproduced at the time of assembling.

Fig. D-31

13. The piston pins are press-fitted. Using ahydraulic press and the jig shown in the Fig. D-32, remove the pins.

NOTES:

1) The connecting rods, pistons, rings and pins must be set aside separately in groups so that the same combination or set as before can be reproduced at the time of assembling.
2) Keep the connecting rods with the nuts fitted in the bearing cap holes.

b. Installation of piston pins

1. When driving in or out the piston pin using a hydraulic press, positively place the piston on the base with the center of the pin in line with that of the base.
2. When pressing out the pin, use a pilot collar on the piston pin in line and apply force on the collar by the ram. When pressing in the pin, put the pilot collar through the piston pin boss and the small end of the connecting rod, place the pin on the collar in line and press it by the ram.

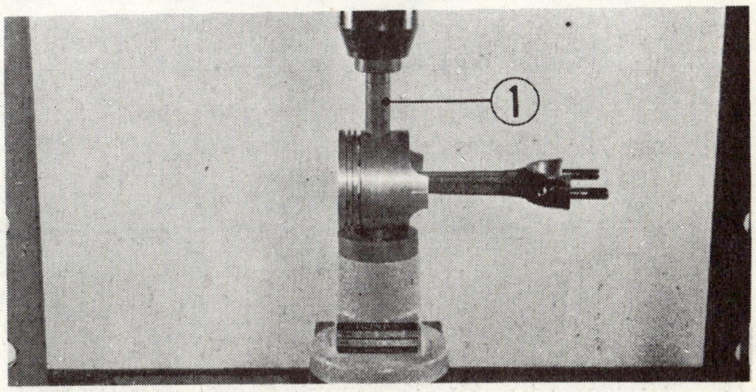

Fig. D-32
① Pilot collar

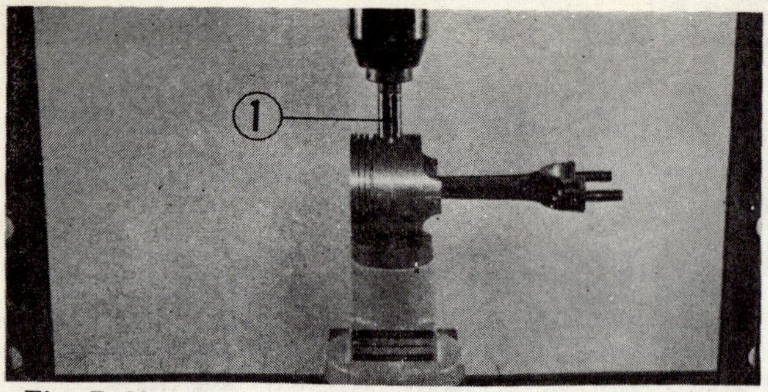

Fig. D-33
① Piston pin

3. When installing, apply a coat of hypoid gear oil to the pin hole and connecting rod small end.
4. Sudden rise of the hydraulic pressure recorded on the meter indicates that the fitting of the pin has been completed.
5. After installing the piston pin, check to be sure its both ends are equally fitted to the piston bosses.

c. Inspection

See Maintenance Standard.

	Inspection item		Remedy
Cylinders	Bore		
	Difference in bore between top and bottom parts		
	Out of round		
	Difference in cylinder bores		
	Warpage of cylinder gasketed surface		
Pistons	Outside diameter of crown		
	Outside diameter of skirt		
	Diameter of piston pin hole		
	Minimum clearance in cylinder		
	Piston ring groove width	Top	
		Second	
		Oil	
Piston rings	Thickness	Top	Replace if service limit is exceeded.
		Second	
	Piston ring clearance	Top	
		Second	
	Ring gap	Top	
		Second	
		Oil	
Piston pins	Piston pin clearance		
Connecting rods	Axial play of big end		
	Radial play of big end		
	Offset of big and small ends		
	Twist of big and small ends		
	Press-in allowance of piston pin		

d. Assembly

To assemble the pistons and connecting rods, reverse the disassembly procedures. Pay attention to the following points:

1. Prior to assembly, decarbon the parts if necessary and wash them thoroughly.
2. When installing the connecting rod to the piston, install the new piston pin with the mark on the piston crown facing the oil jet hole in the connecting rod.

   To install the pin, see page 104.
3. To install the oil ring, first place the spacer and then the rails in position. The spacer and rail gaps must be staggered 2 to 3 cm (0.787 to 1.181 in.).
4. Install the second and top rings in this order in the piston with their markings facing upward.

   NOTE:

   After installing all rings in the piston, hand-rotate them and check to be sure they move smoothly without any sign of binding.

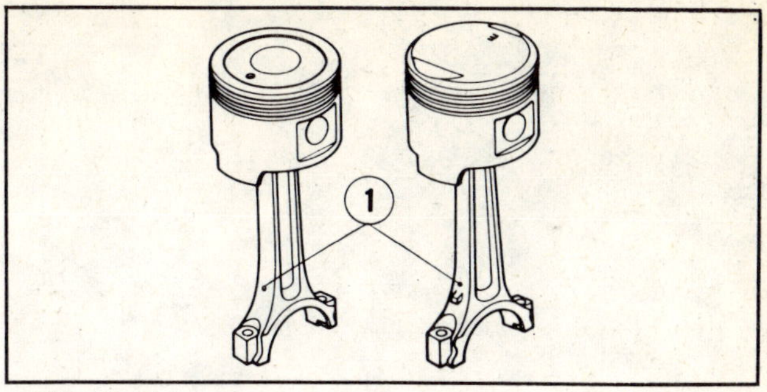

Fig. D-34
① Oil jets

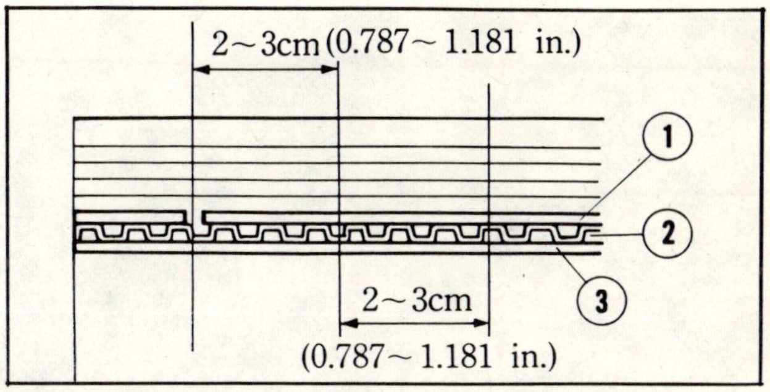

Fig. D-35
① Rail ② Spacer ③ Rail

5. The ring gaps must be staggered 120° or 180° and must NOT be in the direction of the piston pin boss or at right angles to the pin.

   NOTE:

   The gap of the three-piece type oil ring refers to that of the spacer.

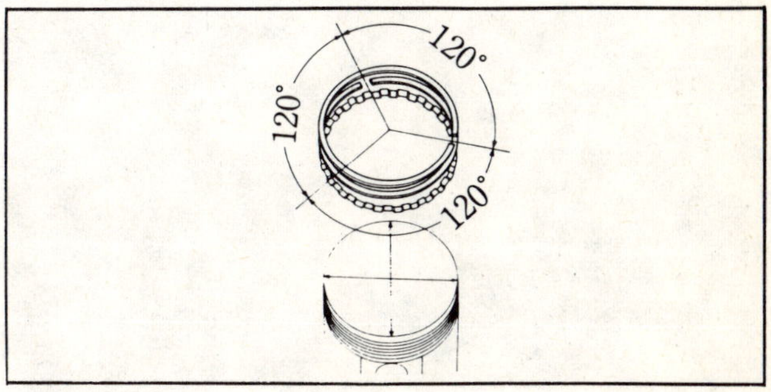

Fig. D-36

● **Oversized pistons and rings**

- The piston and rings are marked or stamped for identification.
- The end face of the piston pin is painted in blue.
- The pistons and rings are available in four oversizes, 0.25 mm (0.010 in.), 0.50 mm (0.020 in.), 0.75 mm (0.030 in.) and 1.00 mm (0.040 in).

  The piston pins are available in one oversize, 0.003 mm (0.0001 in.).

NOTE:

A 0.25 mm oversize piston or ring indicates that its outside diameter is larger 0.25 mm than that of the standard one.

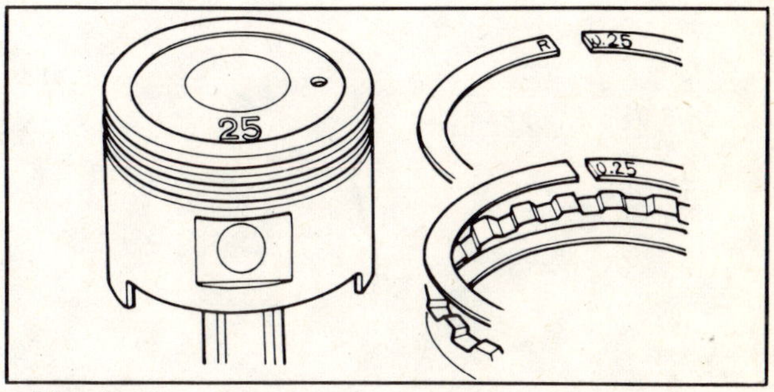

Fig. D-37

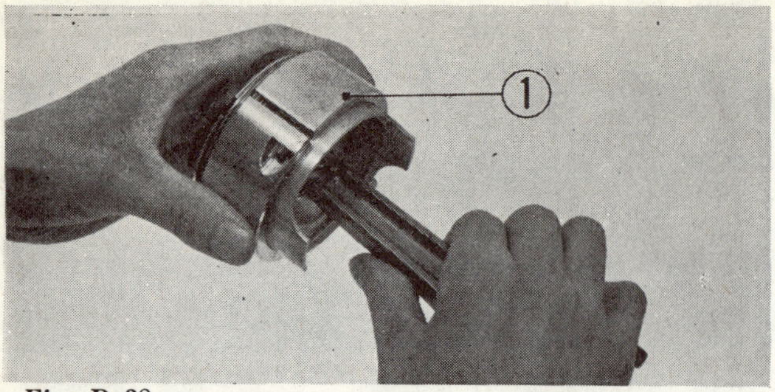

Fig. D-38
① Piston slider

6. Using a piston slider, install the piston in the cylinder with its skirt protruded about 1/3 of the piston height below the slider. Prior to installation, apply a thin coat of oil to the rings and to the cylinder wall.

NOTE:

Direct the connecting rod oil jet hole or the mark on the piston crown toward the intake manifold.

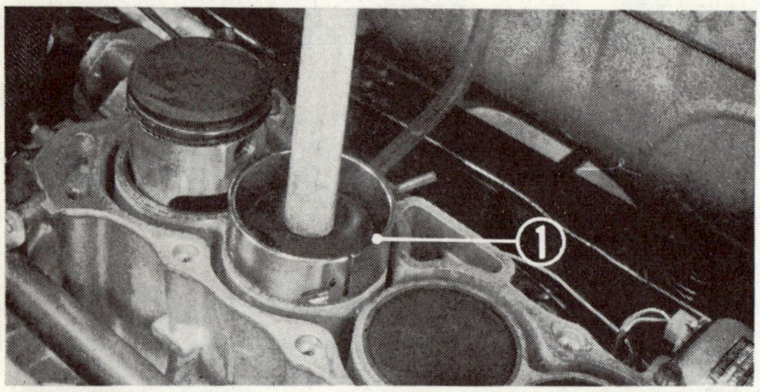

Fig. D-39
① Piston slider

7. Using the wood handle of a hammer, slowly press the piston in the cylinder. Drift the connecting rod until its large end is at right angles to the crankpin or the crankshaft.

NOTE:

Do not change the combination or set of pistons and cylinders.

Fig. D-40

8. Check the oil clearance with a plastigauge and apply a coat of engine oil or molybdenum bisulfide to the connecting rod bearings.

Then, tighten the connecting rod bearing cap bolts to the specified torque.

9. After tightening the bearing cap bolts, rotate the crankshaft and check for smooth rotation.
To check the oil clearance, **see page 110**.

NOTE:

Do not adjust the oil clearance by inserting a shim or shims, or by scraping the end face of the bearing cap.

Crankshaft and Bearings

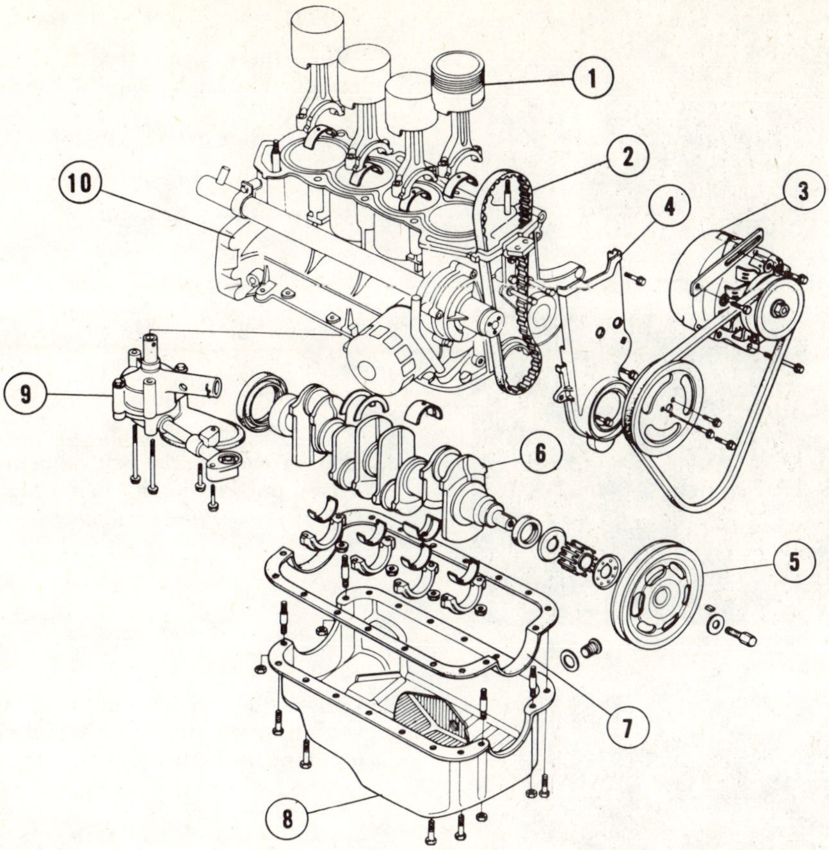

Fig. D-41

① Piston ② Timing belt ③ AC generator ④ Timing belt cover ⑤ Crankshaft pulley
⑥ Crankshaft ⑦ Oil pan gasket ⑧ Oil pan ⑨ Oil pump assembly ⑩ Cylinder block

a. Removal (engine out of car)

1. Remove the engine together with the clutch and transmission cases.
2. Remove the starting motor from the engine.
3. Separate the clutch and transmission cases from the engine.

Fig. D-42

4. Hold the ring gear with a ring gear holder and remove the pressure plate assembly and friction disc.
5. Loosen the crankshaft pulley bolt with the ring gear held with a ring gear holder.
6. Remove the ring gear by loosening its attaching bolts.

Fig. D-43
① Ring gear holder

Fig. D-44
① Distributor ② Fuel pump ③ Carburetor

7. Remove the carburetor.
8. Remove the intake manifold.

NOTE:

When there is no need of disassembling the carburetor, remove the intake manifold together with the carburetor.

9. Remove the exhaust manifold.
10. Remove the fuel pump.
11. Remove the distributor.

Fig. D-45

12. Remove the cylinder head cover and upper timing belt cover.
13. Remove the oil pump gear holder and remove the pump gear and shaft.
14. Loosen the timing belt adjusting bolts and remove the driven pulley with its belt installed on it using a timing belt driven pulley remover.

Fig. D-46

15. Loosen the cylinder head bolts in a criss-cross pattern, beginning on outside bolt, and remove the cylinder head together with the rocker arms and camshaft.

NOTE:

Since the dowels are installed in the cylinder block, remove the cylinder head by lifting it off upward.

16. Remove the water pump pulley.
17. Remove the lower timing belt cover.

Fig. D-47
① Mark

18. Remove the timing belt.

NOTE:

Prior to removal, chalk the direction of the drive on thebelt to facilitate reinstallation.

19. Remove the oil pan. Remove the oil pump and oil pass block.

Fig. D-48
① Bearing cap

20. Loosen the bearing cap bolts in a criss-cross pattern, beginning on outside bolt, and remove the bearing caps.

NOTE:

The main bearings removed must be set aside separately in groups, each marked for cylinder, so that the same combination or set as before can be reproduced at the time of assembling.

21. Remove the connecting rod bearing caps and then the crankshaft.

Fig. D-49

22. Lift off the pistons and connecting rods.

NOTE:

The connecting rods, bearing caps and pistons must be set aside separately in groups, each marked for cylinder, so that the same combination or set as before can be reproduced at the time of assembling.

Fig. D-50

b. Inspection

NOTES:

1) The crankshaft is supported by plain bearings. Keeping the correct bearing (oil) clearance is, therefore, of an extreme importance. Check to be sure it is within the specification. To tightening torque of the bearing cap bolts must be as specified.

2) Do not grind the end face of the connecting rod bearing caps or bearings with a file or an oilstone to decrease the clearance. Since the crush height will be lessened.

3) Do not adjust the clearance with shims or by scraping. Since it may damage the overlay of the bearings.

The "overlay" is a thin lead-tin coating. During break-in period, it provides preferable journal to metal contact, thereby enabling them to carry greater loads even after the overlay has been worn away.

See the Maintenance Standard at the end of this manual.

Inspection item	Remedy
Bend of crankshaft	Replace when service limit is exceeded.
Out of roundness and taper of crankpins and journals	Replace when service limit is exceeded.
Excessive clearance between crankpin or journal and bearing	Replace bearing when service limit is exceeded.
Scores or scratches on sliding surfaces of crankpins and journals	Replace.
Peeling, sticking or any other damage to bearings	Replace bearing.

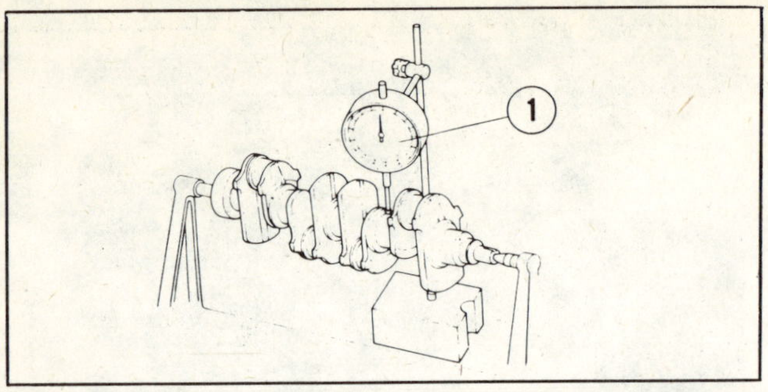

Fig. D-51
① Dial gauge

b-1. Bend of crankshaft

Support the crankshaft on two identical V-blocks placed on a surface plate. Then measure the bend by using a dial gauge at the center journal.

NOTE:

One half of the dial gauge reading is the bend in the crankshaft.

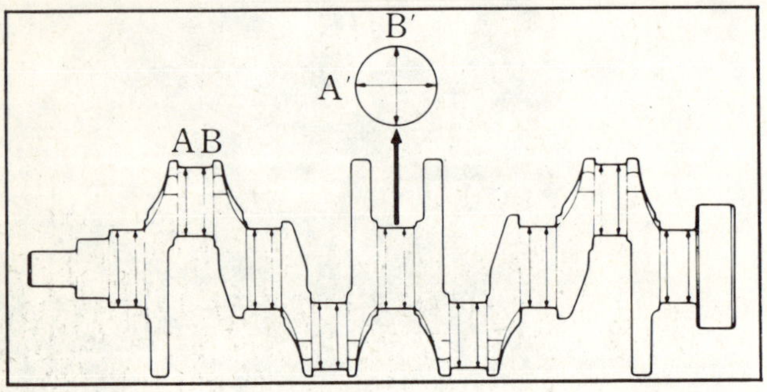

Fig. D-52

b-2. Taper and out of round of crankshaft

Using a micrometer, measure the outside diameters of journals and crankpins at the points shown. The out of roundness is the difference between the longest diameter A′ and the shortest B′. The taper is the difference between the longest diameter A on one of the both ends and the shortest B on the other end.

b-3. Selection of bearings and measurement of oil clearance

NOTES:

1) The sliding surfaces of the bearings are made up of the lead alloy. When wiping off foreign materials, always use polishing agent.
2) The bearings must be placed in position with the locating tab fitted in the recess of the crankcase.
   Improper setting may result in sticking of bearings.
3) Do not recondition the journals and crankpins by grinding. Since their surfaces are specially heat-treated.
4) The crankcase bolt tightening sequence and torque must be as specified.
5) Apply a coat of engine oil or molybdenum bisulfide to the bearings before installing.
6) After tightening the crankshaft center cap and connecting rod bearing caps, hand-rotate the crankshaft to check for smooth rotation.
7) After assembling the engine, run the engine at idle speed for 10 to 20 minutes. Keep in mind that, reviving up the engine suddenly may result in bearing failure.

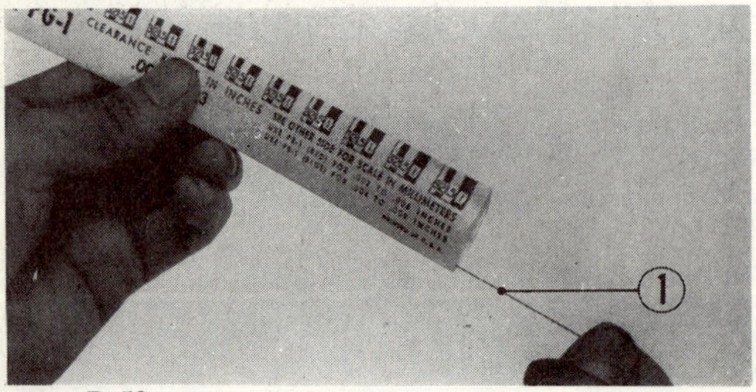

Fig. D-53
① Plastigauge

Measurement of oil clearance (crankshaft)

1. Select the right bearings in accordance with the fit classification table on page 112 and by seeing the fit markings on the crankshaft journal and on the cylinder block. When any bearing must be replaced, use a new bearing identical in marking (color) to the original one.
2. Install the bearings to the crankcase and to the bearing caps by aligning the oil holes and by fitting the locating tab in the recess.

3. Install the crankshaft in place. Cut a 0.025 to 0.076 mm (0.001 ~ 0.003 in.) plastigauge to a proper length and put it on the journal as shown. Be careful not to restrict the oil holes with plastigauge when measuring.
4. Tighten the bearing caps to the specified torque.

   NOTE:

   Tighten the cap bolts evenly in a criss-cross pattern, beginning on center cap, with the crankshaft held stationary.

5. Remove the bearing cap. Measure the amount of increase in the width of the plastigauge as it flattens out, with the scale printed on the bag of the gauge. Take the reading at the widest part of the flattened plastigauge as the oil clearance.

   NOTE:

   For the specified oil clearance, see specification.

6. When the oil clearance exceeds the specification, select the proper bearing according to the fit classification table. If excessive oil clearance exsists even after the bearing has been replaced, install a undersized bearing which is one step smaller than the original bearing in inside diameter. However, excessive oil clearance due to the deformation of the crankcase can not be adjusted. In such a case, replace the crankcase.

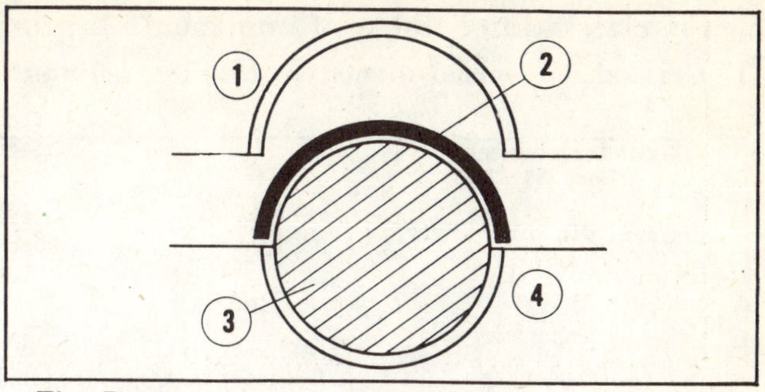

Fig. D-54
① Cap
② Plastigauge
③ Crankshaft journal
④ Case

Fig. D-55

Measurement of oil clearance (connecting rods)

1. The crankshaft main bearings and connecting rod bearings are marked in color for identification.
2. Measure the oil clearance of the connecting rods in the same manner as for the crankshaft.
3. When installing the connecting rod bearing caps, take care of the installation direction.

   NOTE:

   The cap bolts must be evenly tightened in a criss-cross pattern with the crankshaft held stationary.

Fig. D-56

c. Fit classification table of crankshaft bearings

1. Crankshaft journal diameter × cylinder counterbore × main bearing thickness

Unit: mm (in.)

Cylinder counterbore 54 (2.13) / Journal dia. 50 (1.97)	A 0∼+0.006 (+0.0002)	B +0.006 ∼+0.012 (+0.0002∼0.0005)	C +0.012∼+0.018 (0.0005∼0.0007)	D +0.018∼+0.024 (0.0007∼0.0009)
2 0∼−0.006 (−0.0002)	**Red** −0.002 ∼−0.005 (−0.0001∼−0.0002)	**Pink** +0.001∼−0.002 (0.00004∼−0.0001)	**Yellow** +0.004 ∼+0.001 (+0.0002∼0.00004)	**Green** +0.007 ∼+0.004 (+0.0003∼+0.0002)
2 −0.006 ∼−0.012 (−0.0002∼−0.0005)	**Pink** +0.001 ∼−0.002 (0.00004∼−0.0001)	**Yellow** +0.004 ∼+0.001 (+0.0002∼0.00004)	**Green** +0.007 ∼+0.004 (+0.0003∼+0.0002)	**Brown** +0.010 ∼+0.007 (+0.0004∼+0.0003)
3 −0.012 ∼−0.018 (−0.0005∼−0.0007)	**Yellow** +0.004 ∼+0.001 (+0.0002∼0.00004)	**Green** +0.007 ∼+0.004 (+0.0003∼+0.0002)	**Brown** +0.010 ∼+0.007 (+0.0004∼+0.0003)	**Black** +0.013 ∼+0.010 (+0.0005∼+0.0004)
4 −0.018 ∼−0.024 (−0.0007∼−0.0009)	**Green** +0.007 ∼+0.004 (+0.0003∼+0.0002)	**Brown** +0.010 ∼+0.007 (+0.0004∼+0.0003)	**Black** +0.013 ∼+0.010 (+0.0005∼+0.0004)	**Blue** +0.016 ∼+0.013 (+0.0006∼+0.0005)
	Main bearing center stock thickness: 2.0 (0.079)			

1. The figures 1, 2, 3, and 4 and alphabet A, B, C, and D are stamped respectively near the crankshaft journals and cylinder counterbores as shown in the figures.
2. Oil clearance includes an increase of 20μ which is produced by "crushing" the bearing. The crankshaft and main bearings are selective-fitted so that 0.024 to 0.042 mm (0.009 to 0.0017 in.) is obtained in the oil clearance.

2. Crankpin diameter × connecting rod big end hole diameter × connecting rod bearing thickness

Unit: mm (in.)

Connecting rod dia. 43 (1.69) / Crankpin dia. 40 (1.57)	1 0∼+0.006 (+0.0002)	2 +0.006 ∼+0.012 (+0.0002∼+0.0005)	3 +0.012 ∼+0.018 (+0.0005∼+0.0007)	4 +0.018 ∼+0.024 (+0.0007∼+0.0009)
A 0∼−0.006 (−0.0002)	**Red** −0.005 ∼−0.008 (−0.0002∼−0.0003)	**Pink** −0.002 ∼−0.005 (−0.0001∼−0.0002)	**Yellow** +0.001∼−0.002 (0.00004∼−0.0001)	**Green** +0.004 ∼+0.001 (+0.0002∼0.00004)
B −0.006 ∼−0.012 (−0.0002∼−0.0005)	**Pink** −0.002 ∼−0.005 (−0.0001∼−0.0002)	**Yellow** +0.001∼−0.002 (0.00004∼−0.0001)	**Green** +0.004 ∼+0.001 (+0.0002∼0.00004)	**Brown** +0.007 ∼+0.004 (+0.0003∼+0.0002)
C −0.012 ∼−0.018 (−0.0005∼−0.0007)	**Yellow** +0.001∼−0.002 (0.00004∼−0.0001)	**Green** +0.004 ∼+0.001 (+0.0002∼0.00004)	**Brown** +0.007 ∼+0.004 (+0.0003∼+0.0002)	**Black** +0.010 ∼+0.007 (+0.0004∼+0.0003)
D −0.018 ∼−0.024 (−0.0007∼−0.0009)	**Green** +0.004 ∼+0.001 (+0.0002∼0.00004)	**Brown** +0.007 ∼+0.004 (+0.0003∼+0.0002)	**Black** +0.010 ∼+0.007 (+0.0004∼+0.0003)	**Blue** +0.013 ∼+0.010 (+0.0005∼+0.0004)
	Connecting rod bearing center stock thickness: 1.5 (0.059)			

1. The figures 1, 2, 3, and 4 and alphabet A, B, C, and D are stamped respectively near the big end of the connecting rods and the crankpins as shown in the figures.

2. Oil clearance includes an increase of 10μ which is produced by "crushing" the bearing. The crankpins and connecting rod bearings are selective-fitted so that 0.020 to 0.038 mm (0.0008 to 0.0015 in.) is obtained in the oil clearance.
3. Bearings are identified with colors, blue, black, brown, green, yellow, pink and red, each being tolerated 0.003 mm (0.0001 in.).

How to use the table

Knowing a set of fitting marks on the connecting rod and crankpin, or the crankshaft and cylinder counterbore, which are a alphabetical letter or a figure respectively, pick up the collar coding which corresponds to them. A bearing with the collar coding is to be installed.

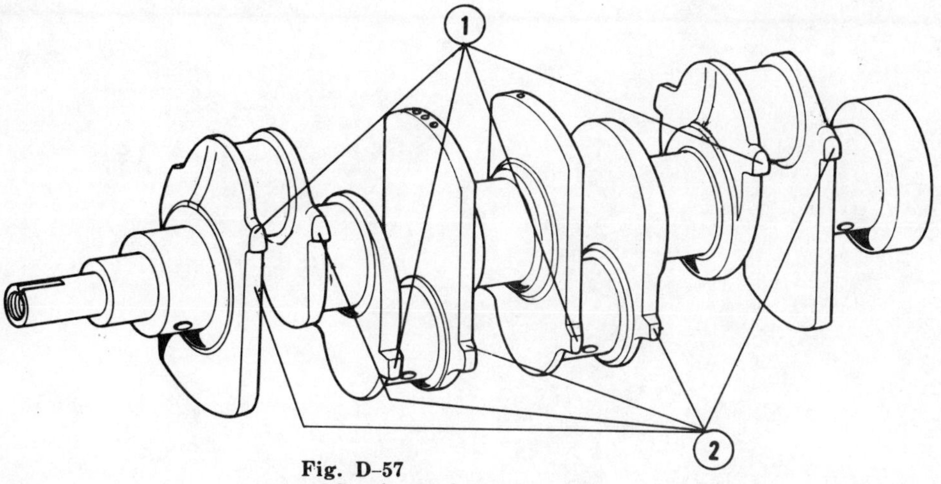

Fig. D-57
① Crank pin fit markings
② Crankshaft journal fit markings

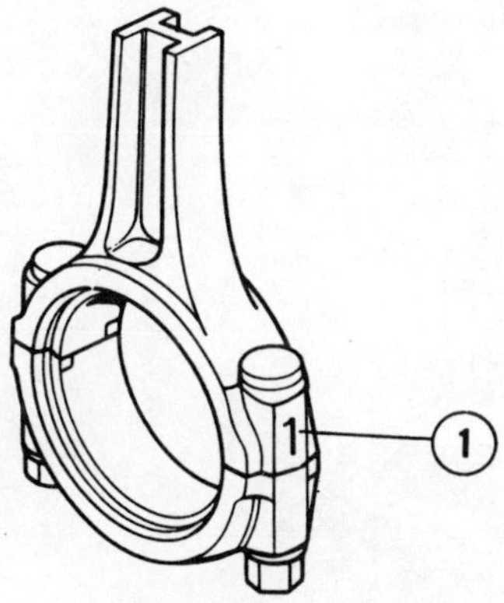

Fig. D-58
① Fit marking

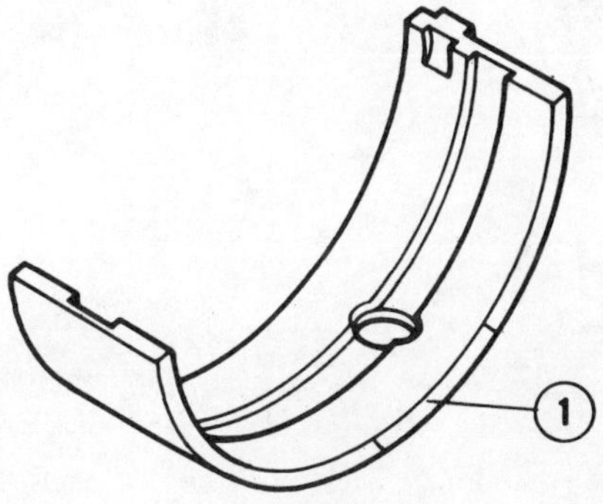

Fig. D-59
① Identification color

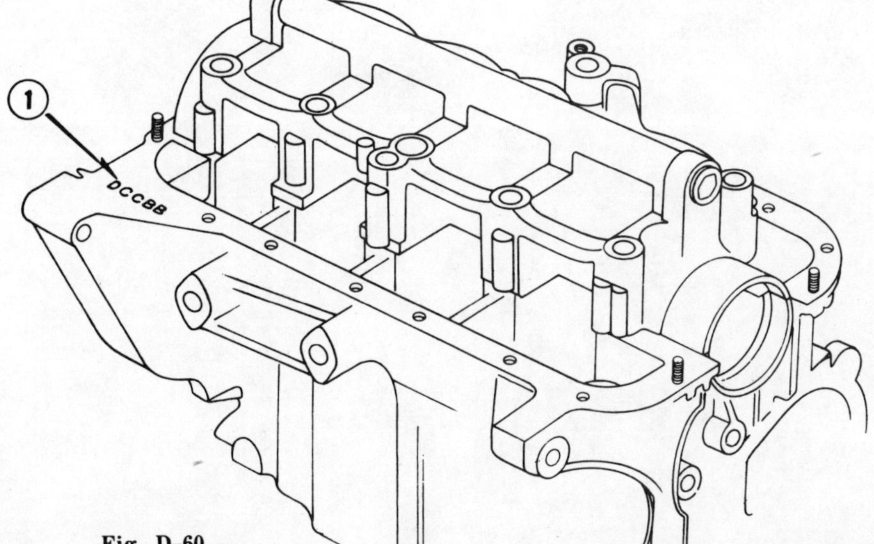

Fig. D-60
① Fit markings

2. Oil clearance includes an increase of 10μ which is produced by "crushing" the bearing. The crankpins and connecting rod bearings are selective-fitted so that 0.020 to 0.038 mm (0.0008 to 0.0015 in.) is obtained in the oil clearance.
3. Bearings are identified with colors, blue, black, brown, green, yellow, pink and red, each being tolerated 0.003 mm (0.0001 in.).

How to use the table

Knowing a set of fitting marks on the connecting rod and crankpin, or the crankshaft and cylinder counterbore, which are a alphabetical letter or a figure respectively, pick up the collar coding which corresponds to them. A bearing with the collar coding is to be installed.

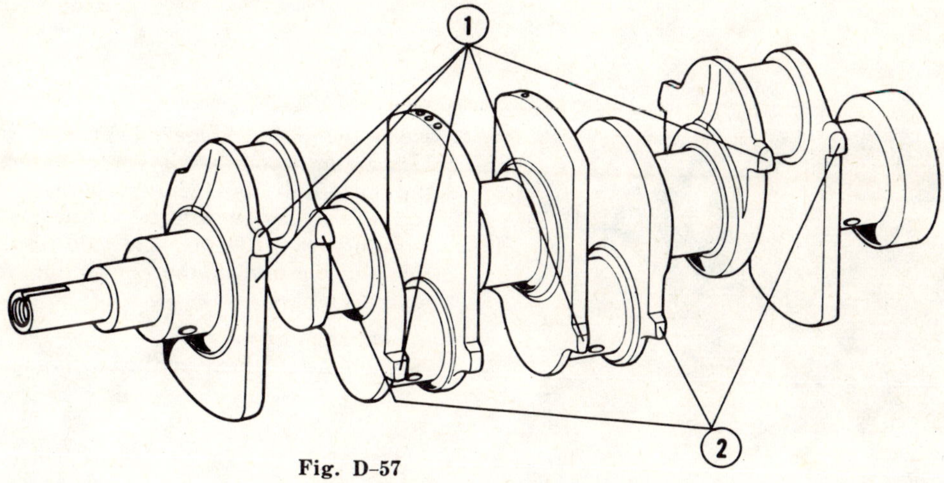

Fig. D-57
① Crankshaft journal fit markings
② Crank pin fit markings

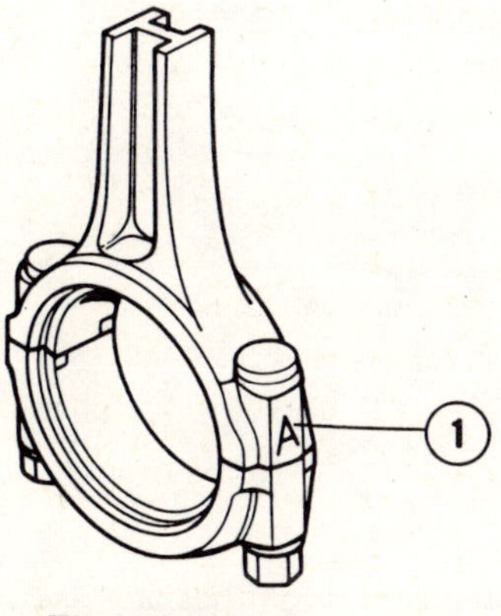

Fig. D-58
① Fit marking

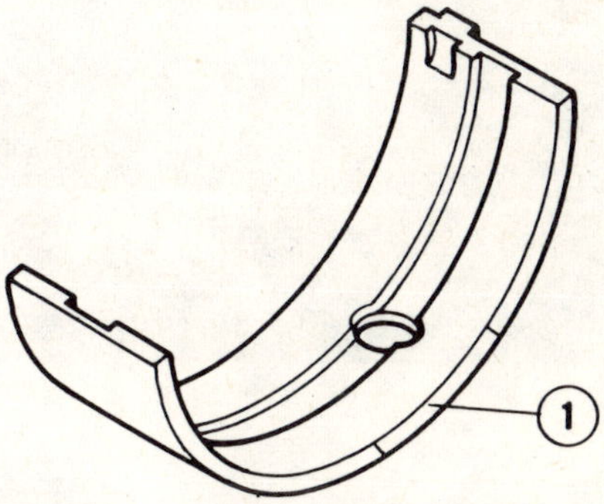

Fig. D-59
① Identification color

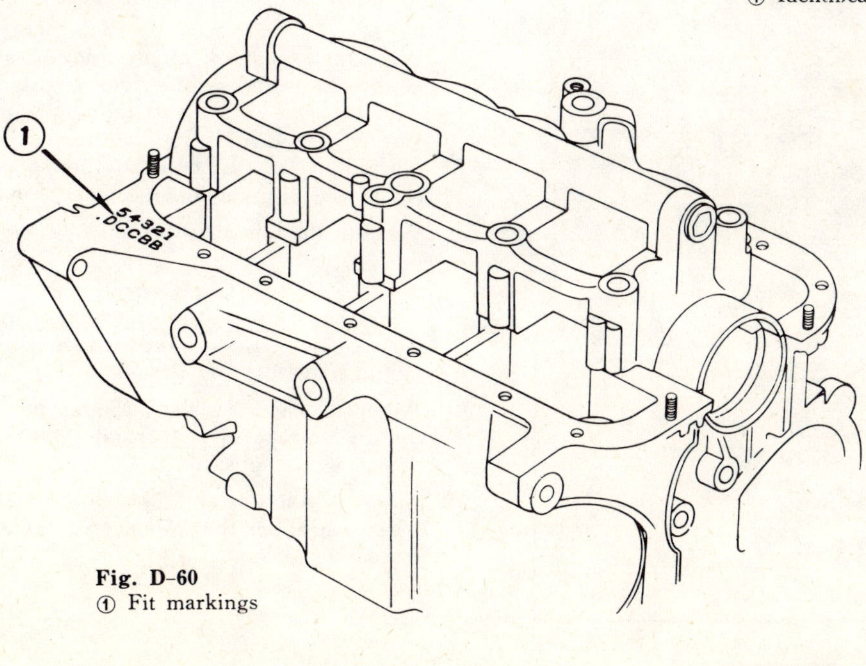

Fig. D-60
① Fit markings

d. Installation

To install the crankshaft and bearings, reverse the removal procedures. Pay attention to the following points:

1. Before installing the piston in the cylinder, apply a coat of engine oil to the inside wall of the cylinder and to the piston ring grooves.

Fig. D-61 Piston slider

2. Keep the piston rings compressed with a piston slider. Using the wood handle of a hammer, when pressing the piston in the cylinder bore, guide the connecting rod which has been fitted to the piston with one hand, working from the opposite side of the engine block so that the big end of the connecting rod can be installed to the crankpin.
3. When assembling the piston to the cylinder bores, use care to install them in their original locations.

Fig. D-62
① Thrust plate

4. When installing the thrust washers, face their sides having grooves outward.
5. Apply a coat of engine oil or molybdenum bisulfide to the surfaces of the bearings before installing.

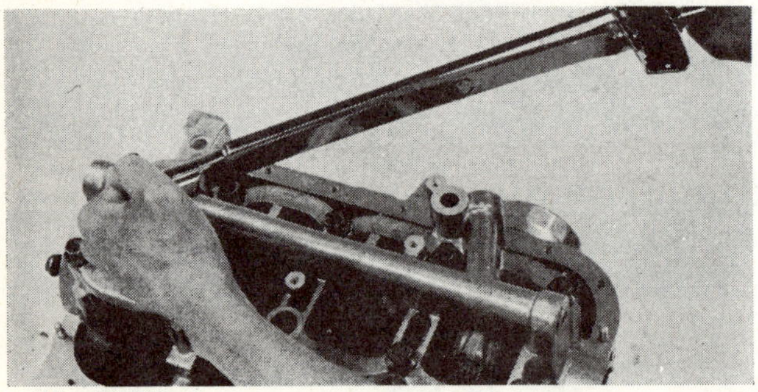

Fig. D-63

6. **Install the connecting rod bearing caps in the correct direction as required. After tightening the cap bolts to the specified torque, check the rods for smooth movement.**
7. Wipe off excess oil on the mating surface between the bearing caps and cylinder block with a clean waste cloth. Apply a coat of sealant (liquid packing) to the mating surface and install the bearing caps in place. After tightening the cap bolts to the specified torque, hand-rotate the crankshaft and check for smooth rotation.
8. When installing the oil pass block, do not forget to install its O-ring.

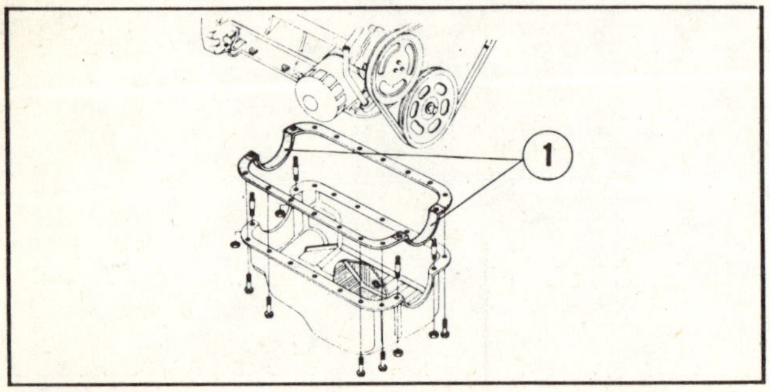

Fig. D-64
① Apply a coat of sealant to these parts

9. Check the oil pan attaching surface of the cylinder block for damage. Reface the surface with a fine oilstone, if damaged.
10. Apply a coat of sealant (liquid packing) to the entire mating surface of the cylinder block except the crankshaft oil seal position.
11. First tighten the oil pan bolts and nuts temporarily. Then retighten them in a criss-cross pattern, beginning on center bolt.

12. Hold the flywheel ring gear with a ring gear holder and tighten the ring gear bolts to the specified torque.

13. To install the drive pulley, install the timing guide plate and guide plate B properly in this order.

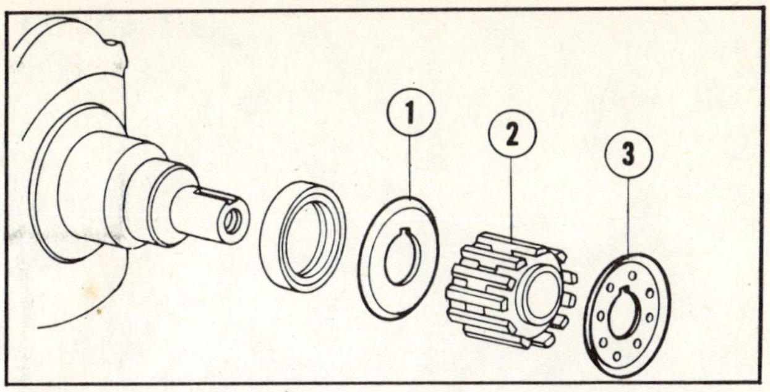

Fig. D-65
① Timing guide plate ② Timing belt drive pulley
③ Timing guide plate B

14. Install the timing belt as indicated by the marking put on the belt. Improper installation will result in a rapid wear of the belt.

Fig. D-66
① Mark

Precautions for handling the timing belt:

- When installing, check the belt for damage and teeth for peeling. Replace the belt if damaged.
- Do not allow oil to come in contact with the belt to prevent the rubber from swelling.
- Do not bend the belt within 13 mm (0.512 in.) radius since it is made up of glass fiber.
- When removing or installing the belt, do not use any tool having a sharp edge such as a screwdriver to prevent damage to the belt.

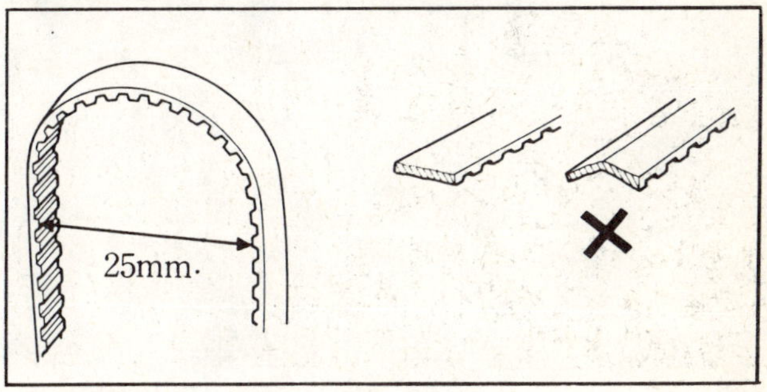

Fig. D-67

15. After installing the timing belt, adjust the belt tension.

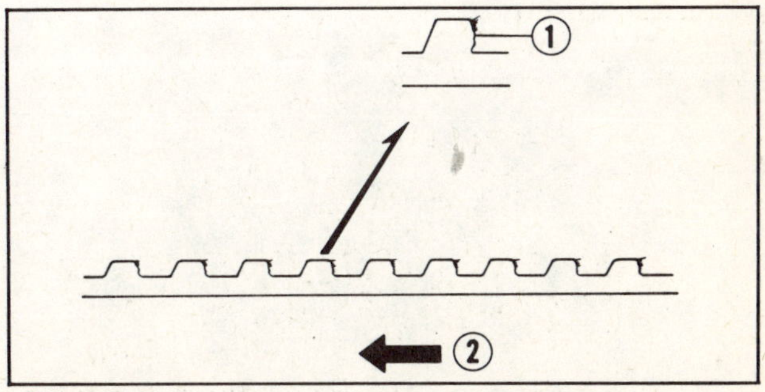

Fig. D-68
① Jacket peeled ② Direction of belt drive

Oil Pump and Fuel Pump

a. Oil pump

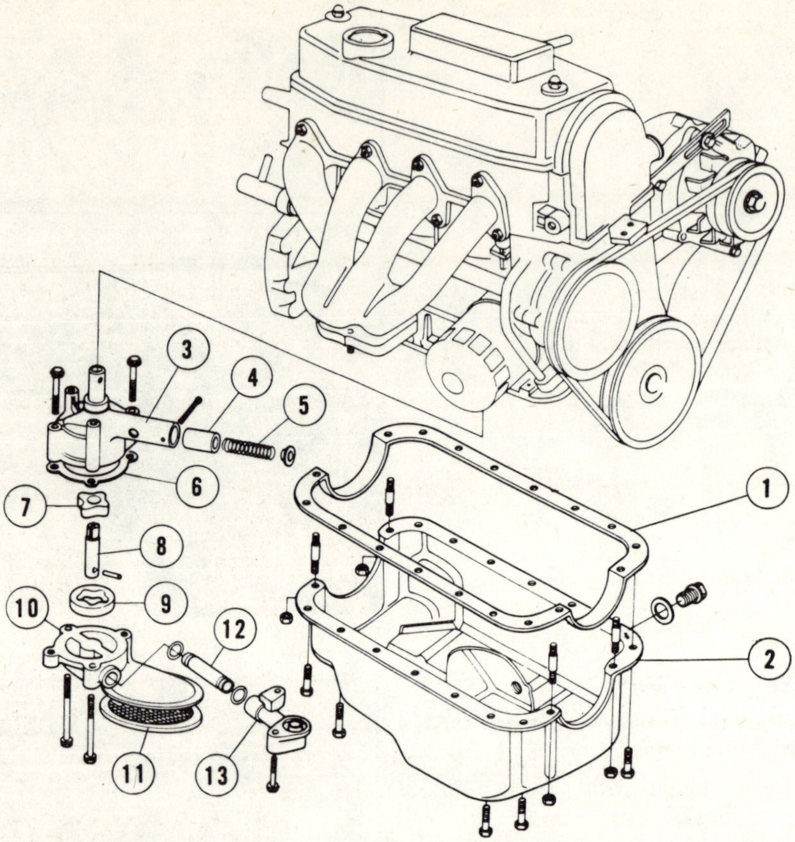

Fig. D–69

① Oil pan gasket
② Oil pan
③ Oil pump body
④ Relief valve
⑤ Relief valve spring
⑥ Pump body gasket
⑦ Pump inner rotor
⑧ Oil pump shaft
⑨ Pump outer rotor
⑩ Oil strainer body
⑪ Oil pump filter screen
⑫ Oil pass pipe
⑬ Oil pass block

a–1. Removal (on car)

1. Pull up the parking brake lever fully. **Jack up** the front of the car and support it with safety **stands.**
2. Attach a sling to the engine and lift it with a chain block just far enough for the chain block to take load.
3. Drain the engine oil pan by removing the drain **bolt.**
4. Remove the center beam and engine mount **rubber A.**
5. Remove the oil pan and clutch case cover.
6. Remove the oil pass block and oil pump assembly.

a–2. Disassembly

Remove the oil relief valve cotter pin. Remove the valve seat, spring, and valve in this order and remove two pump body bolts. Then the pump can be disassembled as shown.

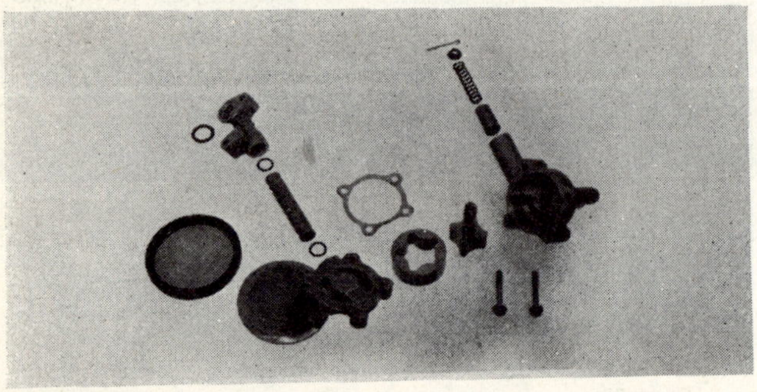

Fig. D–70

a-3. Inspection

See the Maintenance Standard at the end of this manual.

Inspection item	Remedy
Radial clearance between inner and outer rotors	Replace.
Side clearance between rotor and oil strainer body	Replace.
Clearance between outer rotoı and pump body	Replace.
Wear or damage of pump gear	Replace.
Fit of relief valve and damage of its sliding surface	Replace.

● Inspect the pump

After assembling the pump, check it for condition:

Pour engine oil into the oil pan and immerse the strainer side of the pump in the oil. Rotate the pump shaft with a standard-type screwdriver tip placed in the shaft groove, until the oil comes out of the delivery side.

After verifying that the oil comes out sufficiently, plug the oil outlet of the pump with a finger. Check to feel if the oil pressure rises when rotated further, and if the oil comes out when the finger pressure in slightly released.

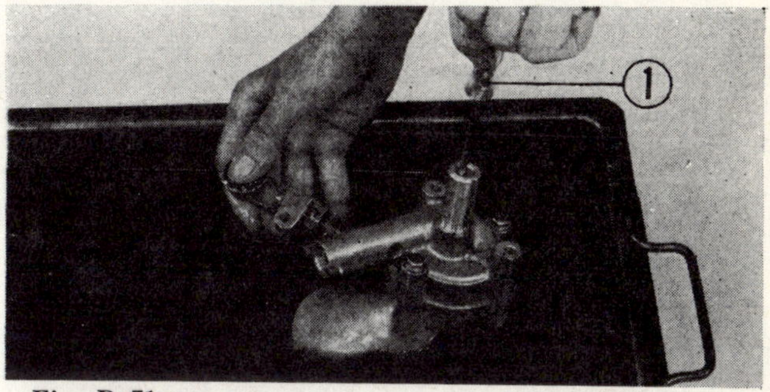

Fig. D-71
① Screwdriver

● Replacement of oil filter

1. Securely put oil filter socket on the filter head, and remove the filter using a ratchet wrench.

   NOTE:
   A small amount of oil will come out. Wipe it with a waste cloth.
2. Replace the filter as an assembly since it is of a cartridge type.
3. Check the filter attaching surface of the cylinder block and the gear holder for damage.
4. Use a new filter gasket when reinstalling.
5. Clean the filter attaching surface of the cylinder block with a clean waste cloth. Apply a thin coat of oil to the filter gasket and install it to the cylinder block.
6. First tighten the filter bolts with fingers, and then re-tighten them with a wrench.
7. After installing, run the engine and check for any sign of oil leak.

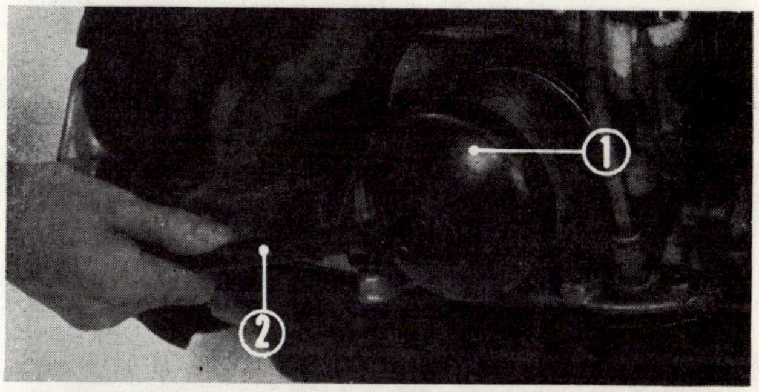

Fig. D-72
① Filter socket ② Wrench

Fig. D-73
① Holder ② Attaching surface ③ Gasket

a-4. Installation

To install the oil pump, reverse the removal procedures.

After installing, run the engine at idle speed and check to see if the oil is pumped up when the oil filter cap is removed.

b. Fuel pump

b-1. Removal

1. Disconnect the inlet and outlet fuel tubes at the pump. Loosen the pump nuts and remove the pump.

   NOTES:

   1) When the fuel tube is disconnected or when the pump is removed, fuel will come out. Wipe the fuel with a waste cloth.

   2) Do not disassemble the pump unless necessary. This disassembly may be the cause of fuel or oil leak. If the pump is defective, replace it as an assembly.

b-2. Inspection

See the Maintenance Standard at the end of this manual.

Inspection item	Remedy
Looseness of pump connector	Retighten.
Looseness of upper and lower body and cover screws	Retighten.
Looseness of rocker arm pin	Retighten or replace.
Contamination or clogging of air hole	Repair or replace.
Improper operation of pump	Replace.

Fig. D-74
① Air hole

Inspect the air hole for condition

Check to see if the air hole is dirty with fuel or oil. If so, the diaphragm is damaged. Replace the diaphragm. Also check the hole for clogging. If necessary, direct a blast of compressed air to the hole and then check to be sure fuel or oil does not come out.

Fig. D-75

Inspect the pump for operation

1. Disconnect the fuel tube at the carburetor. Connect a pressure gauge to the delivery side of the pump. Start the engine and measure the pump delivery pressure.

   Delivery pressure specification:

   0.18 kg/cm^2 (2.56 lb/in^2)

2. After measuring, stop the engine and check to see if the gauge pointer drops suddenly. If so, fuel or oil may leak from the diaphragm or from the valves.

Cam speed, rpm	Delivery pressure, kg/cm^2 (lb/in^2)	Suction pressure, mmHg	Displacement, cc (in^3)/min
300	0.18 (2.56)	450	450 (27), min.
1500	0.18 (2.56)	250±50	700 (43), min.
3000	0.18 (2.56)	250±50	750 (46), min.

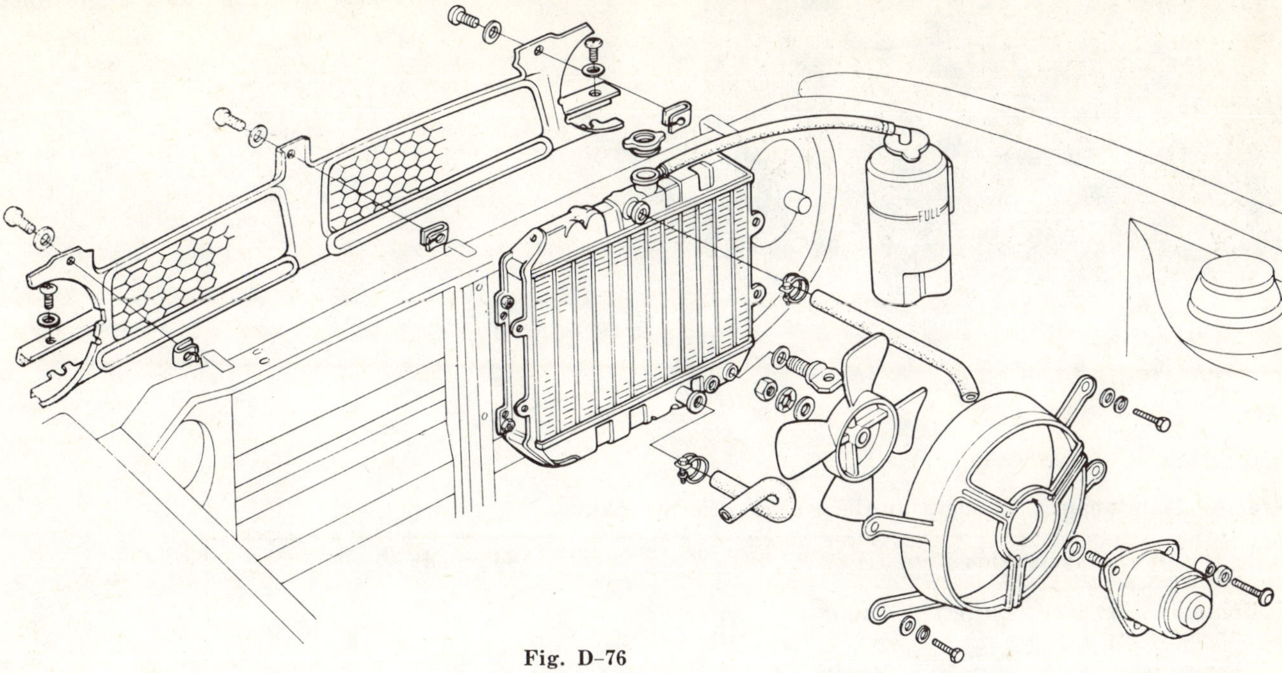

Fig. D-76

NOTES:

1) When inspec ting or servicing the radiator, be sure to turn the ignition switch off and disconnect the radiator fan motor wire at the connector, since the radiator mounted on the Honda CIVIC is so designed that the fan motor automatically starts when the cooling water temperature exceeds 90°C (194°F).
2) When removing the radiator, take care not to damage the core and fins.

a. Removal (on car)

Radiator

1. Drain the radiator from the lower radiator tank by loosening the drain cock.
2. Remove the fan motor and disconnect the thermoswitch cord.
3. Disconnect the upper water hose at the upper tank and the lower hose at the connecting pipe.
4. Remove the front grille.
5. Remove the radiator and spare tank by loosening the radiator bolts.

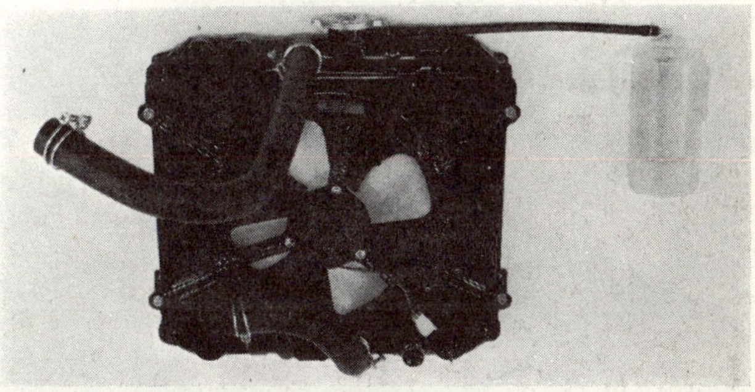

Fig. D-77

Water pump

1. Drain the radiator.
2. Loosen the generator bolts and adjusting bolt. Move the generator toward the cylinder block and remove the drive belt.
3. Loosen the pump bolts and remove the pump together with the seal rubber C.

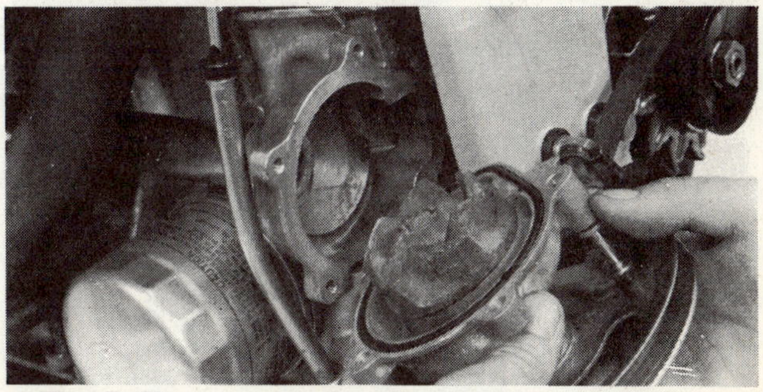

Fig. D-78

Fig. D-79 Thermostat

Thermostat

1. Remove the thermostat cover and then the thermostat.

b. Inspection

See the Maintenance Standard at the end of this manual.

Inspection item	Remedy
Contamination and incorrect operation of thermostat	Replace.
Contamination of radiator and clogging of core	Repair or replace.
Leak from radiator	Repair or replace.
Bend or other damage of radiator core	Repair or replace.
Damage of hoses	Replace.
Damage or contamination of cylinder head and carburetor attaching surfaces of intake manifold	Repair or replace.
Contamination or damage of water pump impeller	Replace.
Malfunction of thermoswitch	Replace.

• See the inspection of the cooling system.

c. Installation

To install, reverse the removal procedures.

CLUTCH

NOTE:

When removing the transmission, clutch and differential from the engine, remove the transmission and clutch cases as a unit.

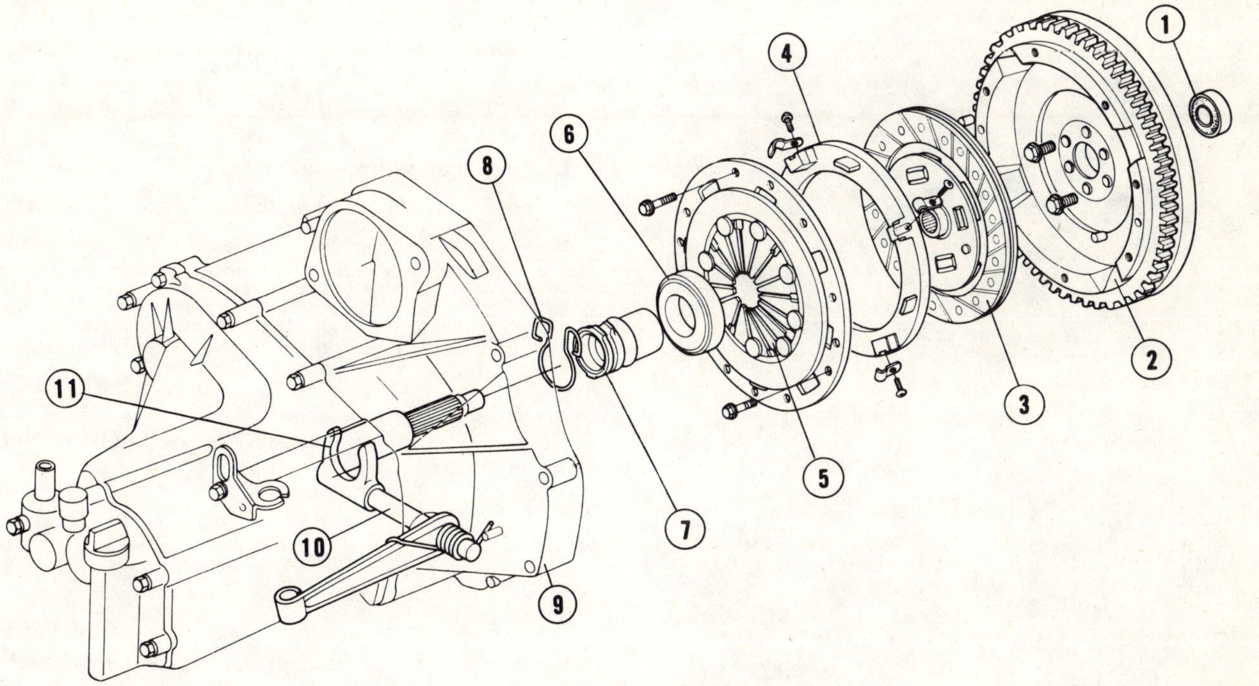

Fig. E–1

① Clutch pilot bearing
② Flywheel
③ Friction disc
④ Pressure plate
⑤ Diaphragm spring
⑥ Release bearing
⑦ Release bearing holder
⑧ Release arm clip
⑨ Clutch case
⑩ Release shaft complete
⑪ Clutch release arm A

a. Removal (on car)

1. Drain the transmission by removing the drain bolt.
2. Disconnect the cord at the positive terminal of the battery and the ground cable at the transmission case.
3. Disconnect the following cables and cord:
   - Clutch control cable
   - Starting motor control cable
   - Speedometer drive cable
   - Back up light switch cord
4. Loosen the right and left wheel clip nuts.
5. Fully pull up the parking brake lever. Jack up the front of the car and support it with safety stands. Remove the wheels and place them under both side sills in back of the safety stands.
6. Remove the right engine mount rubber B.

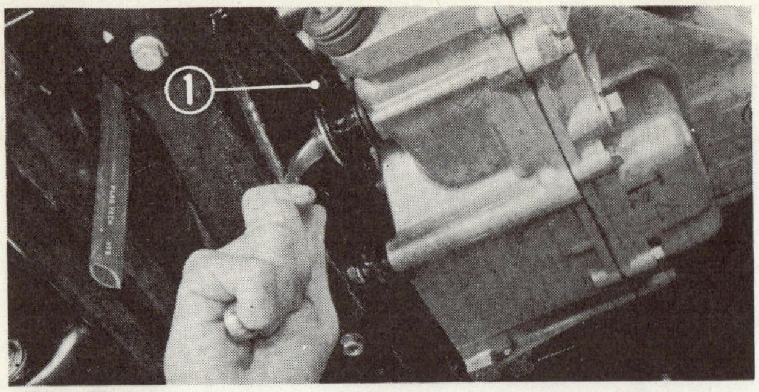

Fig. E–2
① Gearshift extension

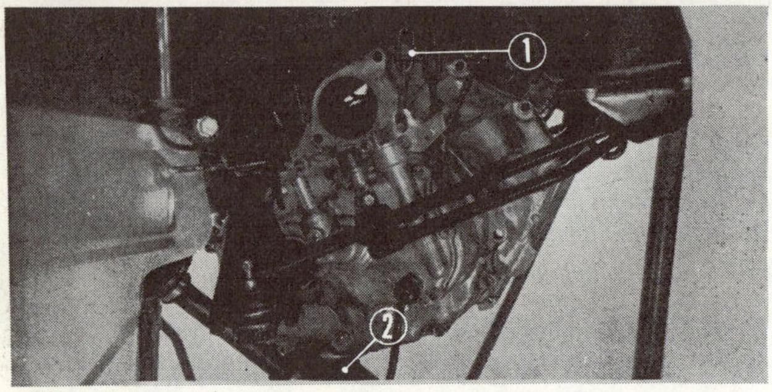

Fig. E–3
① Chain ② Jack

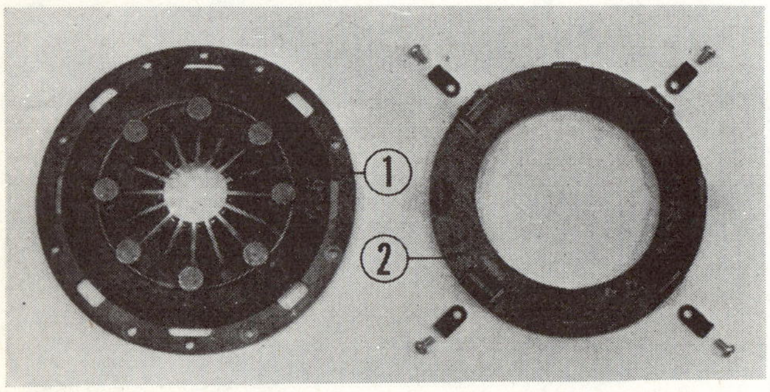

Fig. E–4
① Diaphragm spring ② Pressure plate

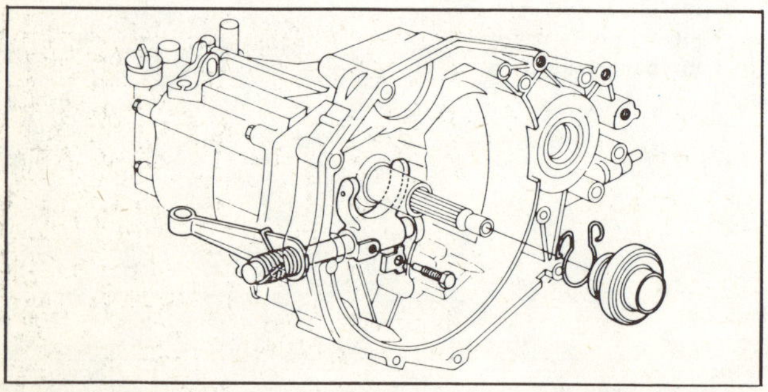

Fig. E–5

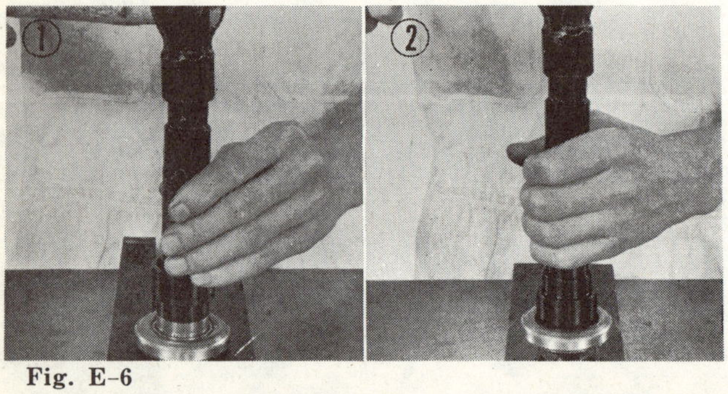

Fig. E–6
① Removing ② Installing

7. Remove the lower arm and tie rod end ball joints. Then pull out the right and left drive shafts.
8. Pull out the gearshift rod 8 mm (0.315 in.) spring pin and disconnect the rod at the transmission case.
9. Disconnect the gearshift extension at the engine side joint.
10. Attach a chain to the clutch control cable sling and lift it with a chain block just far enough for the chain block to take load.
11. Remove the three center mount bolts.

    NOTE:

    Do not remove the center beam.
12. Support the bottom of the transmission case with a garage jack.
13. Remove the starting motor.
14. Remove four clutch case bolts.
15. Slowly lower the chain block and garage jack, and lower the transmission and clutch cases as a unit.

    NOTE:

    To facilitate removal of the cases, incline the transmission cover slightly.
16. Hold the flywheel ring gear with a ring gear holder and remove the pressure plate and flywheel in this order from the engine.

b. Disassembly

1. Remove the clutch stopper springs and disassemble the pressure plate as shown.
2. Remove the release arm clip from the release arm A with a screwdriver and pull the release bearing backward.
3. Pull the release shaft out of the clutch case by straightening the lug of the 8 mm (0.315 in.) lock washer and loosening the bolt. Then remove the release arm A..
4. Remove the release bearing using a bearing driver.

   NOTE:

   When removing or installing the release bearing, use a bearing driver and an attachment. At this time, reverse the attachment.

c. Inspection

See Maintenance Standard at the end of this manual.

Inspection item	Remedy
Wear of contact surface of pressure plate	Replace.
Wear of friction disc (rivet sinkage, thickness, condition of surface)	Replace.
Wear of main shaft serration	Replace.
Wear of friction disc serration	Replace.
Wear of contact surface of diaphragm spring release bearing	Replace.
Face runout of friction disc	Replace.

c-1. Release bearing

Press the bearing toward the holder and rotate it to check for humming or erratic rotation.

NOTES:

1) The bearing is filled with grease and, therefore, it must not be washed in gasoline.
2) When installing a new bearing, install it with the machined (radius) side facing in the direction opposite to the holder.

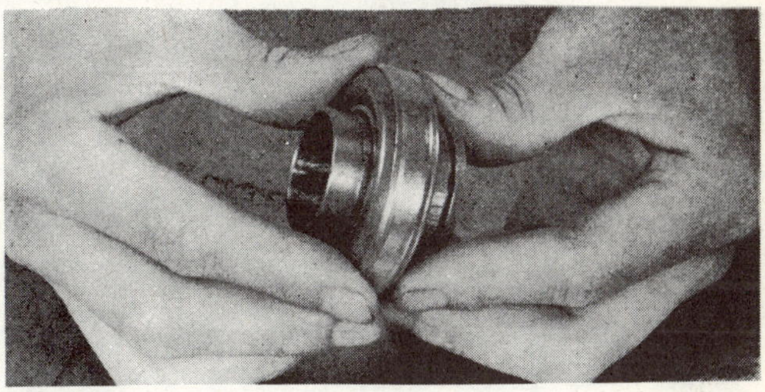

Fig. E-7

c-2. Clutch friction disc

Check the disc for wear, judging from the rivet sinkage. Replace the disc, if it is wet with oil or grease, it is excessively hardened, or if the torsion rubber is deteriorated. Install the friction disc on the main shaft and check for face runout of the disc and rattle of the shaft spline with a dial gauge.

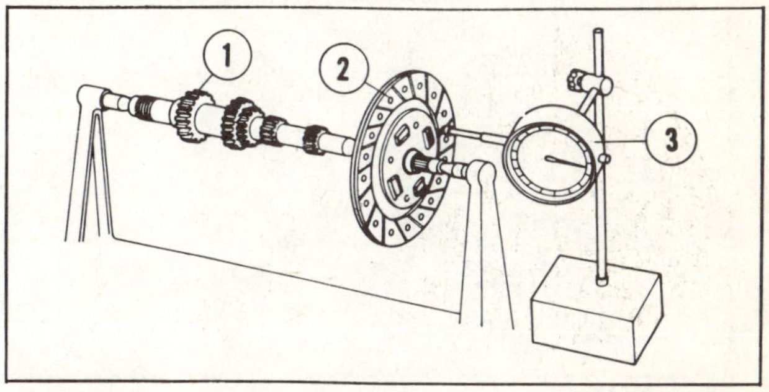

Fig. E-8
① Main shaft ② Friction disc ③ Dial gauge

c-3. Pressure plate and flywheel

Check the pressure plate and disc contact surface of the flywheel for cracks or wear and replace them if excessively damaged.

If they are slightly damaged, reface them with a fine (#500 or #600) emery paper.

Check the clutch pilot bearing for humming or erratic rotation, and replace it if necessary.

Fig. E-9

d. Assembly

To assemble the clutch, reverse the disassembly procedures. Pay attention to the following points:

1. Wipe oil or grease off the flywheel and end face of the crankshaft with a clean waste cloth. Properly fit the flywheel on the end face of the crankshaft and tighten their attaching bolts temporarily.

   Hold the ring gear with a ring gear holder and then retighten the attaching bolts to the specified torque.

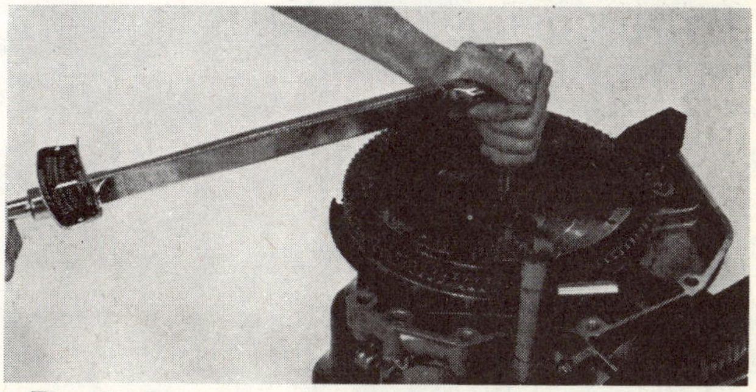

Fig. E-10

Fig. E-11

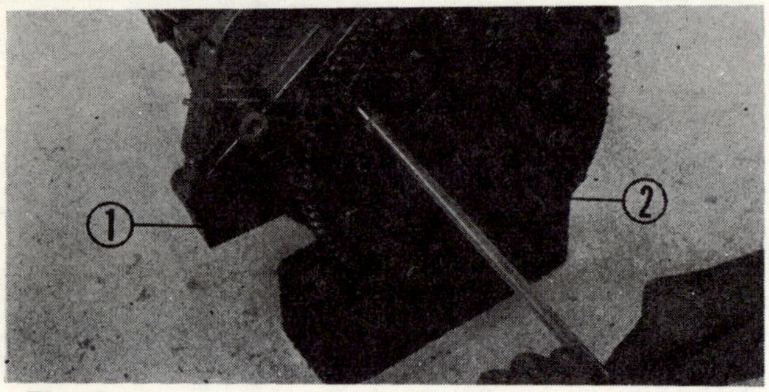

Fig. E-12
① Ring gear holder ② Center pin

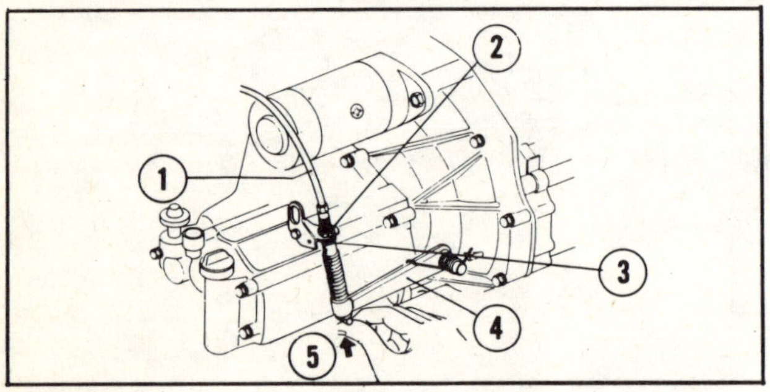

Fig. E-13
① Clutch control cable ② Adjusting nut ③ Lock nut
④ Clutch release lever ⑤ Clutch release lever play

2. Install the friction disc and pressure plate in this order. To install, align the matching marks on the outer periphery of the flywheel and on the clutch cover.

   NOTE:

   A failure to align the matching marks will disturb the balance of the weight of the clutch assembly.

3. When tightening the pressure plate bolts, insert the clutch center pin in place to bring the friction disc spline hole to the center. Then tighten the bolts evenly in a criss-cross pattern.
4. When installing the clutch case to the engine, first insert the main shaft into the friction disc spline hole properly. Then, properly fit the clutch case to the engine securely and tighten the attaching bolts to the proper specification.

   NOTE:

   If the case bolts are tightened with the case and engine fitted loose, the friction disc takes load and would be damaged.

5. Upon assembling, adjust the clutch. To adjust, turn the adjusting nut until the play of release lever is 2 to 3 mm (0.079 to 0.118 in.) as measured at the tip of the lever. Then lock the adjusting nut with the lock nut.
6. After adjusting, check the clutch for proper operation. To check, run the engine at idle speed and check for abnormal noise or for hard shifting.

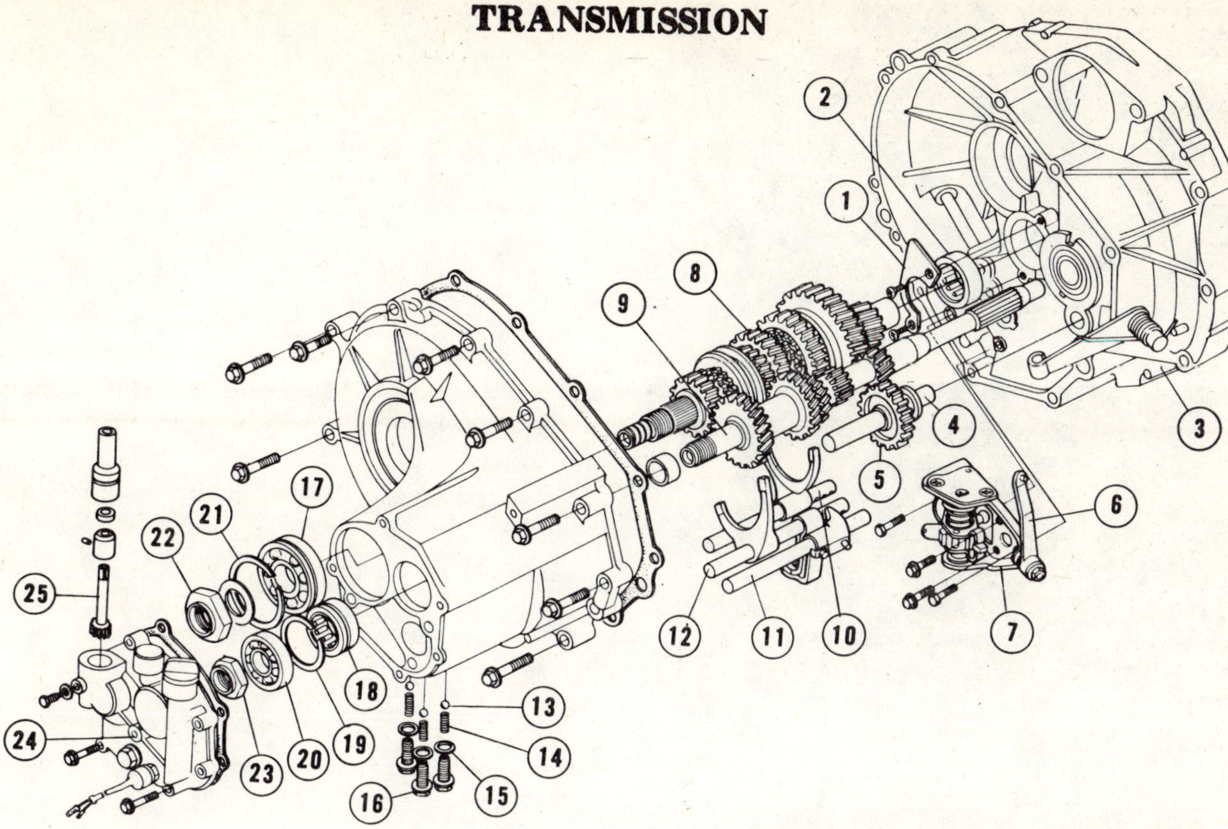

Fig. F–1

① Needle roller bearing set plate
② Needle roller bearing
③ Clutch case
④ Reverse gear shaft
⑤ Reverse idle gear
⑥ Reverse shift fork
⑦ Arm B holder complete
⑧ Countershaft gear assembly
⑨ Main shaft
⑩ Low fork shaft
⑪ Reverse fork shaft
⑫ Top fork shaft
⑬ Steel ball
⑭ Ball set spring
⑮ Drain plug washer
⑯ Set ball spring screw
⑰ Ball bearing
⑱ Needle roller bearing
⑲ 48 mm snap ring
⑳ Ball bearing
㉑ 62 mm snap ring
㉒ 23 mm lock nut
㉓ 20 mm lock nut
㉔ Transmission cover
㉕ Speedometer gear

a. Removal (on car)

1. Remov the transmission case from the engine together with the clutch case.

b. Disassembly

1. Remove the transmission cover.
2. Straighten the lug of the 20 mm lock nut of the main shaft with a standard-type screwdriver and remove the lock nut by using two wrenches as shown.

NOTE:

20 mm lock nut has left-hand threads. Place the gear-shift lever in any speed position, and hold the counter-shaft side nut with a wrench and loosen the main shaft side nut.

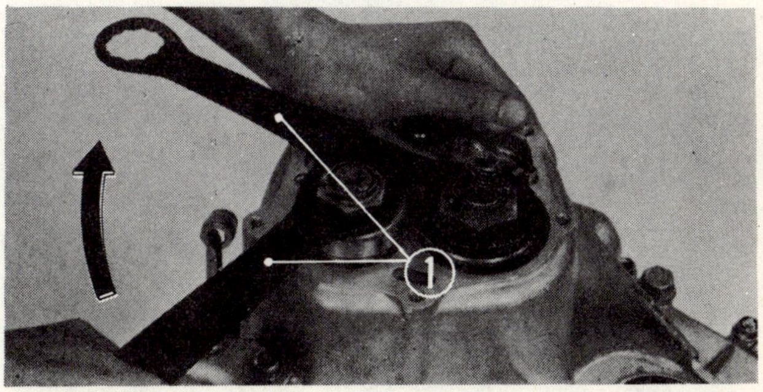

Fig. F–2
① Lock nut wrenches

Fig. F-3

3. Remove the main shaft ball bearing and then two snap rings with a snap ring pliers.

Fig. F-4
① Lock ball screws ② Spring ③ Ball

4. Loosen three lock ball screws and remove the screws, springs and balls.

NOTE:

After removing, set aside the screws, springs and balls in a group.

Fig. F-5
① Transmission case ② Clutch case

5. Loosen the transmission case bolts and lightly tap the case with a mallet to remove it.

NOTES:

1) While removing, keep the clutch case facing downward and the transmission case upward in a horizontal position.
2) Do not insert a screwdriver into between the mating surfaces of the clutch and transmission cases for prying off.

Fig. F-6
① Differential gear holder ② Lock nut wrench

6. Hold the differential gears with a differential gear holder and loosen the countershaft lock nut.

NOTE:

Engage the lugs of both ends of the differential gear holder in the dowel pin holes in the clutch case and the center lug in the hole in the differential case.

The countershaft lock nut has right-hand threads.

Fig. F-7
① Reverse gear ② Reverse shift fork
③ Arm holder

7. Remove the reverse idle gear and shaft.
8. Remove the reverse shift fork.
9. Remove the arm B holder together with the shift arm B.

10. Remove the main shaft and countershaft together with the low and top fork shafts.
11. Remove the shift rod boot. Remove the shift arm A lock washer and bolt and remove the shift rod and shift forks from the clutch case.

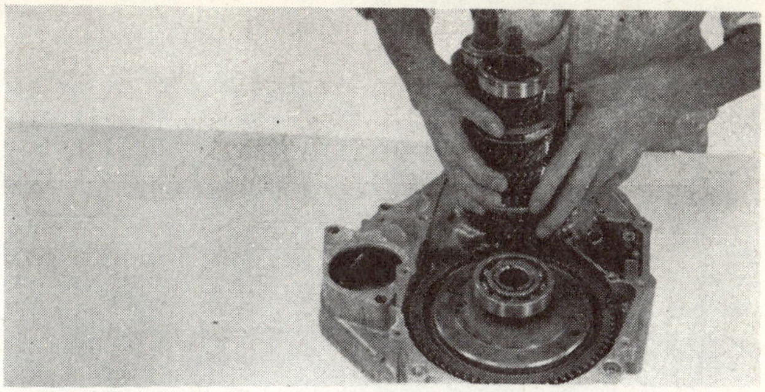

Fig. F-8

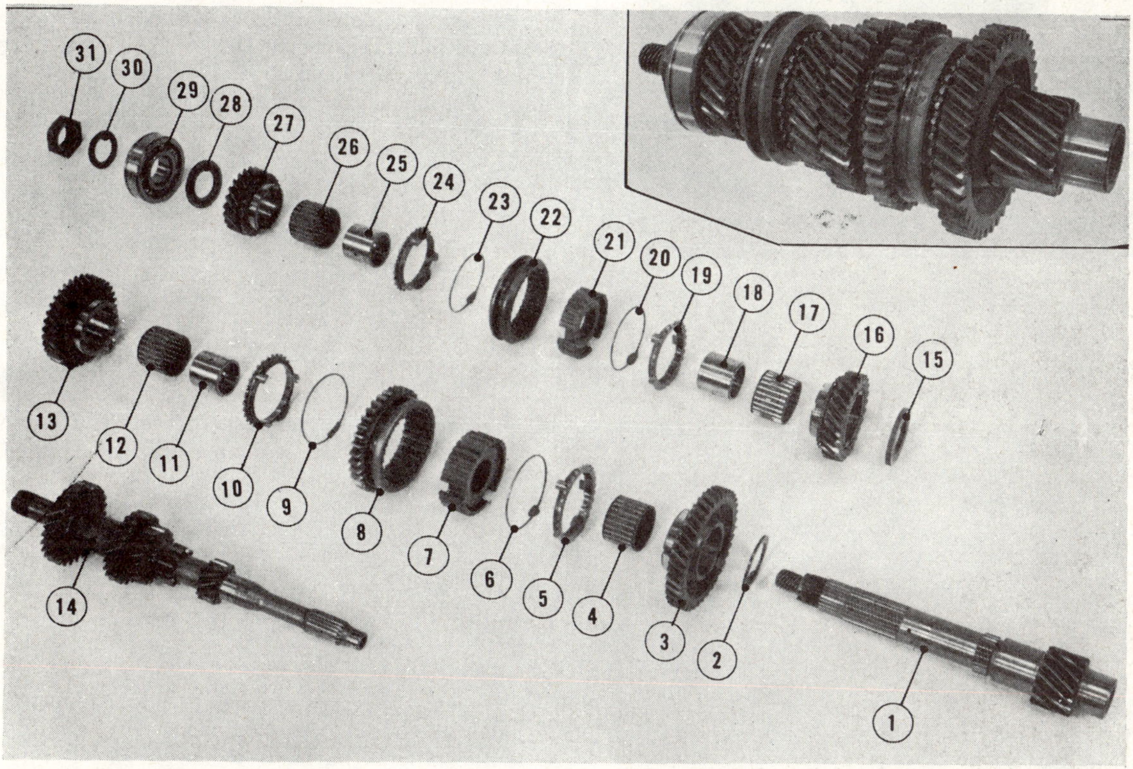

Fig. F-9

① Countershaft
② Thrust plate
③ Countershaft low gear
④ Needle roller bearing
⑤ Blocking ring
⑥ Synchronizer spring
⑦ Synchronizer hub
⑧ Countershaft reverse gear
⑨ Synchronizer spring
⑩ Blocking ring
⑪ Distance collar
⑫ Needle roller bearing
⑬ Countershaft second gear
⑭ Main shaft
⑮ Distance plate
⑯ Countershaft third gear
⑰ Needle roller bearing
⑱ Distance collar
⑲ Blocking ring
⑳ Synchronizer spring
㉑ Synchronizer hub
㉒ Synchronizer sleeve
㉓ Synchronizer spring
㉔ Blocking ring
㉕ Distance collar
㉖ Needle roller bearing
㉗ Countershaft top gear
㉘ Thrust plate
㉙ Ball bearing
㉚ Thrust plate
㉛ Lock nut

c. Inspection

See Maintenance Standard at the end of this manual.

Inspection item	Remedy
Wear or damage of gear teeth	Repair or replace.
Roughness or damage of gear surface (tapered) which comes in contact with blocking ring	Repair or replace.
Wear or damage of gear and shaft sliding surfaces	Repair or replace.
Clearance between shift fork and synchronizer sleeve	Replace.
Wear or damage of shaft and gear splines	Repair of replace.
Clearance between blocking ring and gear (check by pressing blocking ring against gear)	Replace.
Side clearance between gears	Replace thrust plate.

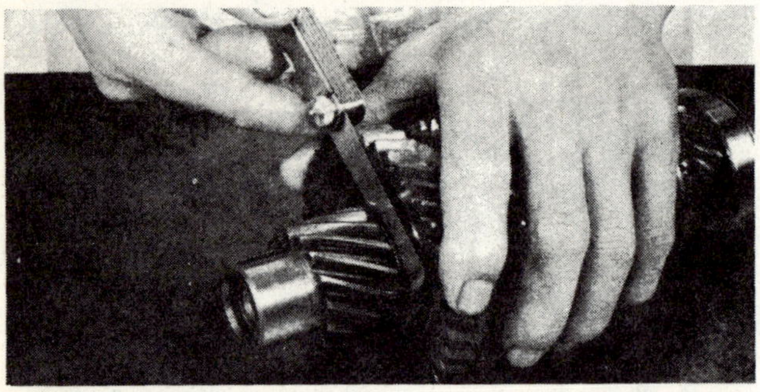

Fig. F-10

c-1. Side clearance of countershaft gears

1. Measure the axial clearance (side clearance) of the low gear with a feeler gauge. If the clearance exceeds the service limit, replace the thrust plate. Similarly measure the clearance of the second, third and fourth gears. If the clearance exceeds the service limit, replace the distance collar.

Unit: mm (in.)

Thrust plate		Distance collar	
Class	Thickness	Class	Length
A	$2.0^{-0.02}_{-0.05}$ (0.077~0.078)	1	$28.0^{+0.09}_{+0.07}$ (1.105~1.106)
B	$2.0^{-0.05}_{-0.08}$ (0.076~0.077)	2	$28.0^{+0.07}_{+0.05}$ (1.104~1.105)
C	$2.0^{-0.08}_{-0.11}$ (0.074~0.076)		

2. The thrust plates and distance collars for adjusting the side clearance are available in several thicknesses as shown in the table.

c-2. Synchronizer sleeves

1. Check the tapered surfaces of each blocking ring and gear for roughness or damage. Apply a thin coat of oil to the tapered surfaces of the blocking ring and gear, rotate the ring and gear while pressing them together with hand and check for slippage.

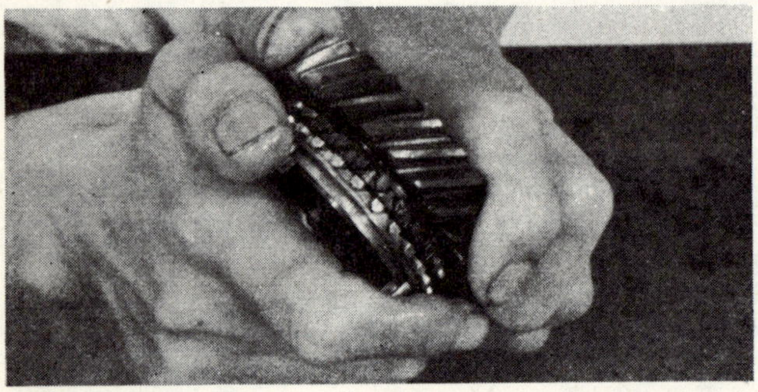

Fig. F-11

2. If the tapered surface of the blocking ring is badly defective, replace the ring. If the defect is slight, apply a thin coat of fine lapping compound to the ring and lap it together with the gear. The contact surface must be over 80 %.

NOTES:

1) Do not lap the ring with the gear excessively.
2) After lapping, wash the compound off the surface with dry cleaning solvent and check for slippage again.

3. Press the blocking ring against the tapered surface of gear and measure the clearance between the ring and gear. If it is too small, replace the ring.

   NOTE:

   Press the ring against the gear evenly, and measure the clearance on the periphery of the ring.

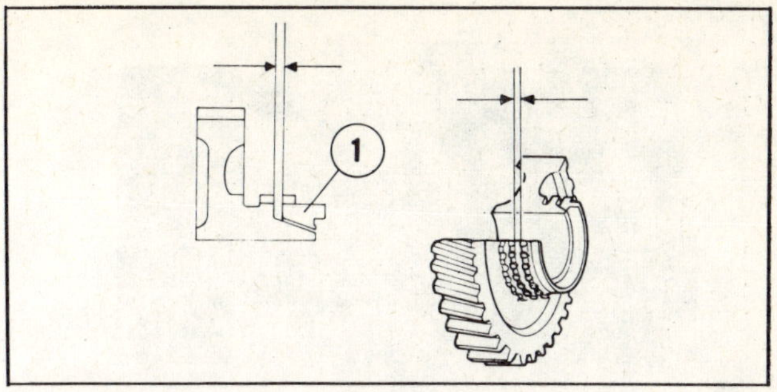

Fig. F-12
① Blocking ring

4. Measure the clearance between each shift fork and synchronizer sleeve.

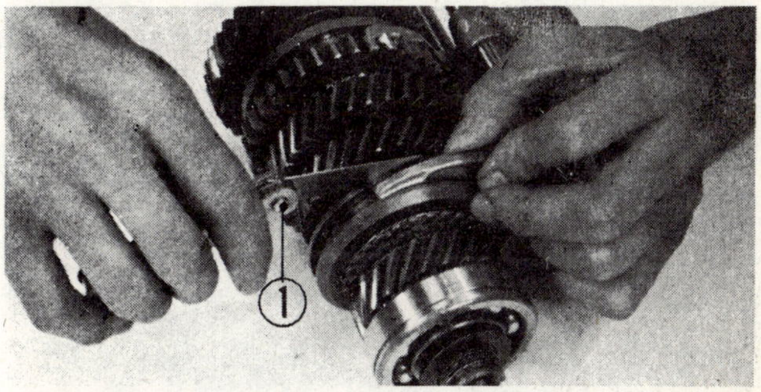

Fig. F-13
① Feeler gauge

5. Check the oil holes in the each gear and countershaft for clogging and shaft spline for wear or damage.

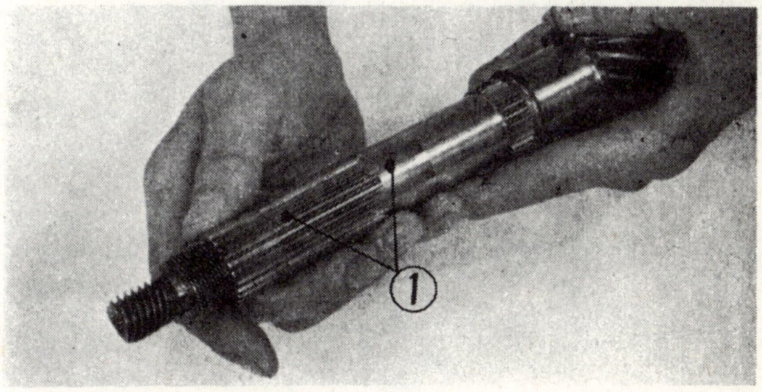

Fig. F-14
① Oil holes

d. Assembly

To assemble the transmission, reverse the disassembly procedures. Pay attention to the following points:

1. Install the synchronizer hub and sleeve.

   The synchronizer hub has three sets of two teeth cut deeper than others, equally spaced on its periphery; the sleeve has teeth cut to match in meshing.
2. Apply a coat of oil to the gears and needle roller bearings before installing.

Fig. F-15

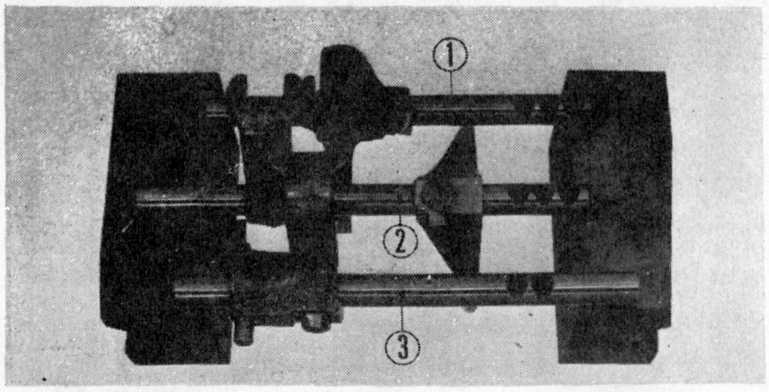

Fig. F-16
① Low fork shaft ② Top fork shaft
③ Reverse fork shaft

3. A correct set or combination of shift fork and shaft must be assembled in the correct direction.
4. Bend the lug of the lock washer securely.

Fig. F-17

5. Install the top and low fork shafts. Fit the shift forks in the grooves in the gear sleeves and install the fork shafts together with the main shaft and countershaft. Then, install the reverse gear shaft in place.

Fig. F-18

6. Use a new transmission case packing. Check the transmission and clutch case packings for damage. Then, install the transmission case to the clutch case by aligning their dowels. At this time, make sure the main shaft, countershatt, shift forks and reverse gear shaft properly engage.
7. Tighten the case bolts evenly in a criss-cross pattern and check the main shaft and countershaft for smooth rotation.

Fig. F-19

8. Put three lock ball spring screws in place. Apply a thin coat of grease to the balls and to the spring screws to stick them for ease of assembly.

Fig. F-20

9. Tighten the main shaft and countershaft lock nuts to the specified torque. Then lock the nuts by upsetting with a chisel.
10. After assembling the transmission, move the shift rod and check to see if the gears are shifted without any sign of binding.

Transmission (with Engine)

The engine can be removed or installed in the same manner as for a car equipped with manual transmission.

Following are the points of procedures different from those for a manual transmission type car.

a. Removal

1. Remove the center console and then control rod pin A.
2. Remove the front floor center mat and remove the control cable bracket nuts.
3. Jack up the front of the car and support it with safety stands.
4. Remove two selector lever bracket nuts at front side.

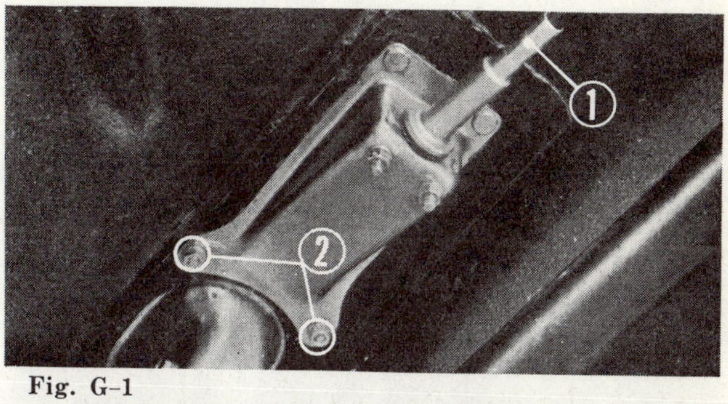

Fig. G–1
① Control cable ② Disconnect here

5. Loosen the bolts securing the control cable holder and support beam and disconnect the control cable.

Fig. G–2
① Control cable holder

6. Disconnect the automatic transmission fluid cooler hose at the transmission.
7. Remove the transmission together with the engine from the body. Then remove the engine mount rubber A and torque converter case cover.
8. Remove the starting motor. Separate the transmission case from the engine by removing the four 8 mm bolts.

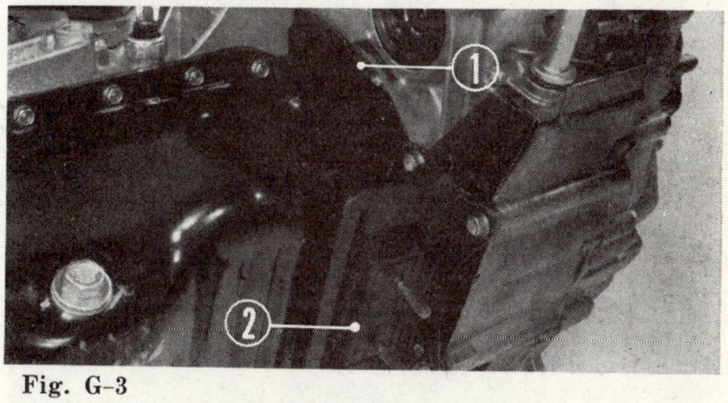

Fig. G–3
① Torque converter case cover
② Engine mount rubber A

b. Installation

To install the transmission, reverse the removal procedures. Pay attention to the following points:

1. Before installing the transmission, check to be sure the torque converter stator hub is correctly located and rotates smoothly. (Use the stator shaft for this purpose.)

Fig. G–4
① Stator shaft ② Torque converter

Fig. G-5

2. Install the transmission in place by fitting the stator and stator shaft serrations and the transmission main shaft and torque converter turbine serrations. To facilitate installation, hand-rotate the ring gear.
3. Before installing the engine to the body, test it for abnormal noise or oil leaks. Also measure the hydraulic line pressure and check to be sure it is within the specification.
4. After installing the engine, inspect the following items:
   Stall rpm
   Selector lever positions
   Pressures of "L" and "D" clutches.

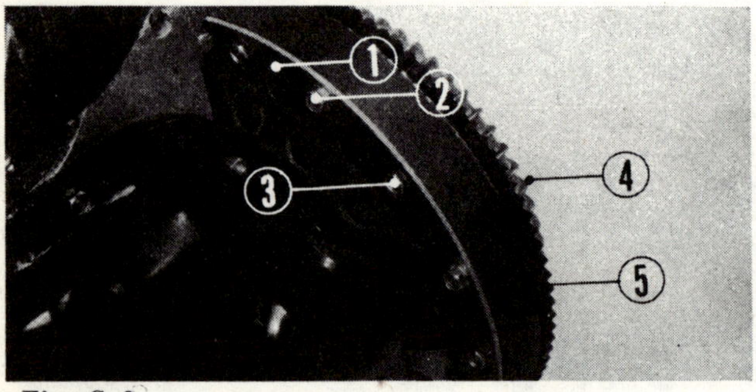

Fig. G-6
① Drive plate ② Torque converter assembly bolt
③ Ring gear bolt ④ Ring gear ⑤ Torque converter

Torque Converter (Transmission off Engine)

a. Removal and disassembly

1. Remove the engine together with transmission case from the body.
2. Separate the transmission case from the engine.
3. Remove the eight 6 mm bolts securing the torque converter assembly to the drive plate and remove the torque converter assembly.
4. Remove the drive plate. (When the torque converter only is removed, it is unnecessary to remove this cover.)
5. The torque converter can be disassembled as shown below.

· **Disassembled view**

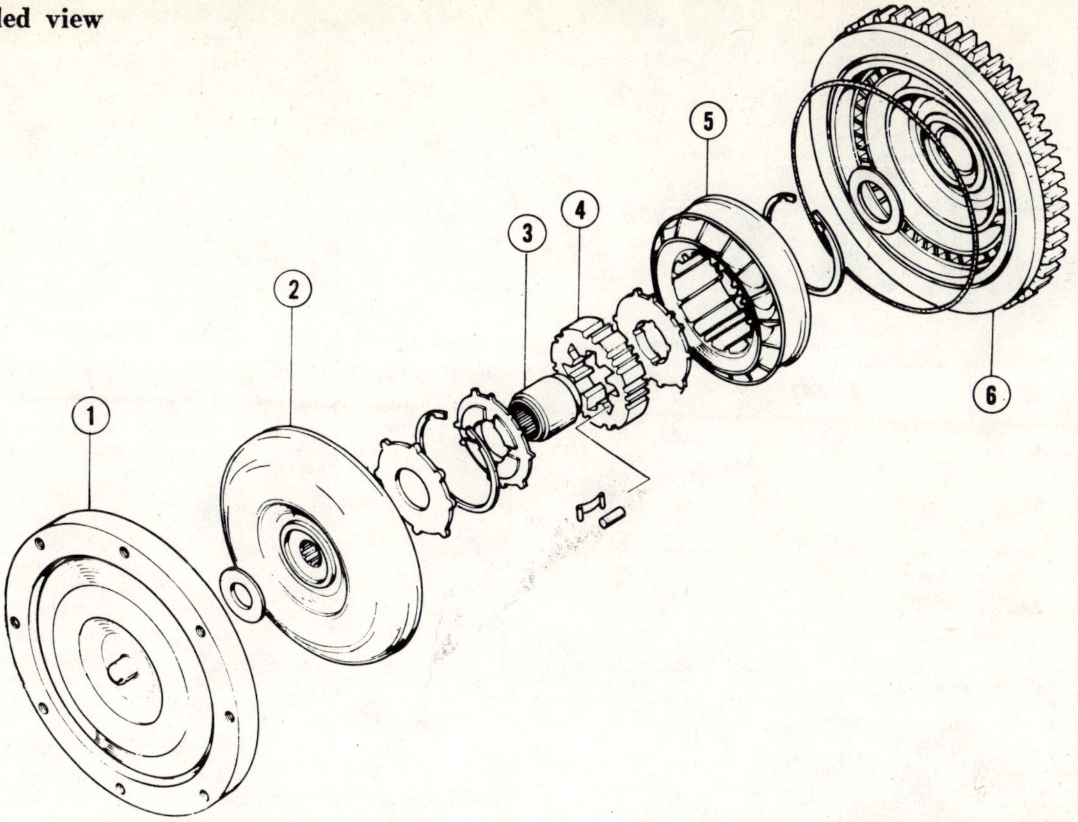

Fig. G-7

① Torque converter pump cover
② Torque converter turbine
③ Stator hub
④ Startor cam
⑤ Stator
⑥ Torque converter pump

b. Inspection

See Maintenance Standard at the end of this manual.

Inspection item	Remedy
Wear or other damage of inside and outside surfaces of stator hub	Replace if it exceeds service limit.
Wear or other damage of stator side plates	Replace if it exceeds service limit.
Damage of roller attaching part of stator cam	Replace.
Wear of inside surface of cover boss bushing	Replace if it exceeds service limit.
Wear or other damage of torque converter turbine washer attaching surface	Replace if it exceeds service limit.
Wear or other damage of thrust washers	Replace if it exceeds service limit.

Fig. G-8

1. Measure the inside diameter of the cover boss bushing. When it exceeds the service limit, replace the torque converter cover.

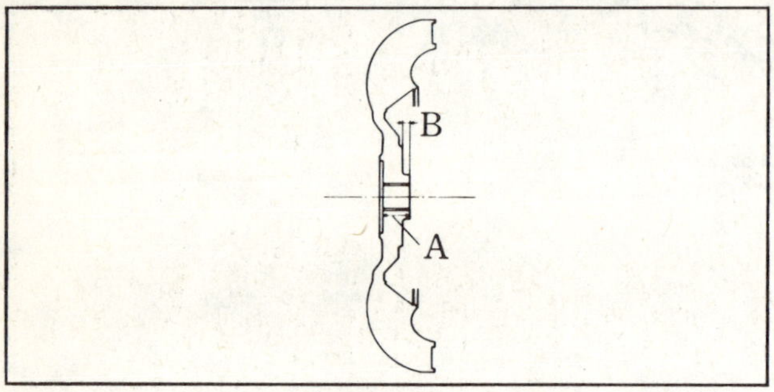

Fig. G-9

2. Measure the thickness of the washer attaching part of the torque converter turbine. Compute the difference between the two measurement points A and B in Fig. G-9.

Fig. G-10
① Stator ② Stator hub ③ Stator cam
④ Roller ⑤ Spring

c. Assembly

To assemble the torque converter, reverse the disassembly procedures. Pay attention to the following points:

1. Assemble the stator, hub, cam, rollers and springs as shown.

Fig. G-11
① Lock ② Turn

2. Temporarily install the stator shaft to the stator as shown in Fig. G-11 and, viewing from the pump side (the side with the thinner vane), confirm that the stator is locked when turned to the right and rotated to the left by fixing the stator with one hand and attempting to rotate the stator shaft with other hand.

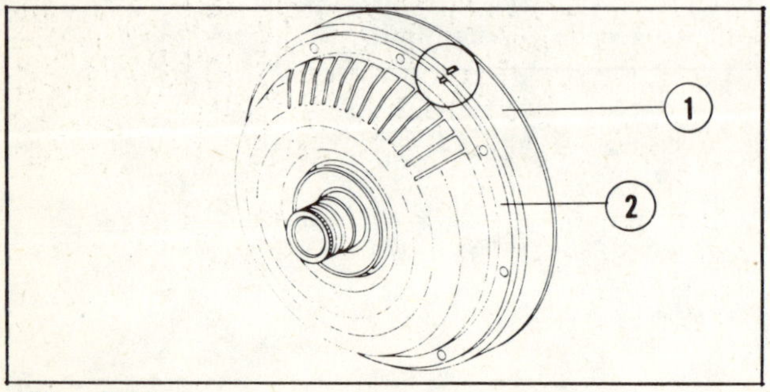

Fig. G-12
① Torque converter pump cover
② Torque converter pump

3. Assemble the torque converter pump and the torque converter cover so that the marks on the pumps and the cover should be apart within the attaching bolt pitch. (Fig. G-12).
   If it is necessary to replace either part of them, select the part of which balancing mark can be within the bolt pitch from the mark on the corresponding part.
4. Apply a coat of automatic transmission fluid to the thrust washer and install it in place.
5. Install the ring gear on the torque converter with the chamfered side of the gear facing toward the converter body.

d. Installation

To install, reverse the removal procedures. Pay attention to the following point:

1. Tighten the drive plate, ring gear and torque converter bolts to the specified torque.
See the Maintenance Standard.

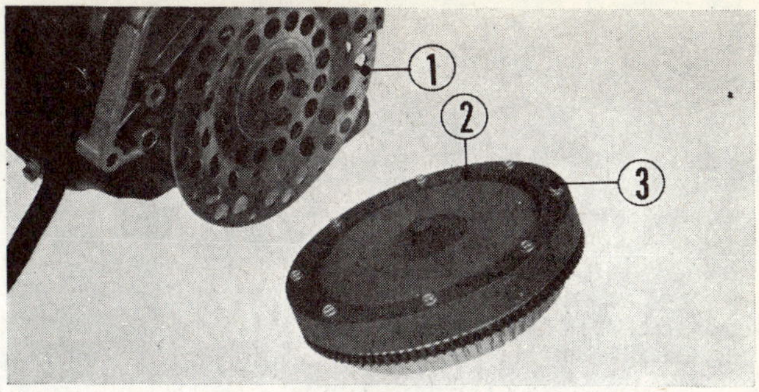

Fig. G-13
① Drive plate ② Torque converter assembly
③ Ring gear bolt

Right Side Cover (Transmission off Engine)

a. Removal and disassembly

1. Remove the engine from the body.
2. Remove the automatic transmission fluid level gauge.
3. Remove the right side cover.
4. The right side cover can be disassembled as shown.

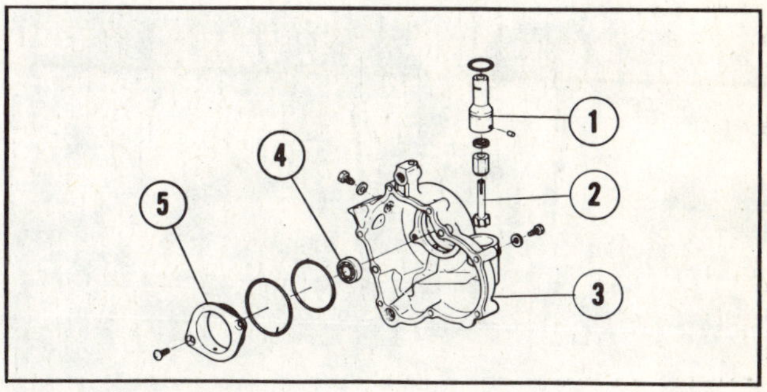

Fig. G-14
① Speedometer gear holder ② Speedometer gear
③ Right side cover ④ 6001 ball bearing
⑤ Sealing guide

b. Inspection

See the Maintenance Standard at the end of this manual.

Inspection item	Remedy
Abnormal noise or rattle of bearing	Replace if necessary.
Breakage of speedometer gear	Replace if necessary.
Restrictions of automatic transmission fluid lines	Clean if necessary.
Deformation, scratches or other damage of case attaching surface	Replace if necessary.
Wear, scratches or other damage of "L" clutch journal	Replace if it exceeds service limit.

Fig. G-15
① Ball bearing ④ Case attaching surface
② Speedometer gear ⑤ Sealing guide
③ Oil hole

Fig. G-16
① Countershaft ② Seal ring ③ "L" clutch
④ Dowel ⑤ O-Ring

c. Installation

To install the right side cover, reverse the removal procedures. Pay attention to the following points:

1. Do not forget to install the two dowels in place.
2. Before installing, check to be sure that the nuts on the ends of the main shaft and countershaft are respectively staked properly as instructed in the pertinent section and that the "L" clutch seal ring rotates smoothly.
3. Take care not to damage the "L" clutch seal ring when installing.
4. Do not forget to position an O-ring on the mating surface at the oil passage.

Fig. G-17
① 16 mm lock nut wrench ② "L" clutch

"L" Clutch, "L" Gear and Counter-Shaft "L" Gear (Transmission off Engine)

a. Removal

1. Separate the transmission case from the engine.
2. Remove the right side cover.
3. Straighten the 20 mm lock washer and remove the 16 mm lock nut from the main shaft.

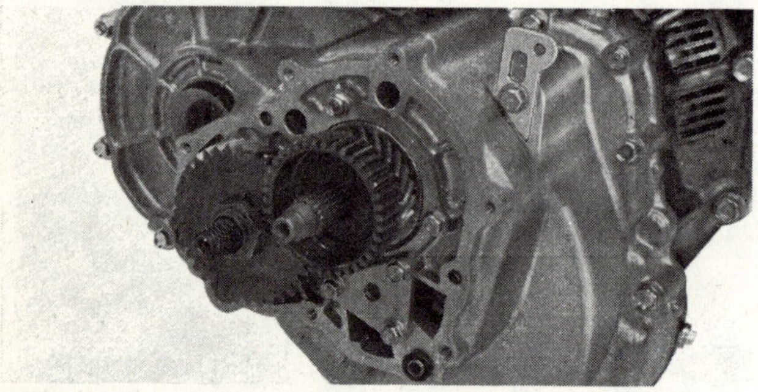

Fig. G-18

4. Remove the "L" clutch assembly.

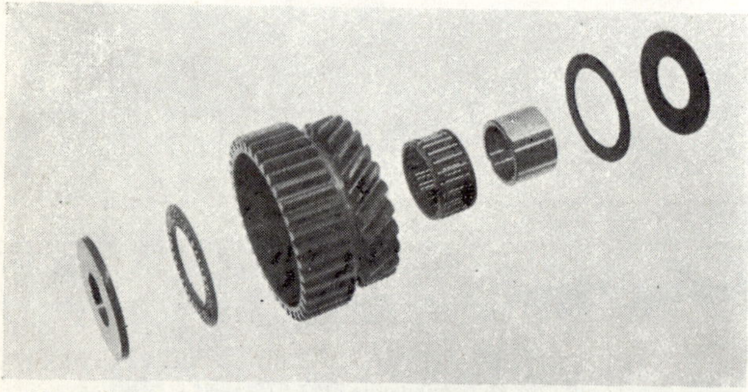

Fig. G-19

5. Remove the "L" gear, thrust washers and needle roller bearing.

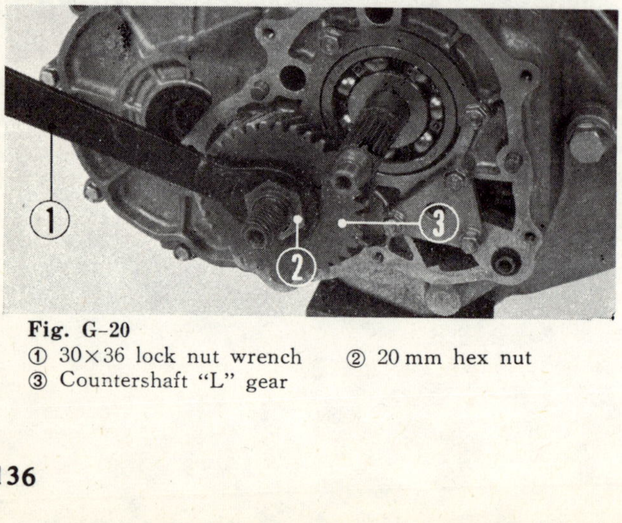

Fig. G-20
① 30×36 lock nut wrench ② 20 mm hex nut
③ Countershaft "L" gear

6. Move the selector lever to Ⓟ postion. Straighten the 20 mm hex. lock nut and remove it.
7. Remove the countershaft "L" gear and thrust washer.

b. Disassembly

1. Disassemble the clutch assembly in the sequence as shown in Fig. G-21.

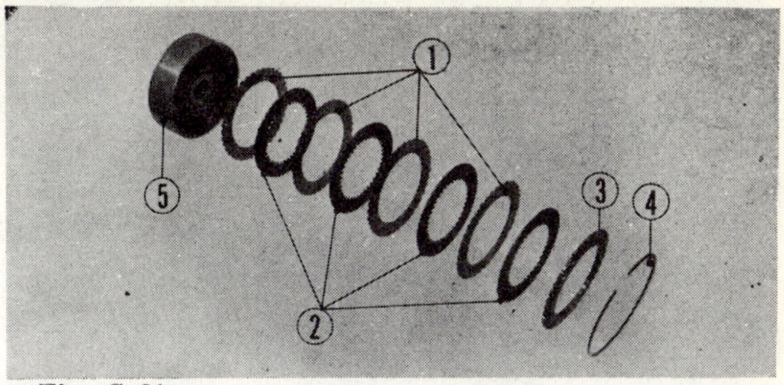

Fig. G-21
① Clutch plates
② Clutch discs
③ Clutch end plate
④ Retaining ring
⑤ Clutch piston guide

2. Remove the clutch return spring.

   Using a spring compressor, compress the spring and remove the retaining ring. Then remove the spring.

Fig. G-22
① Spring compressor

3. Remove the clutch piston. (To remove the clutch piston easily, blow compressed air in the oil passage.)

Fig. G-23

c. Inspection

See the Maintenance Standard at the end of this manual.

Inspection item	Remedy
Breakage of oil seal ring	Replace.
Fatigue of clutch return spring	Replace if it exceeds service limit.
Wear or other damage of clutch plates, discs and end plate	Replace if it exceeds service limit.
Wear or other damage of gear teeth	Replace if it exceeds service limit.
Wear or other damage of inside surface of "L" gear	Replace if it exceeds service limit.
Thickness of thrust washer	Replace if it exceeds service limit.
Abnormal noise or rattle of needle roller bearing	Replace if it exceeds service limit.
Axial play of "L" gear	Replace if it exceeds service limit.

Fig. G-24
① Clutch piston guide ③ Oil holes
② Clutch piston ④ Check valve

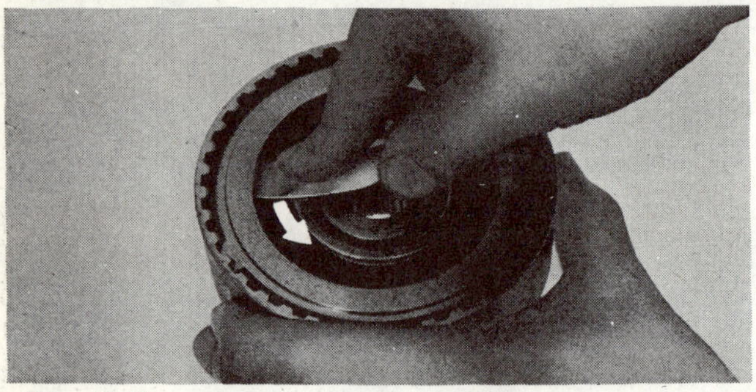

Fig. G-25

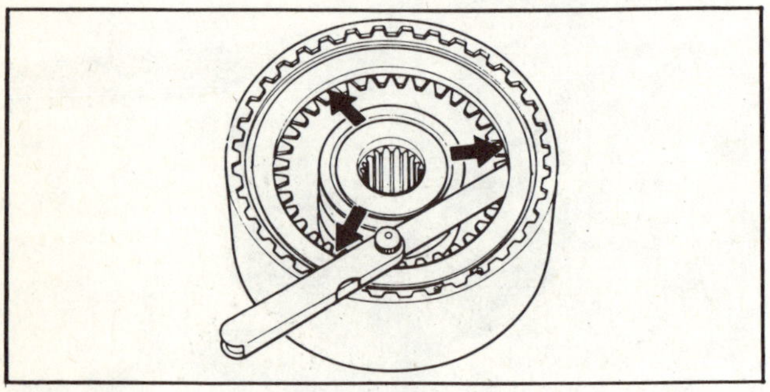

Fig. G-26

Fig. G-27

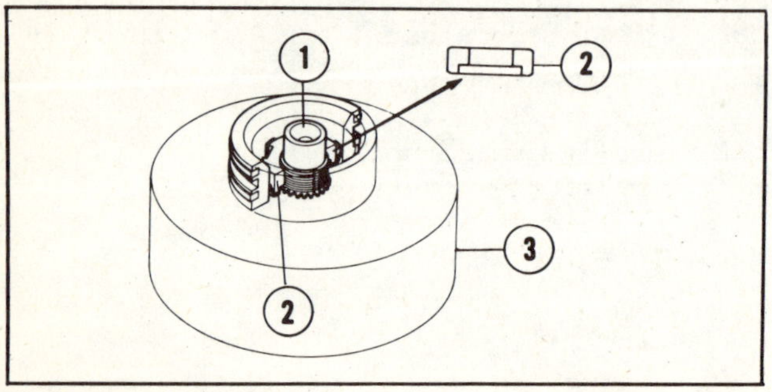

Fig. G-28
① Main shaft ② 16 mm lock nut ③ "L" Clutch

d. Assembly and installation

To assemble and install, reverse the disassembly and removal procedures. Pay attention to the following points:

1. Before installing the piston, check to be sure its check valve does not rattle and the clutch piston guide oil hole is not restricting.

2. After assembling the clutch, measure the initial clutch clearance and adjust it with an end plate of right thickness, if necessary. See the Maintenance Standard at the end of this manual.

   To measure, insert a thickness gauge into between the pressure end plate and the top disc, and check all the way around the clutch to find the minimum clearance.

   • Find and measure the minimum clearance and other two points which are approx. 120° apart from the minimum clearance point.

   • To make a true measurement, insert a feeler gauge of the same thickness as well in the opposite side of the measuring point to keep the top disc horizontal.

3. Install the 16 mm lock nut on the main shaft with the recessed side downward.

4. Move the selector lever to Ⓟ position and tighten the 20 mm lock nut on the countershaft to the specified torque.
5. Hold the torque converter main shaft and stator shaft splines with the main shaft holder (special tool) and tighten the 16 mm lock nut on the main shaft to the specified torque. See the Torque Specifications.

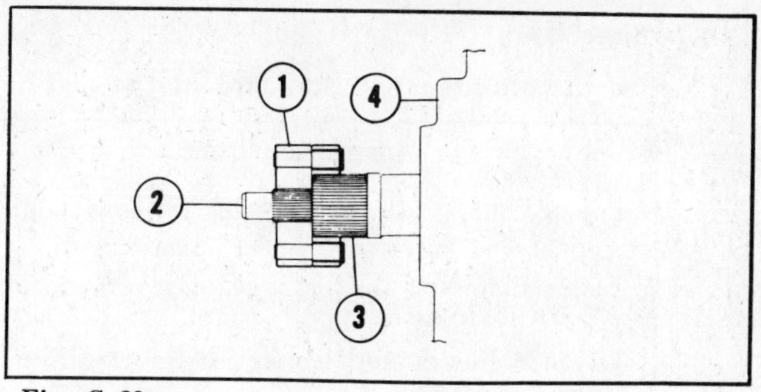

Fig. G-29
① Main shaft holder
② Main shaft
③ Stator shaft
④ Torque converter case

Transmission Case (Transmission off Engine)

a. Removal

1. Remove the transmission case and engine as a unit.
2. Remove the right side cover.
3. Remove the "L" clutch and "L" gear.
4. Remove the bearing retainer and, setting the special tool (Puller transmission case) as shown in Fig. G-30, remove the transmission case.
5. Disassemble the transmission case in the sequence as shown below.

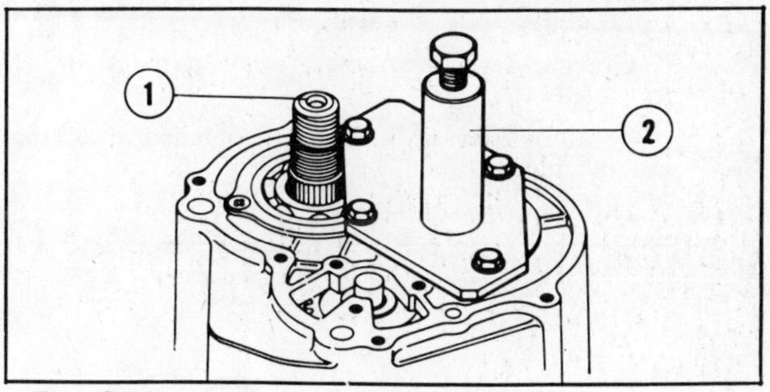

Fig. G-30
① Counter shaft
② Puller transmission case

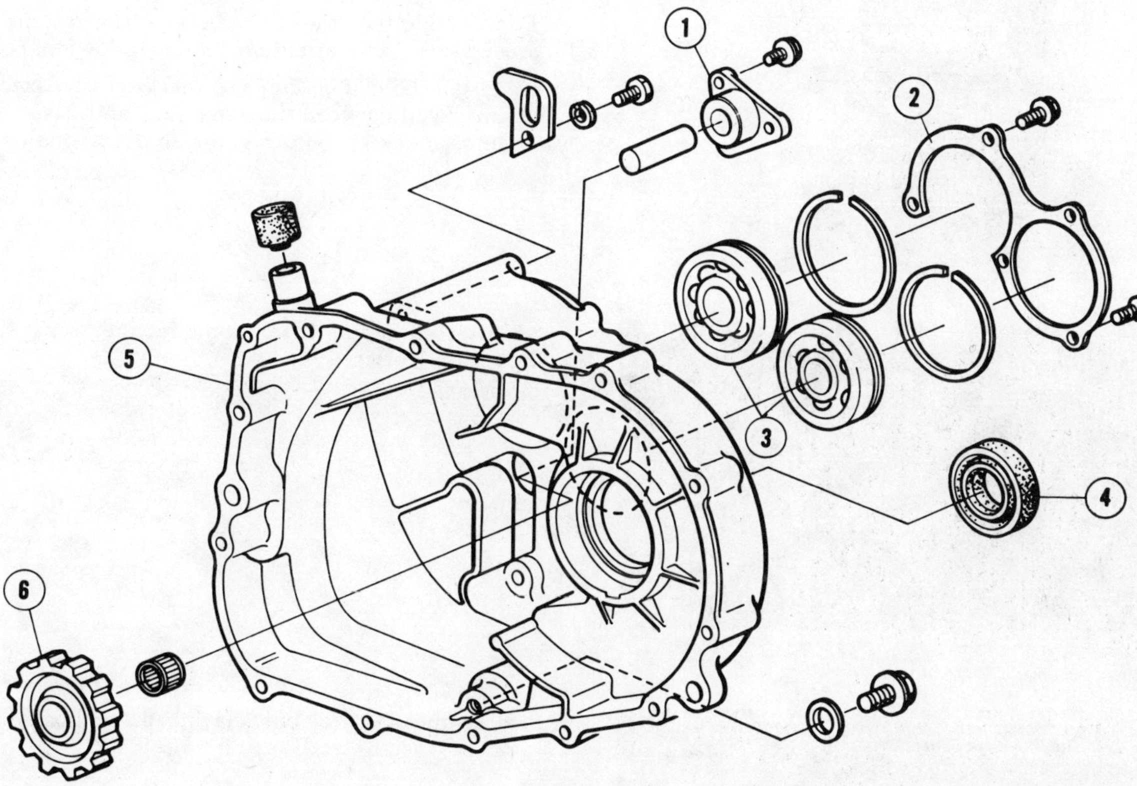

Fig. G-31

① Reverse idle gear shaft holder
② Bearing retainer
③ 16008 ball bearing
④ 35×56×9 oil seal
⑤ Transmission case
⑥ Reverse idle gear

b. Inspection

See the Maintenance Standard at the end of this manual.

Inspection item	Remedy
Outside diameter and damage of reverse idle gear shaft	Replace if it exceeds service limit.
Inside diameter and damage of reverse idle gear shaft holder	Replace if it exceeds service limit.
Inside diameter and damage of reverse idle gear teeth	Replace if it exceeds service limit.
Axial play of reverse idle gear	Repair or replace.
Abnormal noise or rattle of ball bearings and needle roller bearing	Replace if it exceeds service limit.
Cracks, deterioration or other damage of oil seal	Replace.
Scratches or other damage of case attaching surface	Repair or replace.
Deformation of aluminum washers	Replace.

c. Assembly

To assemble the transmission case, reverse the disassembly procedures. Pay attention to the following points:

1. Install the ball bearings so that excessive clearance is not produced between the snap ring and case. After installing, check the bearing for smooth rotation.

Fig. G-32

2. After tightening the bearing retainer screws, stake them with a chisel.

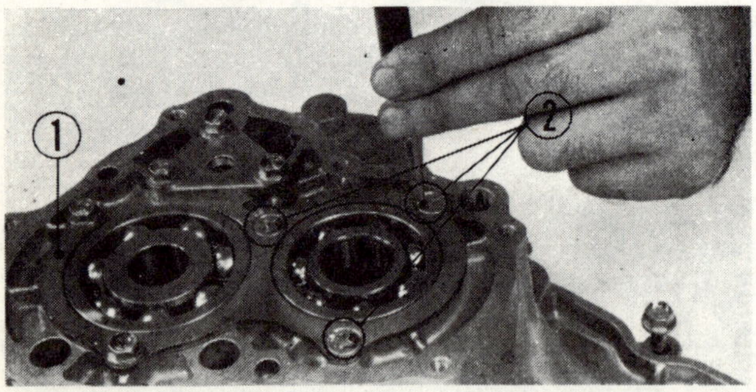

Fig. G-33
① Bearing retainer ② Upset

3. Use the 639 seal only having an embossed marking.

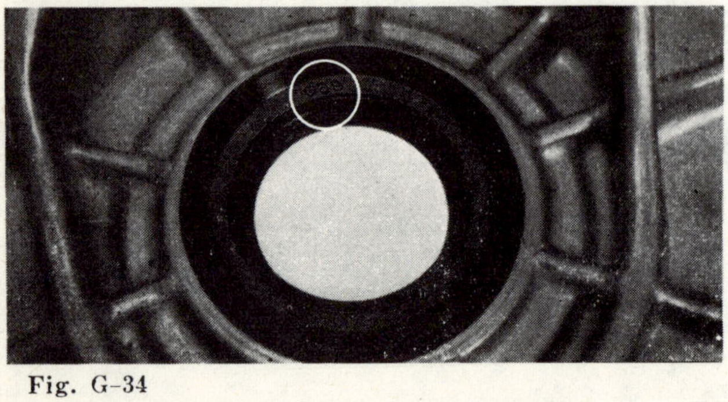

Fig. G-34

4. Align the shaft holes and install the transmission case in place, taking care not to damage the differential side oil seal. After installing, check if the main shaft and countershaft rotate smoothly without any sign of binding.

Fig. G-35

Main Shaft and Countershaft (Transmission off Engine)

NOTE:

Identify the main shaft and countershaft washers so that they can be assembled in their original locations at assembly.

a. Removal

1. Remove the transmission case and engine as a unit.
2. Remove the right side cover and transmission case.
3. Remove the shift fork shaft retaining screw and remove the shift fork shaft, reverse gear selector and countershaft reverse gear.
4. Remove the main shaft and countershaft at the same time.

Fig. G-36
① Countershaft ② Main shaft
③ Countershaft reverse gear ④ Reverse shift fork
⑤ Shift fork shaft set screw

Fig. G-37

5. Slide the "D" gear, "D" clutch, needle roller bearings and thrust washers off the main shaft·

   NOTE:

   To disassemble and assemble the "D" clutch, see the paragraph for the "L" clutch on page 137.

6. Slide the thrust washers, countershaft reverse gear and reverse gear selector off the countershaft.

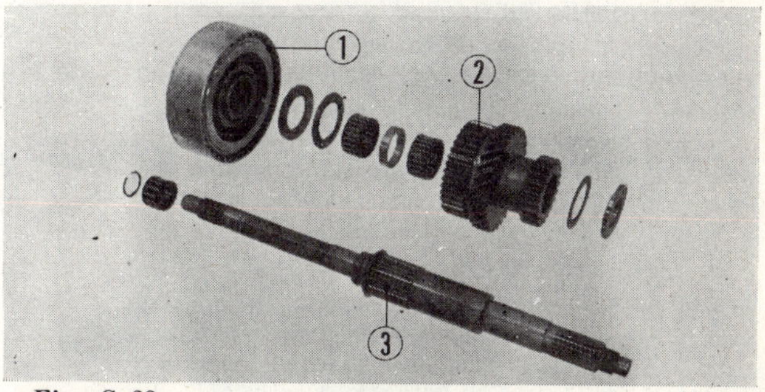

Fig. G-38
① "D" clutch ② "D" gear ③ Main shaft

7. Using a press, remove the reverse gear hub and countershaft "D" gear at the same time.
8. Remove the snap ring and remove the needle roller bearing.

Fig. G-39
① Countershaft ② Reverse gear hub
③ Countershaft "D" gear

b. Inspection

See the Maintenance Standard at the end of this manual.

Inspection item	Remedy
Breakage of oil seal rings	Replace.
Fatigue of clutch return spring	Replace if it exceeds service limit.
Wear or other damage of clutch plates, discs and end plate	Replace if it exceeds service limit.
Wear or other damage of gear teeth	Replace if it exceeds service limit.
Wear or other damage of inside surface of "D" gear	Replace if it exceeds service limit.
Wear or other damage of needle roller bearing journals of main shaft and countershaft	Replace if it exceeds service limit.
Thickness of thrust washers	Replace if it exceeds service limit.
Thickness of reverse shift fork fingers	Replace if it exceeds service limit.
Abnormal noise or rattle of needle roller bearings	Replace if it exceeds service limit.

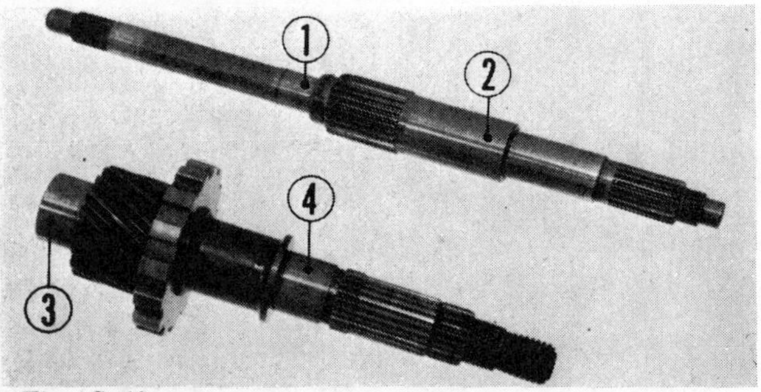

Fig. G-40
① Stator shaft needle roller bearing surface
② "D" gear needle roller bearing surface
③ "L" side needle roller bearing surface
④ Countershaft "D" gear needle roller bearing surface

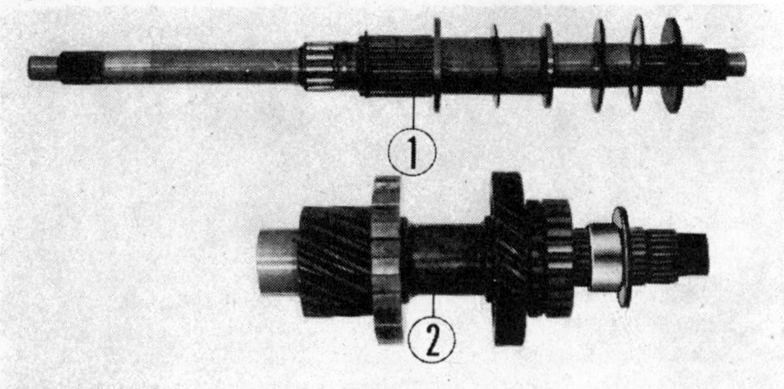

Fig. G-41
① Main shaft ② Countershaft

• Needle roller bearing journals of main shaft and countershaft to be inspected are as shown in Fig. G-40.

c. Installation

To install, reverse the removal procedures. Pay attention to the following points:

1. Use the thrust washers in proper places.
2. Before installing the transmission case, check to be sure the main shaft and countershaft rotate smoothly.

Parking Control (Transmission off Engine)

a. Removal

1. Separate the transmission case from the engine.
2. Disconnect the control lever and cable.

Fig. G-42
① Control lever ② Control cable

3. Remove the right side cover, transmission case and countershaft reverse gear.
4. Fully pull the manual valve backward (to the "[1]" position). Remove the 6 mm (0.236 in.) roller pin and push the valve into the body.

Fig. G-43
① 6 mm roller pin ② Manual valve ③ Valve body

5. Remove the parking brake shaft and shift lever shaft at the same time.

Fig. G-44

6. Disassemble the parking brake shaft assembly as shown in Fig. G-45.

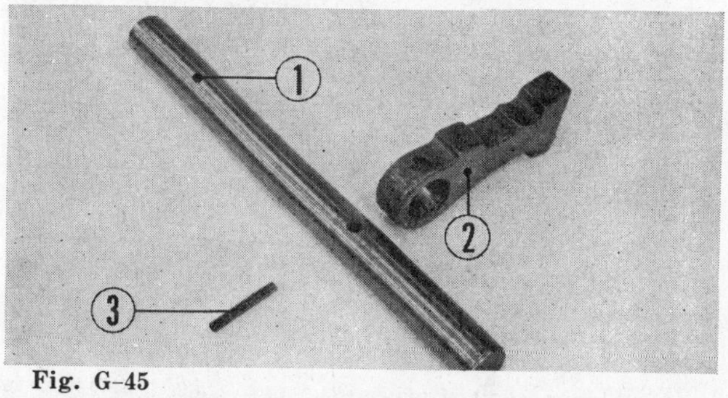

Fig. G-45
① Parking brake shaft ② Parking brake pawl
③ Spring pin

7. Disassemble the shift lever shaft assmbly as shown in Fig. G-46.

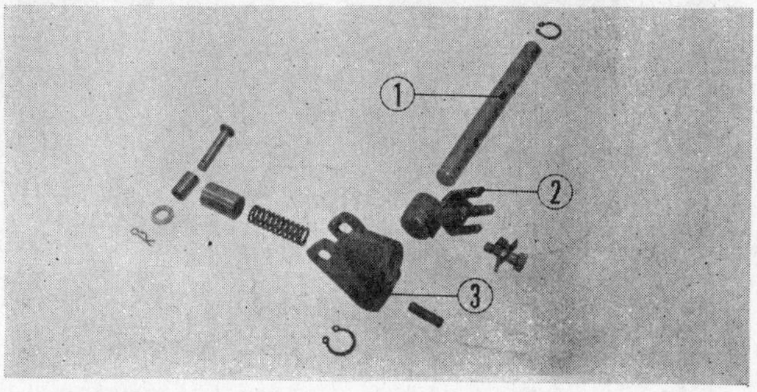

Fig. G-46
① Shift lever shaft ② Manual valve lever
③ Parking brake lever

b. Inspection

See the Maintenance Standard at the end of this manual.

Inspection item	Remedy
Wear of parking brake pawl	Replace if it exceeds service limit.
Bend of parking brake shaft	Replace if it exceeds service limit.
Fatigue or breakage of parking brake spring	Replace if it exceeds service limit.
Fatigue or breakage of release pawl spring	Replace if it exceeds service limit.

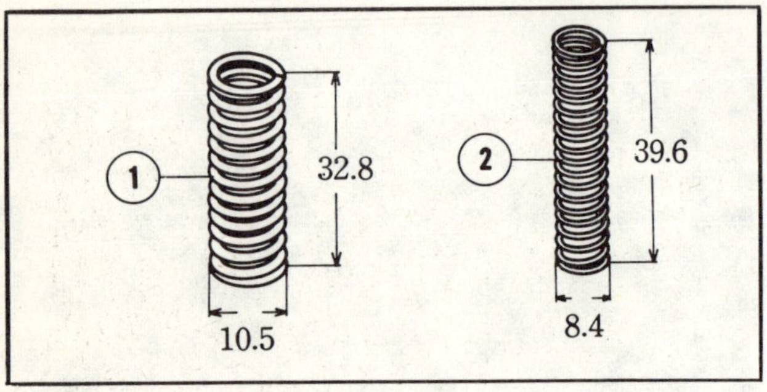

Fig. G-47
① Parking brake spring
② Torque converter check valve spring

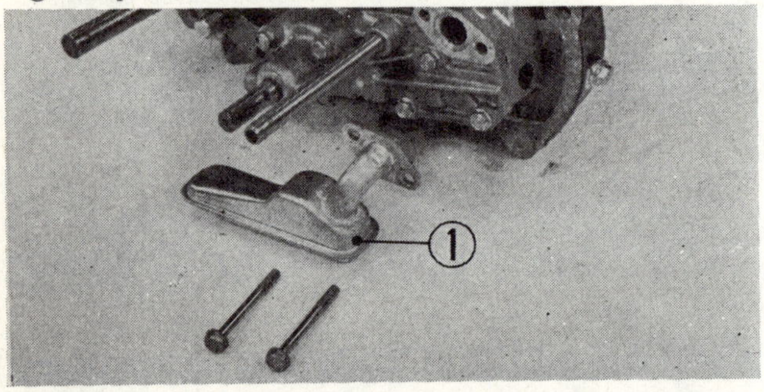

Fig. G-48
① Transmission strainer

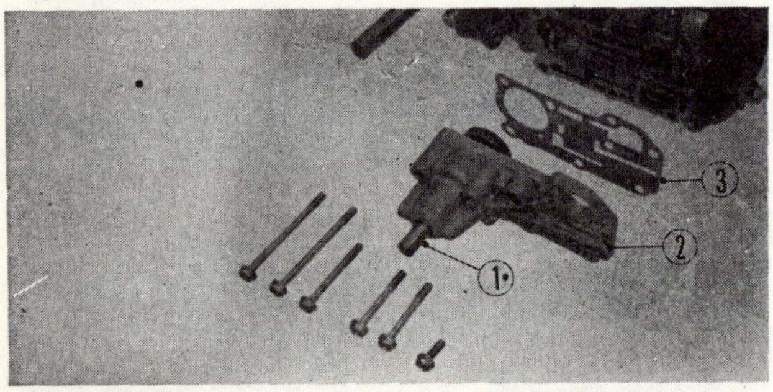

Fig. G-49
① Shift fork shaft ② Serve body complete
③ Servo boby gasket

Fig. G-50
① Stator shaft arm ② Sealing ring guide

c. Assembly

To assemble, reverse the disassembly procedures. Pay attention to the following points:

1. Take care not to transpose the parking brake spring with the torque converter check valve spring when installing.
2. Before installing the transmission case, check the contact of the 6 mm roller pin and parking pawl for condition. Also, check the parking pawl for proper engagement.

Servo Body, Valve Body and Oil Pump (Transmission off Engine)

a. Removal

1. Remove the right side cover.
2. Removo the transmission case.
3. Remove the main shaft and countershaft.
4. Remove the transmission strainer.
5. Remove the servo body assembly.
6. Remove the seal ring guide and then the stator shaft arm.

7. Remove the valve body by lifting its oil pump and servo valve sides evenly.

NOTE :

Take care not to drop the oil pump gear, torque converter check valve and orifice plug.

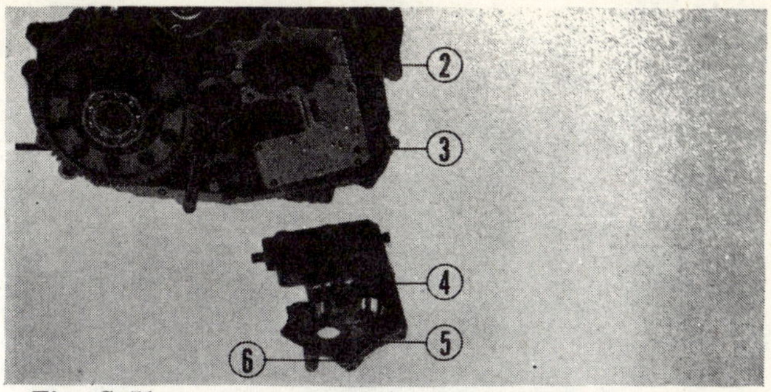

Fig. G-51
① Oil pump drive gear ④ Valve body
② Oil pump driven gear ⑤ Check valve spring
③ Separating plate ⑥ Torque converter check valve

8. Remove the oil pump drive and driven gears.

Fig. G-52
① Torque converter check valve
② Oil pump drive gear ③ Oil pump driven gear

b. Inspection

See the Maintenance Standard.

Inspection item	Remedy
Wear or other damage of inside surface of servo body	Replace if it exceeds service limit.
Gear-to-housing clearance	Replace if it exceeds service limit.
Gear-to-housing thrust clearance	Replace if it exceeds service limit.
Wear of inside surfaces of stator shaft and bushing	Replace if it exceeds service limit.
Fatigue or other damage of check valve spring	Replace if it exceeds service limit.
Restriction of oil passages and holes	Clean.

• Gear-to-housing (valve body) clearance

Measure the clearance with a known feeler gauge. If it exceeds the service limit, replace the pump and gears as an assembly.

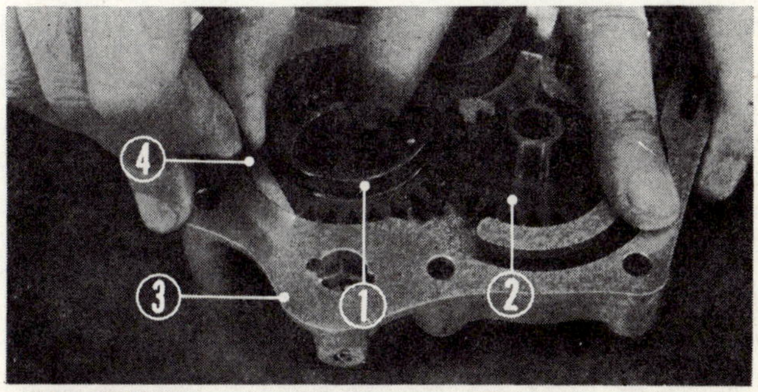

Fig. G-53
① Oil pump drive gear ② Oil pump driven gear
③ Valve body ④ Feeler gauge

Fig. G-54
① Oil pump drive gear ② Oil pump driven gear
③ Valve body ④ Straightedge ⑤ Feeler gauge

- Gear-to-housing thrust clearance

  Measure the distance between the mating surface of the valve body and the side of the gear with a straightedge and a feeler gauge.

  If it exceeds the service limit, replace the pump and gears as an assembly.

c. Installation

To install, reverse the removal procedures. Pay attention to the following points:

1. When installing the valve body, do not forget to install the orifice plug on the servo body side.
2. Tighten the main valve body, seal ring guide, servo valve body and oil strainer bolts to the specified torques.

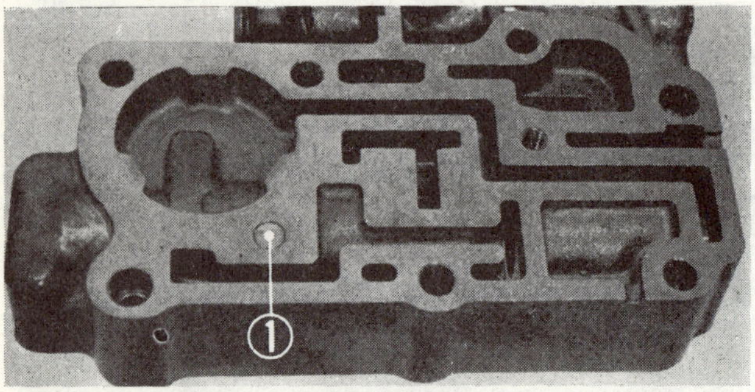

Fig. G-55
① 1.2 mm orifice plug

3. Install the shift fork shaft as shown.

Fig. G-56
① Shift fork shaft

Torque Converter Case (Transmission off Engine)

a. Removal

1. Remove the transmission case and engine as a unit.
2. Remove the right side cover, transmission case, main shaft, countershaft, valve body and parking control.
3. Remove the differential gear assembly.

Fig. G-57
① Differential gear assembly

4. Remove the oil hose joint A and 8 mm sealing bolt.

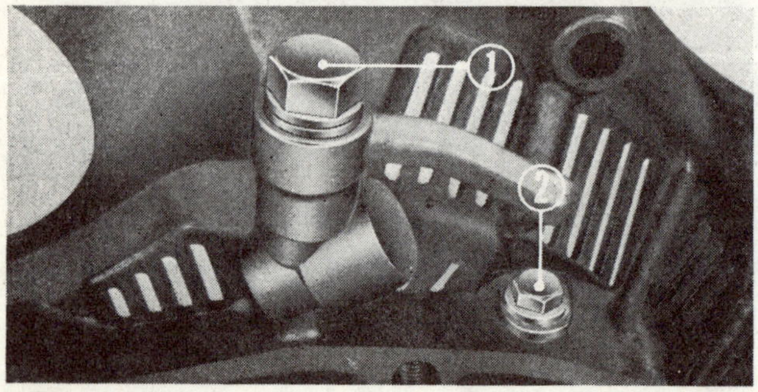

Fig. G-58
① Oil hose joint A ② 8 mm sealing bolt

5. Remove the needle roller bearing from the countershaft with a special tool.

   NOTE:

   Do not reuse the removed bearing.

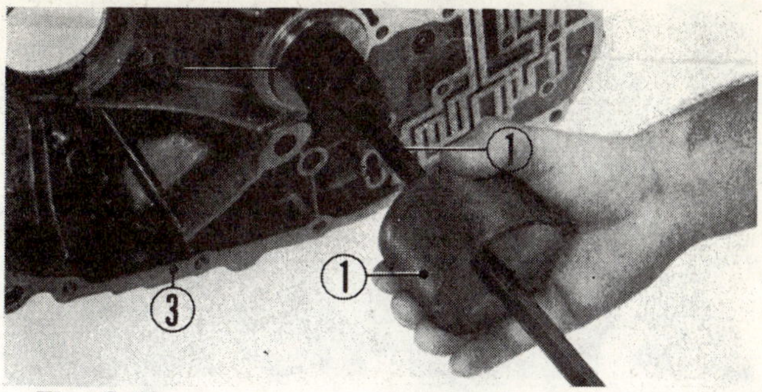

Fig. G-59
① Bearing remover set
② 30×57×17 needle roller bearing
③ Torque converter case

6. Remove the oil seal and ball bearing from the main shaft.
7. Remove the oil seal from the shift lever shaft.
8. Remove the differential side oil seal and 72 mm (2.83 in) retaining ring.

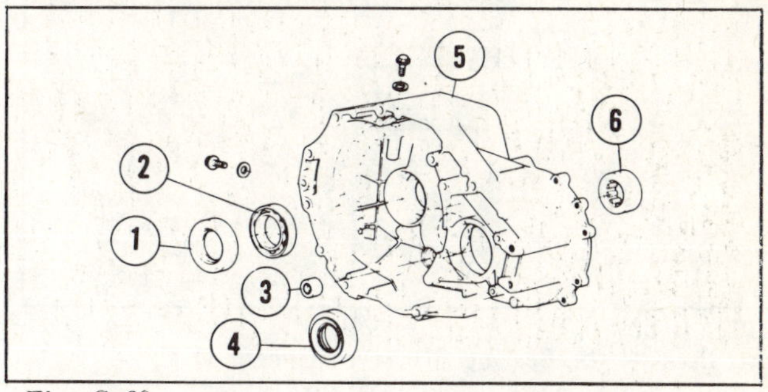

Fig. G-60
① 44×68×8 oil seal ② 16008 ball bearing
③ 12×22×7 oil seal ④ 35×68×9 oil seal
⑤ Torque converter case
⑥ 30×57×17 needle roller bearing

b. Inspection

See the Maintenance Standard.

Inspection item	Remedy
Deformation or other damage of oil seals	Replace.
Abnormal noise or rattle of bearings	Replace.
Restrictions of oil passages and holes	Repair or replace.
Scratches or other damage of case attaching surface	Repair or replace.
Scratches or other damage of valve body attaching surface	Repair or replace.
Deformation of aluminum washers	Replace.

c. Assembly and installation

To install, reverse the removal procedures. Pay attention to the following points:

1. When replacing the torque converter case, transmission case or the differential side bearing, measure the side clearance. If out of specification, adjust by using a 72 mm (2.83 in) retaining ring.
   See the Maintenance Standard.

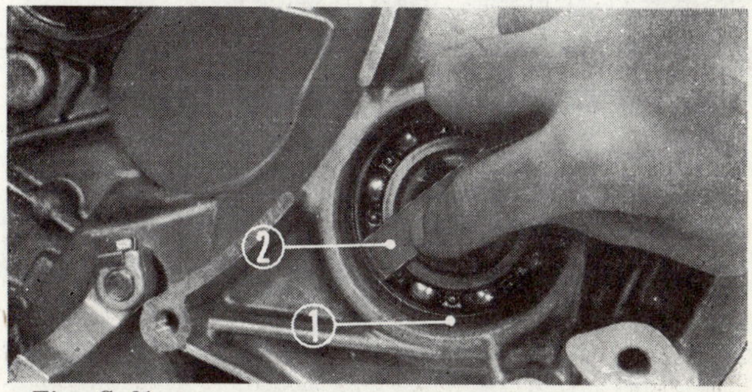

Fig. G-61
① 72 mm set ring ② Feeler gauge

Fig. G-62

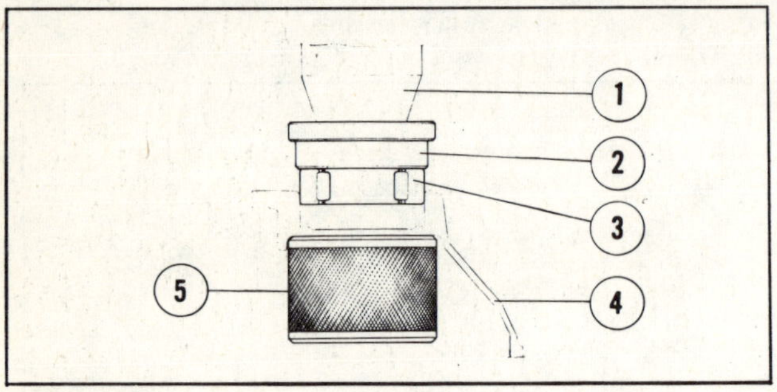

Fig. G-63
① Press ② Bearing driver attachment D
③ 30×57×17 needle roller bearing
④ Torque converter case
⑤ Weight of bearing remover set

2. Install the 72 mm retaining ring and then the differential side oil seal in place.

3. Using a press, install the countershaft needle roller bearing so that it is not higher than the valve body attaching surface.

4. Partially tighten the oil hose joint A in place. After mounting the engine on the car, connect the oil cooler hose and position the oil hose joint A so as not to contact the check bolt or any other part and then retighten the joint A.

Selector Lever

a. Removal

1. Remove the center console.
2. Remove the 8 mm (0.315 in) lock pin and remove the control rod pin A.

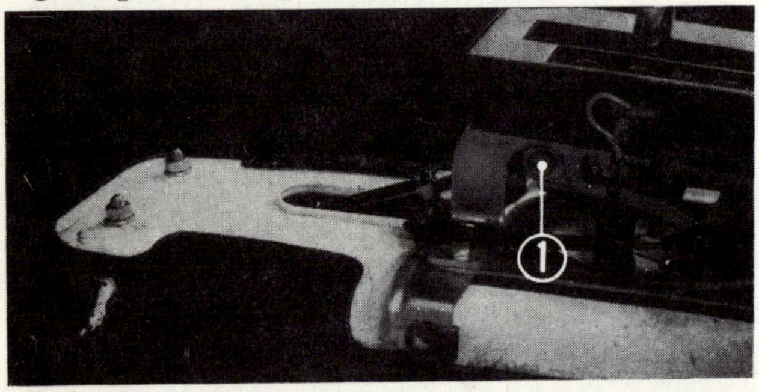

Fig. G-64
① Control rod pin A

3. Jack up the front of the car and support it with safety stands.
4. Remove the selector lever bracket nuts and lift up the selector lever assembly.

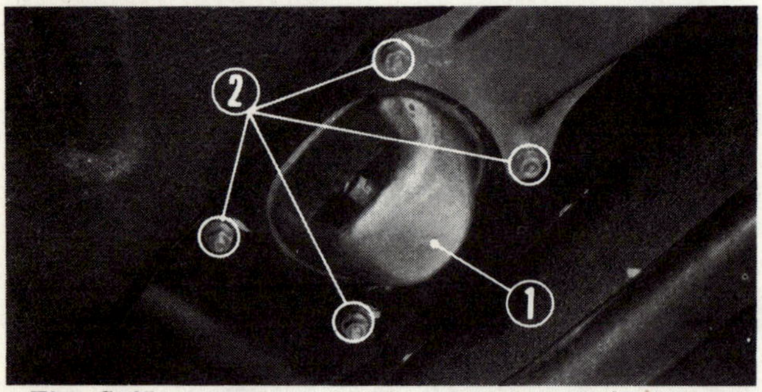

Fig. G-65
① Selector lever bracket ② Attaching nuts

b. Disassembly

1. Remove the 4 mm screws and disassemble the selector lever knob assembly.

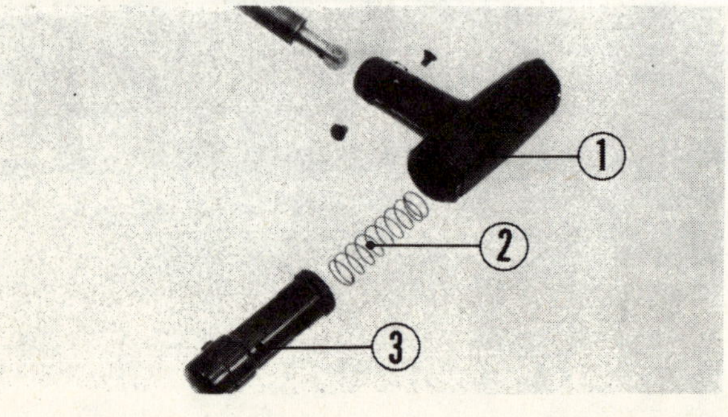

Fig. G-66
① Selector lever knob ② Push knob spring
③ Push knob

2. Remove the 5 mm screws and disassemble the console escutcheon.

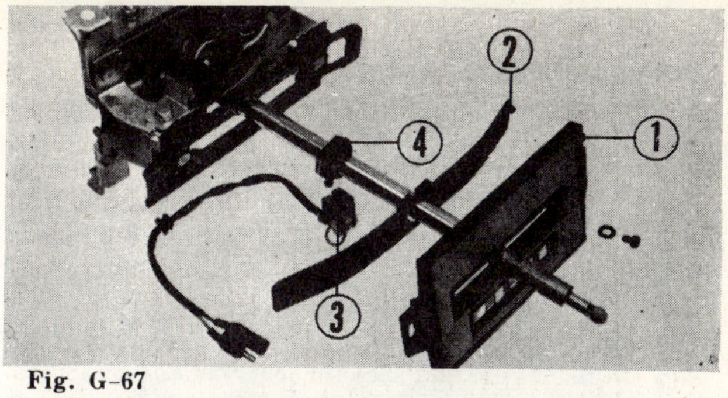

Fig. G-67
① Console escutcheon ② Selector lever cover
③ Indicator ④ Selector lever cover guide

3. Remove the selector lever cover holder and inhibitor/back light switch.

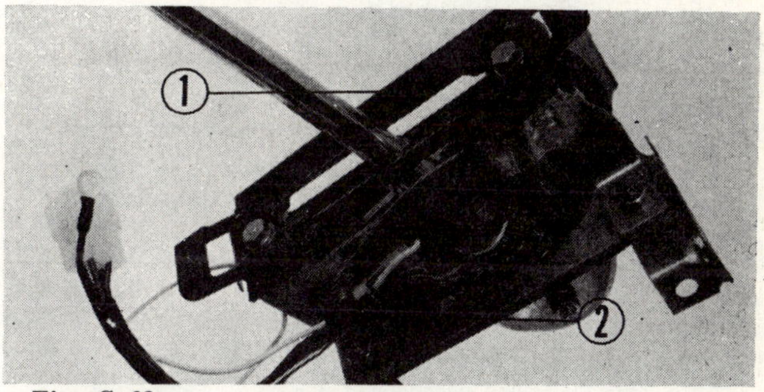

Fig. G-68
① Selector lever cover holder
② Inhibitor/back light switch

4. Remove the control bracket and console front bracket.

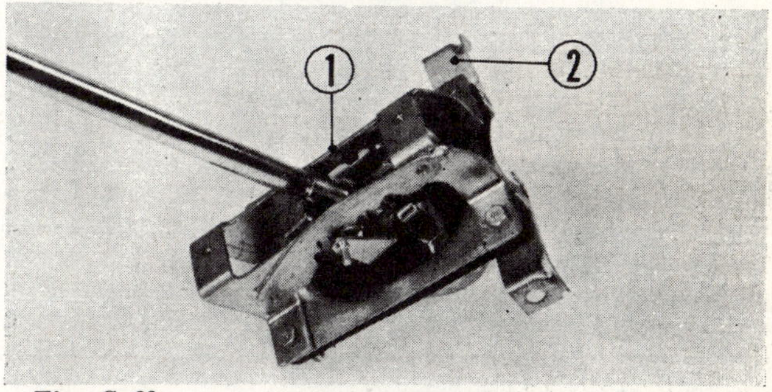

Fig. G-69
① Control bracket ② Console front bracket

5. Remove the selector lever from the bracket.

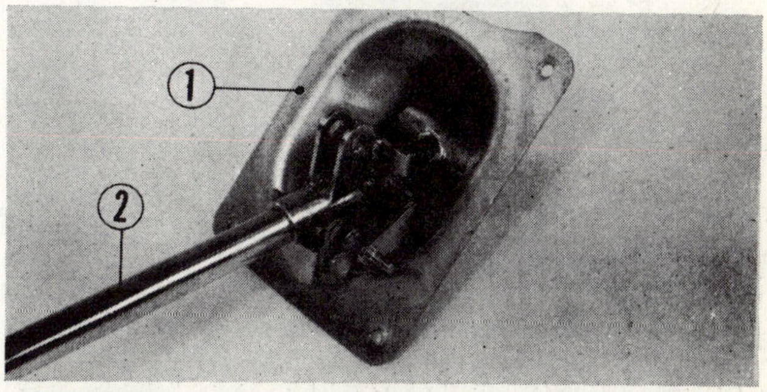

Fig. G-70
① Selector lever bracket ② Selector lever

6. Disassemble the selector lever as shown in Fig. G-71.

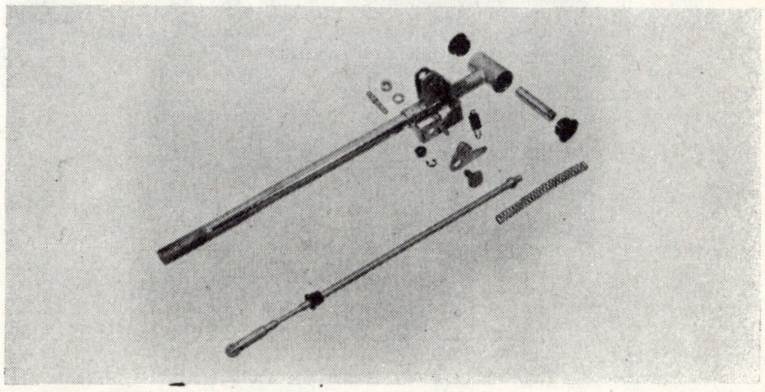

Fig. G-71

c. Inspection

See the Maintenance Standard at the end of this manual.

Inspection item	Remedy
Fatigue or other damage of lock pin spring	Replace if it exceeds service limit.
Fatigue or other damage of push knob spring	Replace if it exceeds service limit.
Fatigue or other damage of selector lever stopper spring	Replace if it exceeds service limit.
Wear or other damage of inside surfaces (on both engine and selector lever sides) of control cable pin	Replace if it exceeds service limit.
Wear or other damage of inside surface of selector lever bushing	Replace if it exceeds service limit.
Wear or other damage of outside surface of control cable pin	Replace if it exceeds service limit.
Wear or other damage of inside surface of selector lever stopper arm	Replace if it exceeds service limit.
Wear or other damage of outside surface of selector lever stopper bolt	Replace if it exceeds service limit.

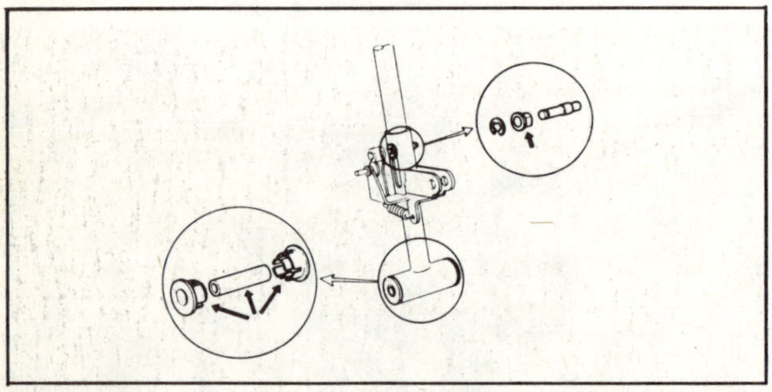

Fig. G–72

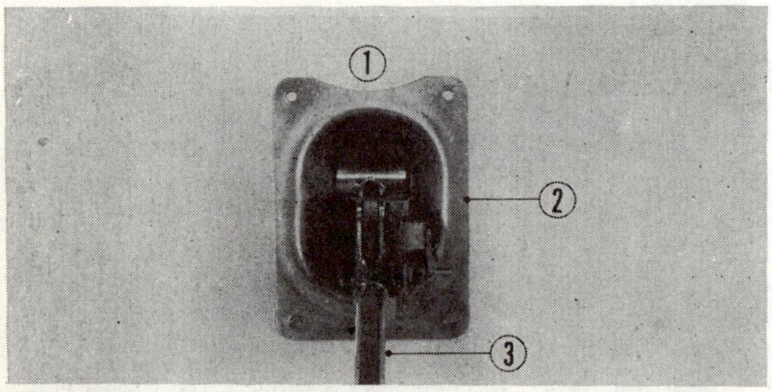

Fig. G–73
① Ⓟ position side ② Selector lever bracket
③ Selector lever

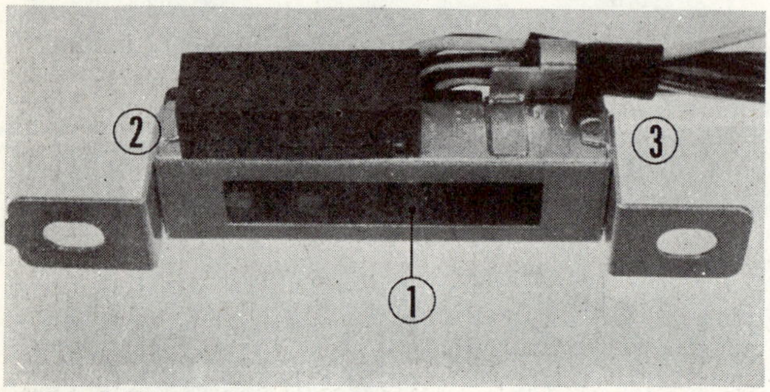

Fig. G–74
① Contact holder ② Ⓟ position side
③ 1 position side

d. Assembly

To assemble the selector lever, reverse the disassembly procedures. Pay attention to the following points:

1. When installing the selector lever bushing, collar and lock pin collar, apply a coat of grease to them.

2. Install the selector lever to the bracket correctly as shown. Incorrect installation of the lever will make it difficult to connect the control cable.

3. Install the inhibitor/back light switch assembly following the instructions given below.
   - Centralize the switch contact holder in the base.

• Move the selector lever to N position. Engage the lever pin in the groove in the contact holder and tighten the pin.
• Be sure that the selector lever assembly is installed in the correct position. (Be careful not to install reversely.)

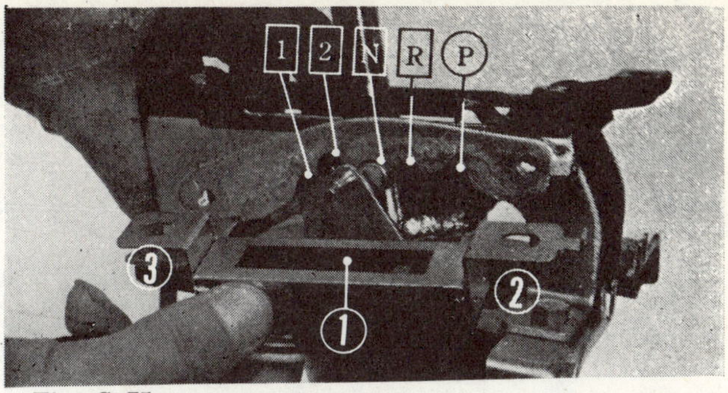

Fig. G–75
① Contact holder ② Ⓟ position side
③ 1 position side

4. Install the selector lever cover holder so that one side having a notch is facing the inhibitor/back light switch.
5. After installing the selector lever, inspect the following items:
   • Operation of inhibitor/back light switch
   • Positions of selector lever

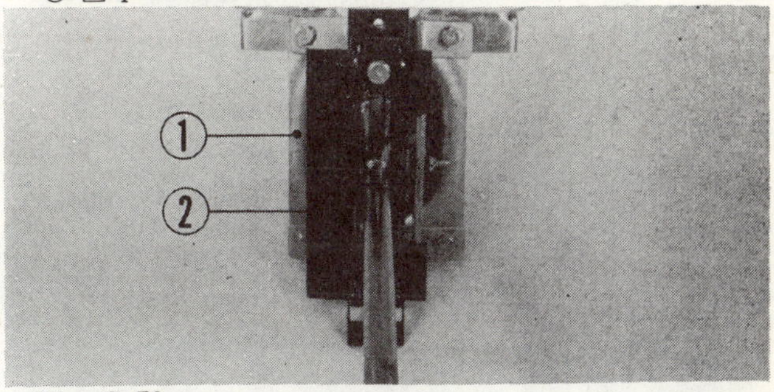

Fig. G–76
① Control bracket ② Selector lever cover holder

Control Cable

a. Removal

1. Remove the center console.
2. Remove the control rod pin A at the selector lever side.
3. Remove the front floor center mat and remove the control cable bracket nuts.

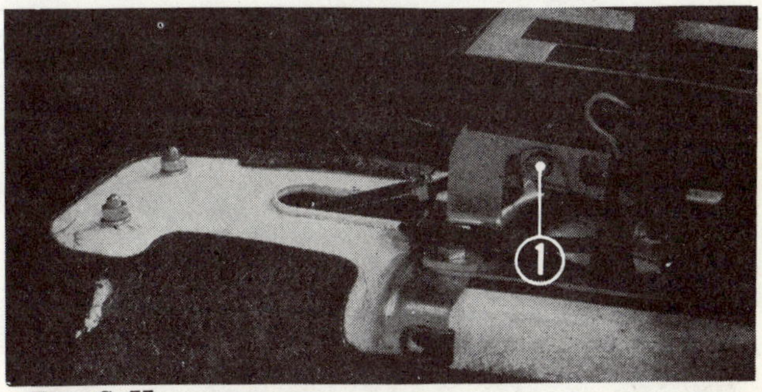

Fig. G–77
① Control rod pin A

4. Jack up the front of the car and support it with safety stands.
5. Remove two selector lever bracket nuts at the front side.

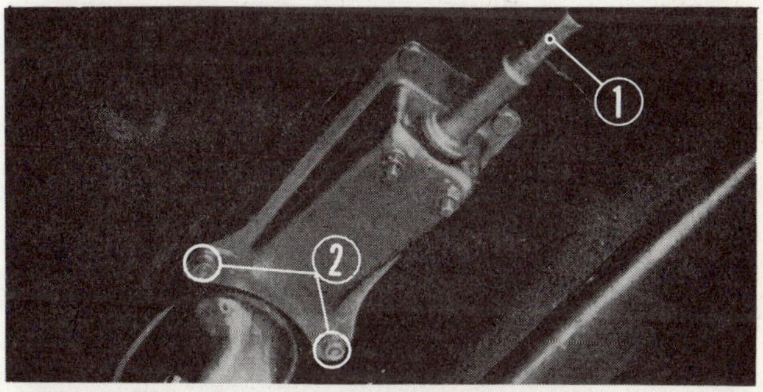

Fig. G–78
① Control cable ② Disconnect here

6. Loosen the control cable holder securing the support beam and disconnect the control cable.
7. Remove the transmission case and engine as a unit from the body, and separate the transmission case from the engine.

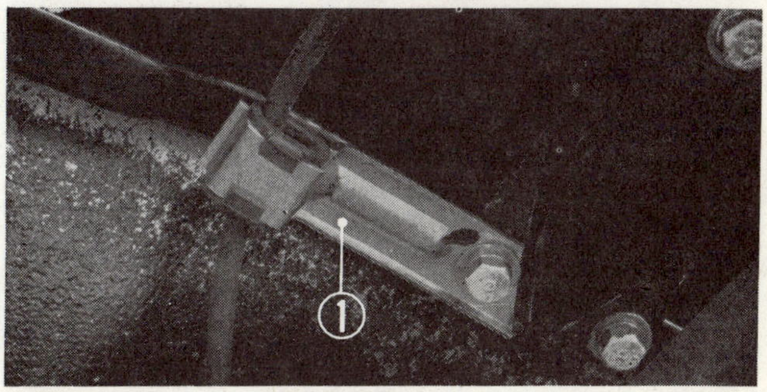

Fig. G–79
① Control cable holder

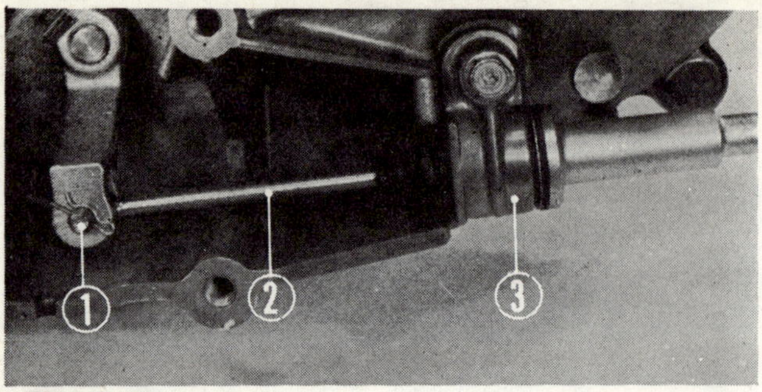

Fig. G-80
① Control rod pin A ② Control cable
③ Cable holder

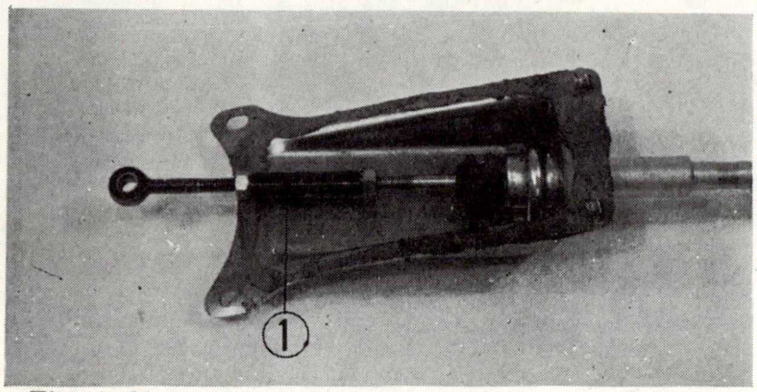

Fig. G-81
① Adjusting turnbuckle

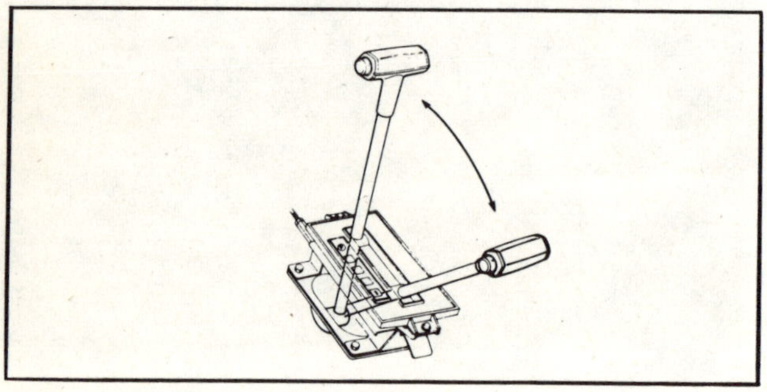

Fig. G-82

8. Remove the cable holder from the torque converter case.
9. Remove the 8 mm lock pin and control rod pin A at the transmission side. Then disconnect the control cable.

b. Installation

To install, reverse the removal procedures. Pay attention to the following points:

1. Check to be sure that the adjusting turnbuckle threads are not damaged.
2. When installing the control rod pin A, cover the selector lever bracket with a waste cloth to prevent the pin from accidentally dropping in the bracket.
3. After installing the rod pin A, adjust the turnbuckle so that the selector lever can travel from Ⓟ to [1] position.
4. Adjust all positions of the selector lever. (See page B-25.)

Trouble Shooting

Complaint			Probable cause	Inspect for :
Engine starts but car does not move with selector lever in any position.	Engine runs up with lever in [2], [1] or [R] position.	"L" pressure is normal.	Defective servo valve	Jamming shift fork shaft Jamming reverse gear selector spline
Remarks : "L" pressure : Line pressure "2" pressure : "D" clutch pressure "1" pressure : "L" clutch pressure		"L" pressure is not produced.	Damage of automatic transmission fluid (ATF) pump drive system	Slippage of pressfitted portion of torque converter pump inner hub Damaged splined portion of ATF pump drive gear
	When accelerator pedal is depressed with lever in [2], [1] or [R] position, car simulates to roll off.	"L" pressure is too low and fluctuates.	Inoperative ATF pump	Insufficient fluid Clogged pump strainer Wear of pump gear
		"L" pressure is too low.	Defective regulator	Sticking regulator valve
Car does not move with selector lever in [2] or [R] position.	Car moves with lever in [1] position. "L" and "2" pressures are normal.	When accelerator pedal is depressed with lever in [2] or [R] position, engine stalls.	Sticking gear	Sticking and jamming reverse idle gear
		When accelerator pedal is depressed with lever in [2] or [R] position, engine runs up.	Defective "D" clutch	Sticking "D" clutch piston Wear of "D" clutch discs
Car does not move with selector lever in [1] position.	Car moves with lever in [2] or [R] position. "L" and "1" pressures are normal.	When accelerator pedal is depressed with lever in [1] position, engine runs up.	Defective "L" clutch	Sticking "L" clutch piston Wear of "L" clutch discs
Car does not reverse.	Car moves with lever in [2] or [1] position.	Engine runs up with lever in [R] position as in [N] position.	Sticking spline	Sticking reverse gear selector spline
		When accelerator pedal is depressed with lever in [R] position, engine stalls.	Maladjusted control cable	Parking brake lever is locked in [R] position
Poor acceleration	[2] or [1] stall rpm is too high.	Stall pressure is too low and fluctuates.	Inoperative ATF pump	Insufficient fluid Clogged pump strainner Wear of pump gear
		Stall pressure is too low.	Defective regulator	Jamming of regulator spring cap and valve body Fatigue or breakage of regulator springs
	[2] stall rpm is too high. [1] stall rpm is normal.	"D" clutch stall pressure is too low.	Defective "D" clutch	Wear of "D" clutch seal ring Breakage of "D" clutch piston O-ring Inoperative "D" clutch piston check valve
	[1] stall rpm is too high. [2] stall rpm is normal.	"L" clutch stall pressure is too low.	Defective "L" clutch	Wear of "L" clutch seal ring Breakage of "L" clutch piston O-ring Inoperative "L" clutch piston check valve
	Stall rpm is normal.	Poor acceleration at higher speeds.	Excessive fluid resistance	Too much fluid
			Torque converter does not function	Sticking torque converter one-way clutch

Complaint			Probable cause	Inspect for :
Poor acceleration (Continued)	Stall rpm is too low.	Poor acceleration in each speed position.	Insufficient engine output	Ignition timing, carburetor, compression pressure, valve timing, etc.
		Acceleration is normal at higher speeds.	Slippage of one-way clutch	Slippage of torque converter one-way clutch.
		Poor acceleration with pedal fully depressed.	Maladjusted throttle control cable	Carburetor throttle valve will not open fully when accelerator pedal is depressed fully.

DIFFERENTIAL

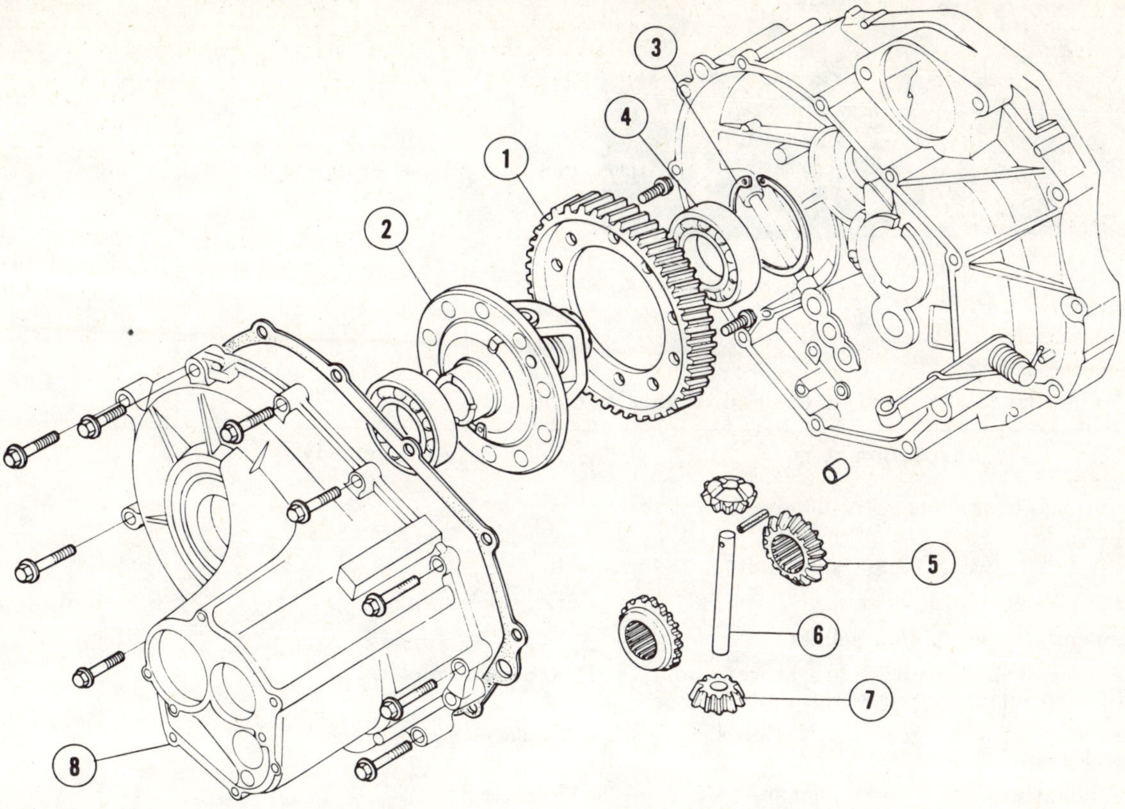

Fig. H-1 ① Final driven gear ② Differential case ③ 72 mm set ring ④ Ball bearing ⑤ Drive pinion ⑥ Pinion shaft ⑦ Differential pinion ⑧ Transmission case

a. Removal (on car)

1. Remove the transmission case from the engine together with the clutch case.

b. Disassembly

1. Separate the transmission case from the clutch case.

2. Remove the differential gears from the clutch case.
   To remove, put a steel plate to the differential case and lightly tap the case with a mallet, as shown.

Fig. H-2

3. Hold the differential case in a vice and loosen 10 mm special bolts in a criss-cross pattern.
   NOTE:
   The special bolts have left-hand threads.

Fig. H-3

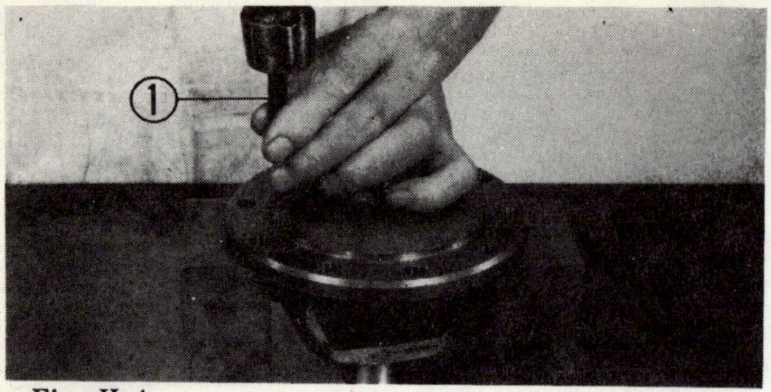

Fig. H-4
① Spring pin driver

4. Remove the 4 mm (0.157 in.) spring pin from the pinion shaft using a spring pin driver, and remove the pinion shaft and gears.

c. Inspection

See Maintenance Standard at the end of this manual.

Inspection item	Remedy
Deformation or damage of differential oil seal lips	Replace.
Clearance between pinion and shaft	Replace.
Wear or damage of gear teeth	Repair or replace.
Damage of drive pinion splines	Repair or replace.
Wear or damage of 6207 ball bearings and bearing attaching surface of case	Repair or replace.
Clearance between gear case pinion shaft hole and shaft	Replace.
Clearance between inboard joint and its opening	Replace.
Outside diameter of 6207 ball bearing attaching part of case	Replace.

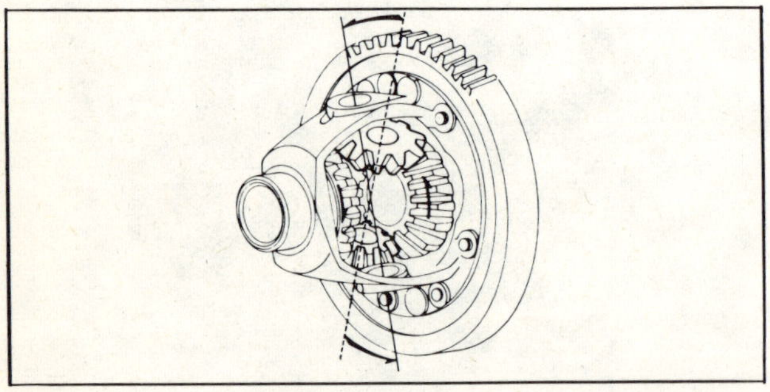

Fig. H-5

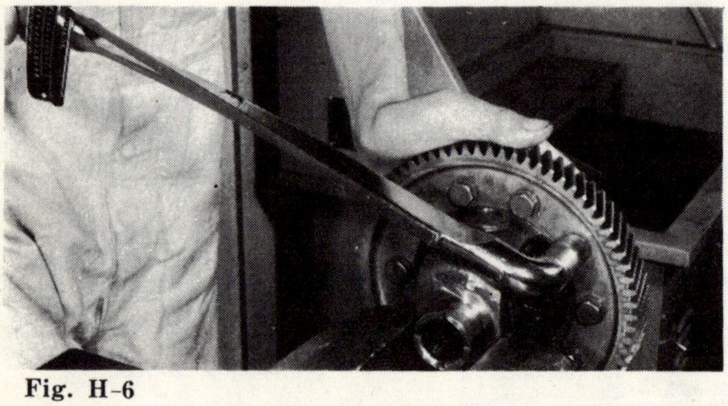

Fig. H-6

d. Assembly

To assemble the differential, reverse the disassembly procedures. Pay attention to the following points:

1. Install the drive pinions and differential pinions. Before putting them together, apply a thin coat of oil or molybdenum bisulfide to the rear sides of the pinion shaft and gears.

   To install, first install the two drive pinions in the differential case at both sides and then mesh the differential pinions with the drive pinions while turning four pinions in the direction indicated by arrows.

2. Install the final driven gear. Hold the differential case in a vice. First tighten the ten special bolts in a crisscross pattern temporarily and then retighten them to the specified torque using a torque wrench.

3. After assembling, check the preload of differential gears.

To check, press the gears against the transmission case and measure the clearance between the case and 72 mm (2.83 in.) set ring by inserting a feeler gauge through the clutch case oil seal hole. If the clearance exceeds the service limit, replace the set ring and measure the clearance again.

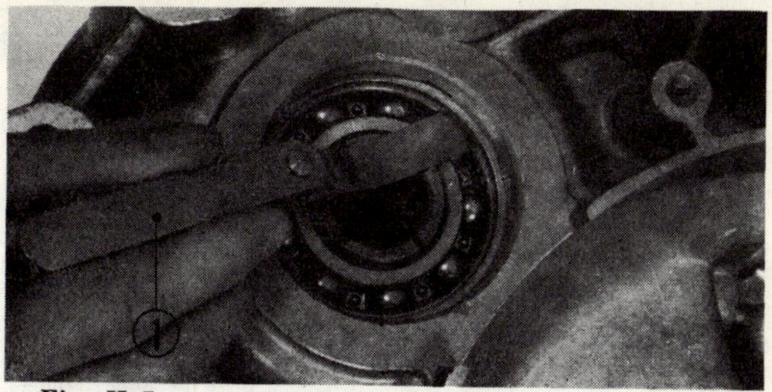

Fig. H–7
① Feeler gauge

The set rings are available in several sizes (thicknesses) as shown in the table.

Unit: mm (in.)

Part No.	Thickness
90414–634	2.45 (0.096)
90415–634	2.55 (0.010)
90416–634	2.65 (0.104)
90417–634	2.75 (0.108)
90418–634	2.85 (0.112)
90419–634	2.95 (0.116)

CARBURETTOR

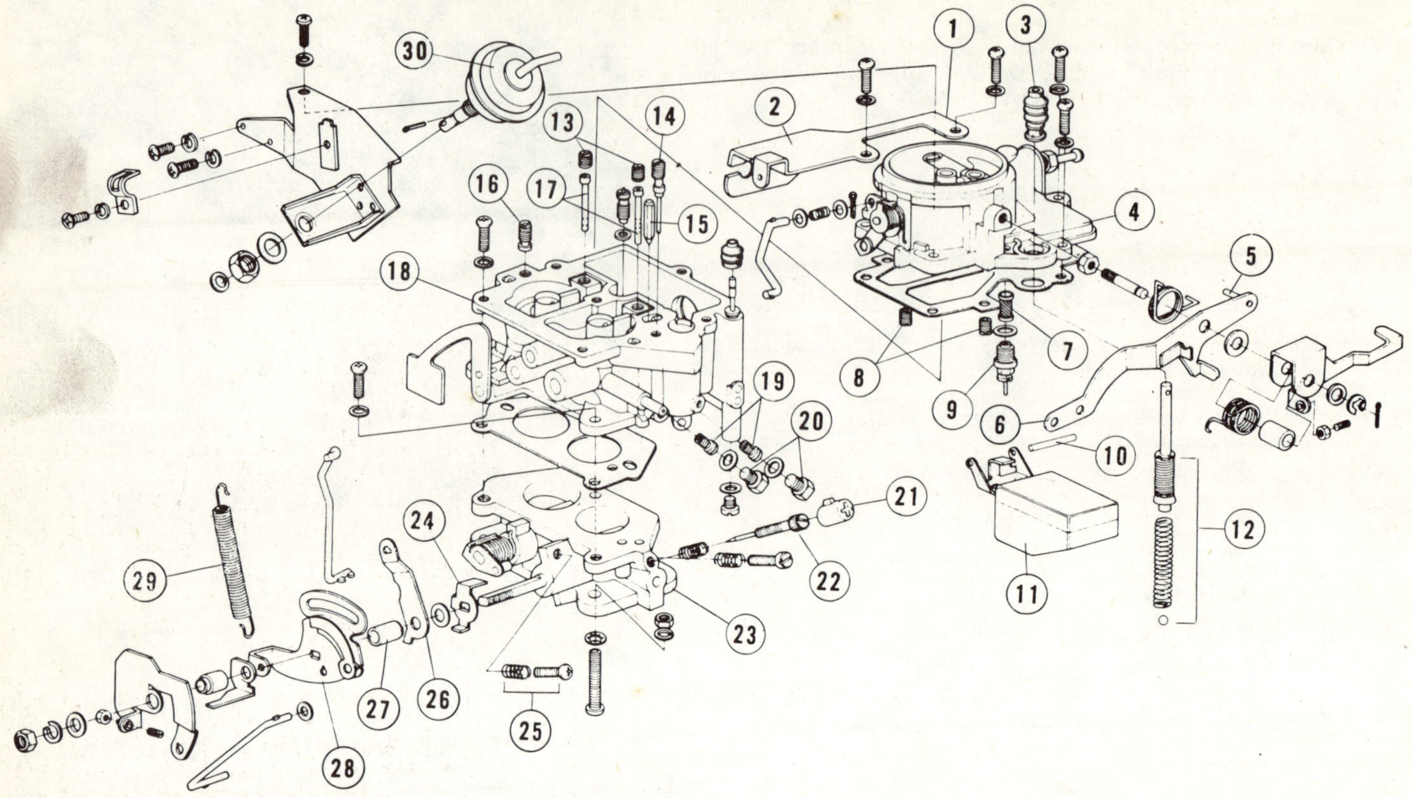

Fig. I–1

① Choke control cable arm	⑪ Float	㉑ Limiter cap
② Spring hanger	⑫ Piston set	㉒ Idle adjusting screw
③ Piston pump cover	⑬ Main air jets	㉓ Throttle chamber
④ Choke chamber	⑭ Slow jet	㉔ Adjusting plate
⑤ Fuel cut-off lever	⑮ Injector weight	㉕ Throttle adjusting screw set
⑥ Pump lever	⑯ Slow air jet	㉖ Choke connecting lever
⑦ Filter	⑰ Emulsion tubes	㉗ Sleeve
⑧ Slow jets	⑱ Float chamber	㉘ Throttle control lever
⑨ Needle valve	⑲ Main jets	㉙ Throttle return spring
⑩ Float shaft	⑳ Drain plugs	㉚ Throttle opener

NOTES:

1) When disassembling the carburetor, use tools of the right size and type to prevent damaging small parts and to have the job done better.
2) The primary- and secondary-side parts must be set aside separately in groups, each identified for its location so that the same combination or installation as before can be reproduced at the time of assembling. Take care to prevent missing of small parts.
3) The carburetor must be disassembled in a clean place. Wash the parts in clean gasoline.
4) After washing all jets, direct a blast of compressed air to them to clean their orifices. Do not use a wire.
5) Use new washers, O-rings, gaskets and cotter pins when reassembling.
6) Do not disassemble the throttle, choke and auxiliary valves and their shafts without valid reason.
7) Replace any assembly part as a unit if necessary.

Fig. I–2
① Throttle control cable ② Choke control cable

a. Removal (on car)

1. Remove the air cleaner case and disconnect the vacuum advance control tube and throttle opener control tube.
2. Disconnect the fuel tube at the carburetor.
3. Disconnect the choke and accelerator control cables.
4. Remove the carburetor by loosening the attaching nuts with a wrench.

NOTES:

1) After removing the carburetor, cover the intake manifold holes with clean waste cloths to keep from foreign materials.
2) When the fuel tube is disconnected, fuel will come out. Wipe fuel off with waste cloths.

b. Disassembly

b–1. Carburetor assembly

1. Remove the throttle return spring.
2. Remove the pump lever and the springs. Disconnect the pump rod.

NOTE:

The pump lever must not be removed unless necessary. When it is removed from necessary, a coat of locking compound must be applied to the threaded surface.

3. Disconnect the choke connecting rod.
4. Remove the spring hanger, choke control cable arm and choke chamber by loosening the five retaining screws.

NOTE:

When removing the choke chamber, take care not to drop the float pin; otherwise the float may be damaged.

5. Separate the float chamber from the throttle chamber by removing the four screws.

Fig. I-3
① Spring hanger ② Throttle return spring
③ Pump lever

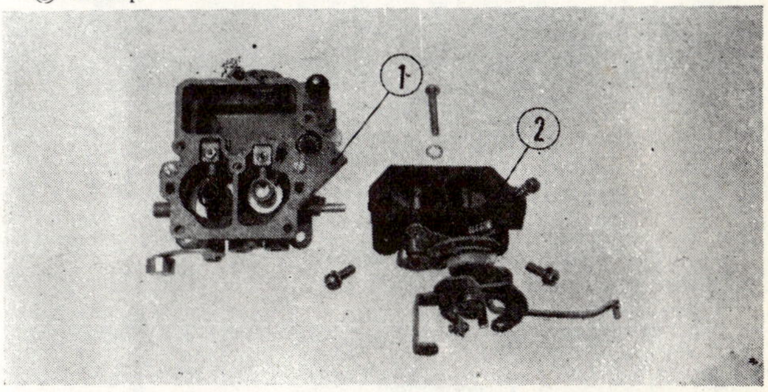

Fig. I-4
① Float chamber ② Throttle chamber

b–2. Choke chamber

1. Remove the piston and pump cover.
2. Pull out the float pin and remove the float, and then the needle valve and filter.
3. Remove the slow air bleed.

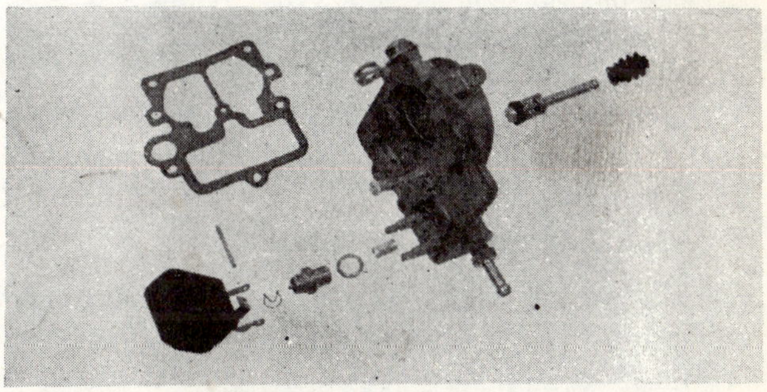

Fig. I-5

b–3. Float chamber

1. Remove the piston return spring and inlet valve.
2. Remove the outlet valve.
3. Remove the main air bleed and disconnect the emulsion tubes.
4. Remove the slow jets.
5. Remove the drain plugs and main jets.

Fig. I-6

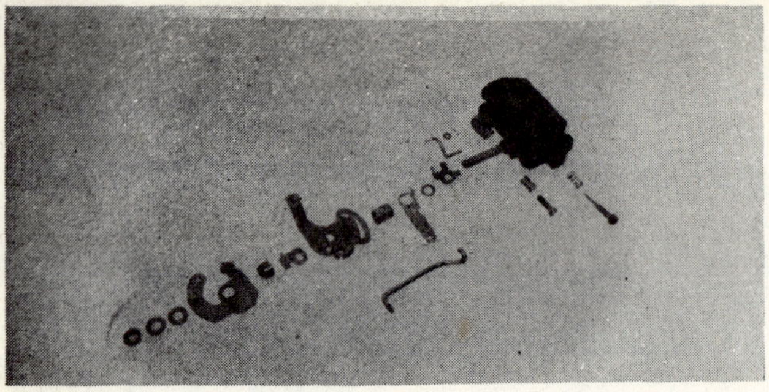

Fig. I-7

b-4. Throttle chamber

1. Remove the throttle adjusting screw and spring.
2. Remove the idle adjusting screw and spring.
3. Loosen the attaching nuts and remove the throttle lever, sleeve, choke connecting lever and adjusting plate in this order.

NOTE:

Do not disassemble the valves unless necessary.

c. Inspection

NOTES:

1) Before checking the parts for condition, wash them in clean gasoline.

   Remove dust and dirt from the jets and orifices in the body with compressed air.
2) Clean the carburetor body with a cleaning brush. Decarbon the throttle, choke and auxilialy valves.
3) When cleaning the jets, do not use a wire; otherwise they may be damaged.

See the Maintenance Standard.

Inspection item	Remedy
Improper contact of needl valve and seat and damage of needle valve	Repair or replace.
Breakage or corrosion of filter	Replace.
Clogging or damage of jets and air bleed	Replace.
Damage or binding of pump piston cup	Replace.
Improper contact of injector weight and seat	Repair or replace.
Improper contact of idle adjusting screw and seat and damage of adjusting screw	Replace.
Rattle or improper operation of throttle shaft	Replace.
Damage or improper operation of diaphragm	Replace.

d. Assembly and adjustment

To assemble the carburetor, reverse the disassembly procedures. Pay attention to the following points:

1. Install the primary- and secondary-side emulsion tubes and jets in correct position.
2. Whenever assembling the carburetor, replace all parts in a gasket, O-ring or cotter pin set regardless of conditions, if it is necessary to replace any of it.
3. To install the piston dust cover, first install it to the choke chamber. Securely engage the cover lip in the chamber.
4. Tighten tne throttle and choke chamber bolts evenly.
5. After assembling, check to be sure each linkage moves smoothly without any sign of binding.

Adjustment

d–1. Float level

The level must be at the mark (●) on the glass.

1. Invert the carburetor body after removing the float chamber and set H distance to 11 mm with the valve stem just contacting the float sheet properly.
2. After adjusting H distance, set the valve stem to float sheet clearance to 1.3 mm to 1.7 mm with the float in the raised position (the lowest position on car) by bending the float stopper properly.

NOTE:

When adjusting the float level, check the needle valve seat for entry of dust and the float for smooth movement.

3. After adjusting, again visually check the float level while the engine is running at idle speed.

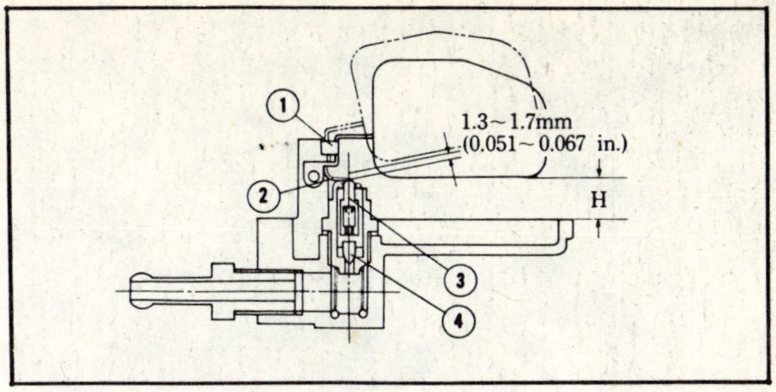

Fig. I-8
① Float stopper ② Float seat ③ Valve stem
④ Needle valve

d–2. Fast idle speed

Using a wire gauge, check the primary throttle valve opening (dimension G1) when the choke valve is fully closed.

G1 specification:

1.48 mm (0.058 in.) for model DCG306–45. DCG306–46
1.22 mm (0.048 in.) for model DCG306–41. DCG306–42

If out of specification, adjust it by bending the adjusting lever stopper. After adjusting, make sure the highest fast idle speed is 2500 to 2800 rpm while the engine is warmed up.

NOTE:

When adjusting the fast idle speed, be sure to return the throttle adjusting screw until it is apart from the stopper.

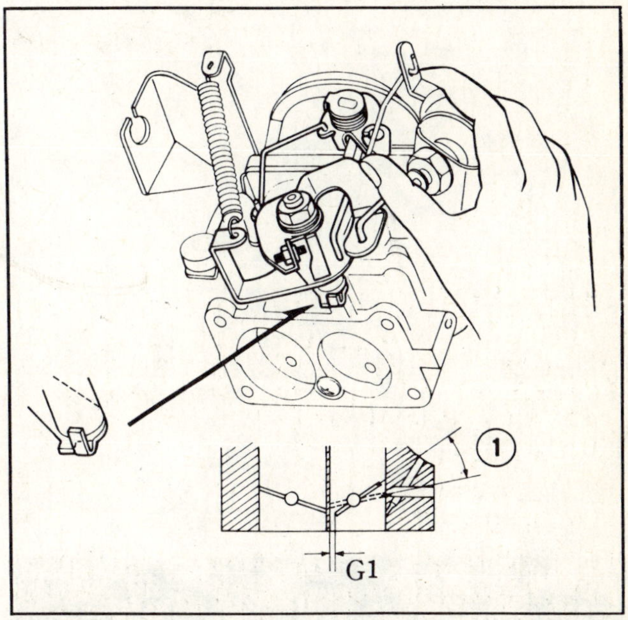

Fig. I-9
① 19.5° for model DCG306–45. DCG306–46
17° for model DCG306–41. DCG306–42

d–3. Throttle valve operation

Check to see if the throttle valve opens fully when the accelerator pedal is depressed fully or when the throttle lever is fully moved in the direction of increasing the engine speed and if the valve closes fully when the pedal or the lever is released. Measure the clearance (G2) between the primary throttle valve and chamber wall at the position where the connecting rod bears on the end of the throttle lever groove.

G2 specification: 5.83 mm (0.230 in.)

If out of specification, adjust it by bending the connecting rod.

NOTE:

After adjusting, operate the throttle lever fully and check for any sign of binding.

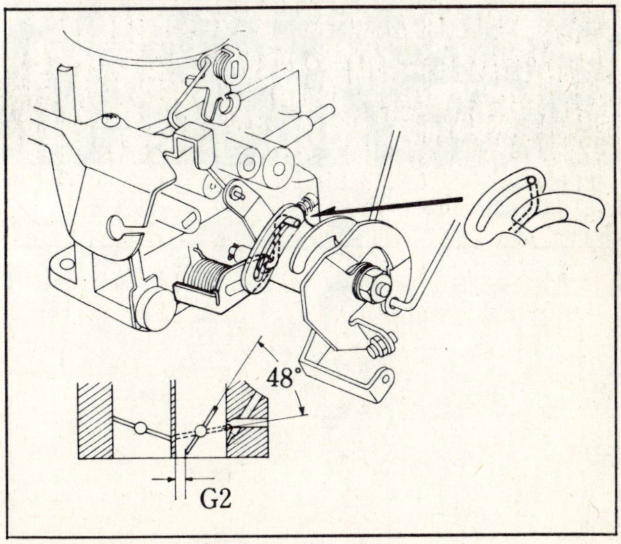

Fig. I-10

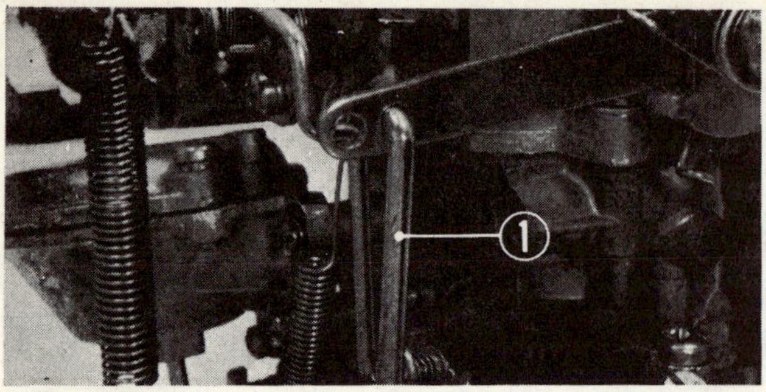

Fig. I-11
① Pump connecting rod

d-4. Accelerator pump displacement

Check the pump for smooth operation. See if fuel spurts out of the pump injector by operating the pump lever or the throttle lever. When the pump is operated slowly, fuel must continuously spurt out until the pump comes to the end of its travel. If the pump is defective, check for clogging or for defective piston. If acceleration is erratic or if starting ability is poor in cold weather, adjust the pump displacement by changing the position of the pump connecting rod.

In hot weather:
Outside position (**decrease**)

In cold weather:
Inside position (**increase**)

d-5. Adjust the idle speep

1. Warm up the engine thoroughly before making this adjustment. In consideration of the load on the generator, turn on the head light high beam.
2. Set the engine rpm to the lowest stead speed by turning the throttle stop screw. Then, bring the engine rpm to the maximum by turning the pilot screw in either direction.
3. Turn the throttle stop screw so that the engine rpm becomes within the specified idle rpm. Finally check to see by turning the pilot screw that the adjusted engine idle speed is still maximum.
4. Then turn the pilot screw 1/16 turn clockwise. Generally adjust the idle speed so that the CO content is within the specification in the working range of the idle adjusting screw limiter cap.

Specification:
1~3% for model DCG306-45, DCG306-46
2~4% for model DCG306-41, DCG306-42

Fig. I-12
① Idle adjusting screw

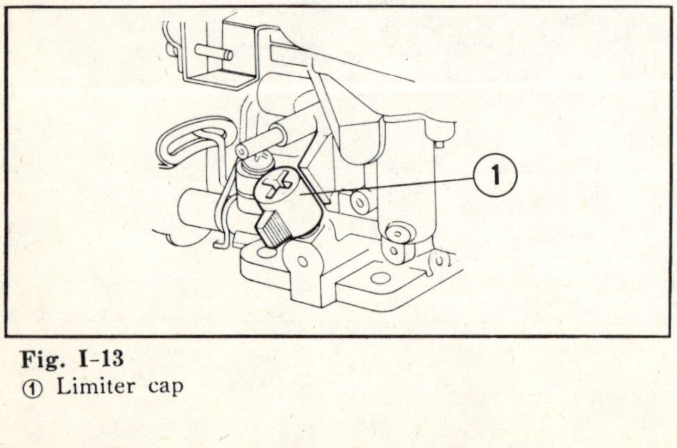

Fig. I-13
① Limiter cap

d-6. Idle adjusting screw

When the adjusting screw limiter cap is removed as for disassembly of the carburetor, adjust the idle adjusting screw as follows:

1. Fully turn the screw in and then out the specified number of turns.
2. Then adjust the screw as in the steps for the idle speed.
3. After adjusting, install the limiter cap in such a way that the tab on the carburetor and the tab on the limiter are in line facing down.

Specified number of returns:
2¾±¼ for model DCG306-45, DCG306-46
1¾±¼ for model DCG306-41, DCG306-42

Precautions for service work

1. If any electrical trouble occurs, trace the trouble in the sequence of the battery, wiring and electrical elements according to the wiring diagram.
2. Check the battery terminals for looseness or contamination. Repair or clean them if necessary.
3. If the fuse is blown, check for problems causing the fuse failure before installing a new fuse. Always install a specified rated fuse.
4. Check the wire harnesses for binding and their connections for looseness. Repair them if necessary.
5. Take care not to damage the wire harnesses and wire leads during service work.
6. After servicing, check to be sure the wire harnesses and wire leads are properly secured with clips and the grommets and covers are securely installed in place.

Engine Electrical System

Ignition System

a. Ignition coil

Specifications

Specified voltage	12 V
Primary coil resistance	3.8±0.38Ω (20°C or 68°F)
Secondary coil resistance	8.0±1.6 kΩ (20°C or 68°F)
Three-point spark length	8 mm (0.315 in.), min. at 8V, 150 rpm * 7 mm (0.276 in.), min. at 12V, 3000 rpm **

* Coil temperature : Room temperature
** Coil temperature : 80°C (176°F)

Fig. J–1 Ignition coil

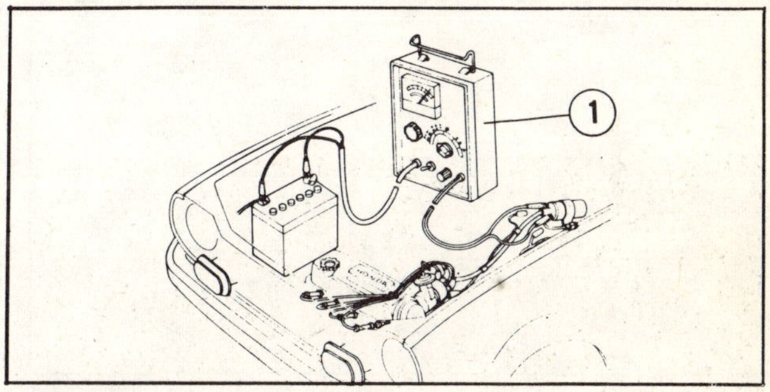

Fig. J–2 Ignition coil output test
① Coil tester

a–1. Coil output

1. Using a three-point spark tester, measure the maximum length of stable spark, together with the distributor having a dwell angle of 49° to 55°.
2. The performance can be tested by a commercially available coil tester.

NOTE :

Heat the coil to a temperature of 80°C (176°F) before testing.

a–2. Coil resistance

1. If the resistance value is below the specification, short circuit is in the coil.
2. If the resistance value is above the specification, the contacts are defective.
3. If there is no continuity, open circuit is in the coil.

NOTE :

The coil must be cooled up to the room temperature because the resistance value varies excessively with changes in temperatures.

b. Distributor

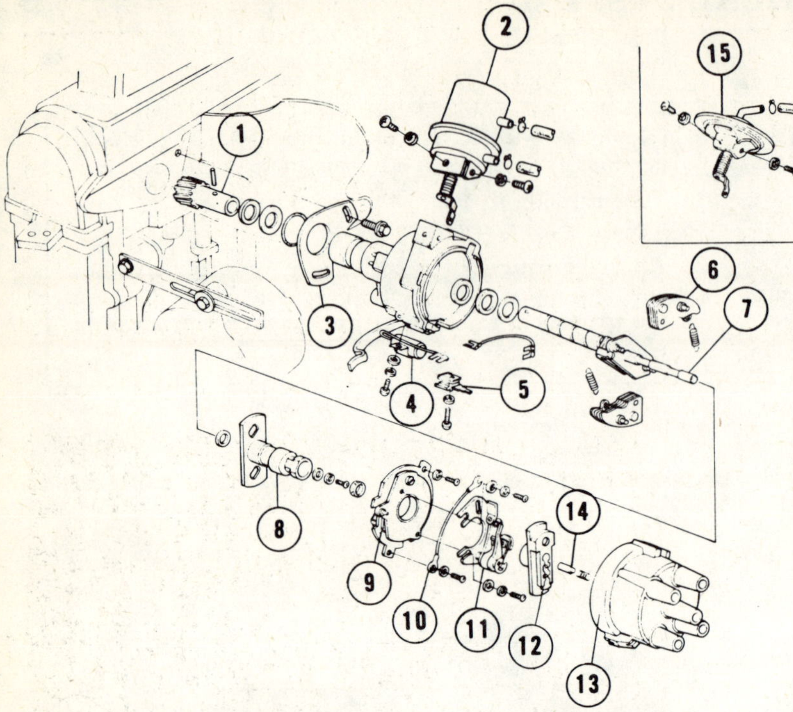

Fig. J-3 ① Pinion ② Vacuum controller ③ Set plate ④ Capacitor ⑤ Terminal ⑥ Governor weight ⑦ Shaft ⑧ Cam ⑨ Breaker plate ⑩ Ground cord ⑪ Contact breaker ⑫ Head rotor ⑬ Cap ⑭ Carbon point ⑮ Vacuum controller

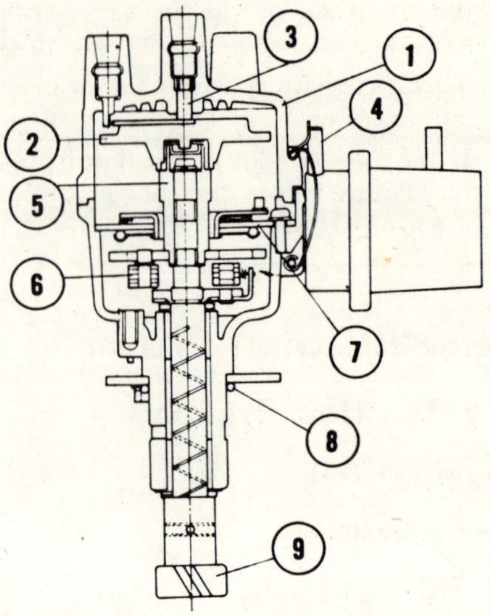

Fig. J-4 ① Distributor cap ② Head rotor ③ Carbon point ④ Vacuum controller ⑤ Cam ⑥ Governor weight ⑦ Breaker plate ⑧ O-ring ⑨ Pinion

Centrifugal governor advancer characteristics

1. For U.S.A. type

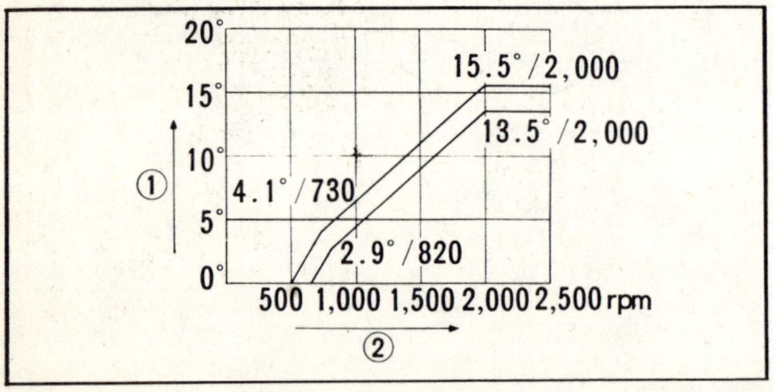

Fig. J-5
① Distributor advance angle ② Distributor rpm

2. Not applicable to U.S.A. type

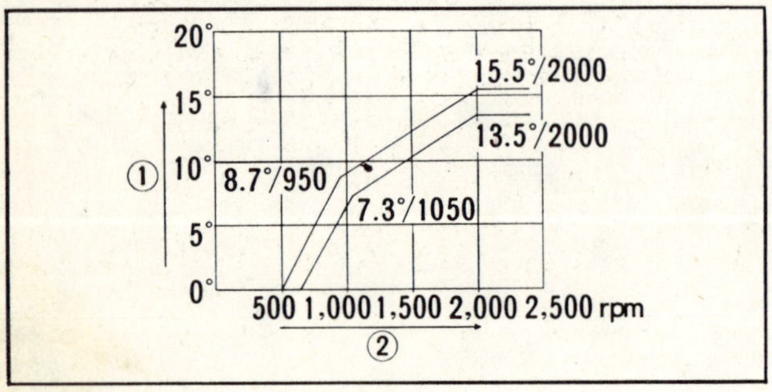

Fig. J-6
① Distributor advance angle
② Distributor rpm.

Specifications

Model		
Direction of rotation		Counterclockwise as viewed from cap side
Firing order		1, 3, 4, 2
Ignition timing (Regarding U.S.A. Type: when idle retarder is not functioning)		5° B.T. D.C.
Dwell angle		49° ~ 55°
Point gap		0.45 ~ 0.55 mm (0.018 ~ 0.022 in.)
Capacity of capacitor		0.22 μF
Advancer characteristics	For U.S.A. type	
	Centrifugal	Starting 575 rpm Maximum (14.5°/2000 rpm)
	Vacuum	Starting 150 mmHg (5.9 inHg) Maximum 7.5°/305 mmHg (12 inHg)
	Not applicable to U.S.A. type	
	Centrifugal	Starting 575 rpm Maximum (14.5°/2000 rpm)
	Vacuum	Starting 120 mmHg (4.7 inHg) Maximum 12°/275 mmHg (10.8 inHg)
For U.S.A. type Retarder characteristic		Starting 300 mmHg (11.8 inHg) Maximum −2.5°/325 mmHg (2.8 inHg)

Vacuum advancer characteristics

1. For U.S.A. type

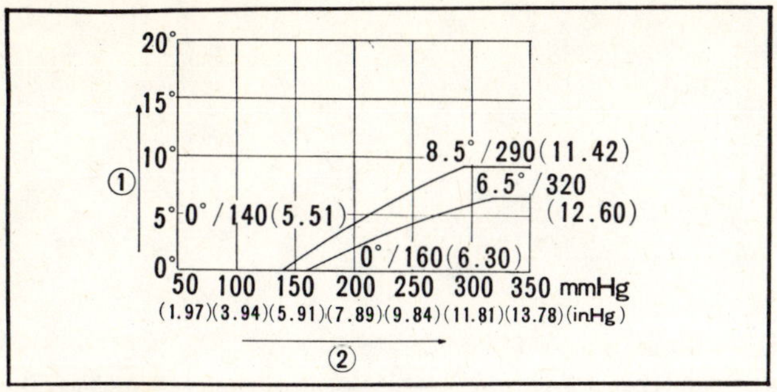

Fig. J-7
① Distributor advance angle
② Vacuum mmHg (in.Hg)

2. Not applicable to U.S.A. type

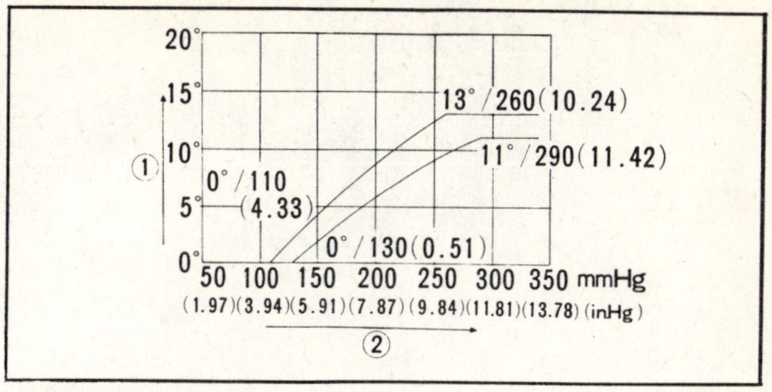

Fig. J-8
① Distributor advance angle
② Vacuum mmHg (in.Hg)

b–1. Removal

1. Disconnect the high tension cords.
2. Disconnect the vacuum tube.
3. Disconnect the ignition coil primary lead.
4. Loosen the distributor bolts and remove the distributor from the engine.

Fig. J-9 Distributor attaching bolt

b–2. Disassembly

NOTE:

The parts removed must be marked or set aside separately in groups, each tagged for its location, so that the same combination or set as before can be reproduced at the time of assembling.

1. Remove the cap and rotor.
2. Remove the vacuum control assembly by removing two set screws and by removing the advancer rod from the breaker plate pin.
3. Loosen the primary lead terminal set screw and remove the terminal.
4. Remove the contact breaker assembly by loosening two set screws. (One set screw is securing the ground cord.)
5. Remove the capacitor by loosening its set screw.
6. Remove the breaker plate by two set screws. (One set screw is securing the ground cord.)

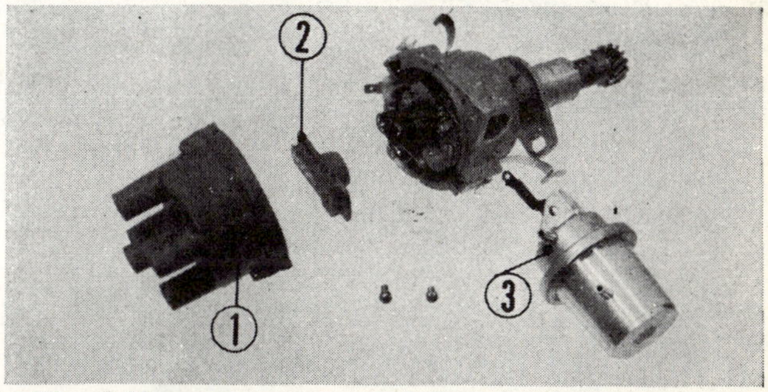

Fig. J-10
① Cap ② Head rotor ③ Vacuum controller

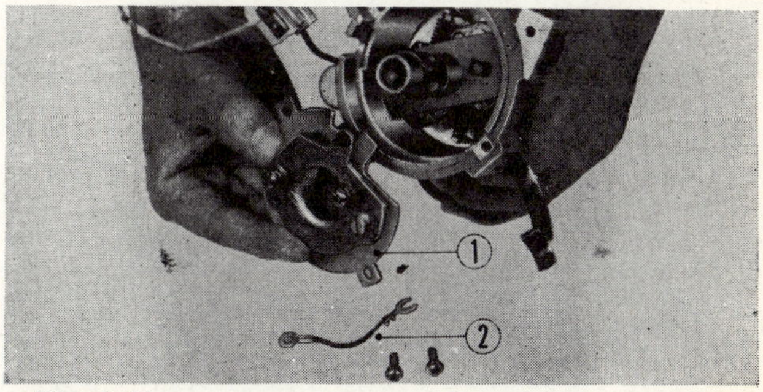

Fig. J-11
① Breaker plate ② Negative ground cord

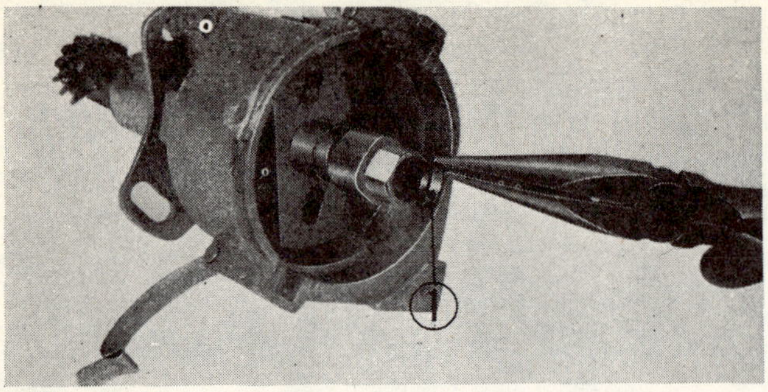

Fig. J-12
① Rubber cap

7. Remove the cam by removing the rubber cap located on the camshaft end and the cam set screw. The cam set screw is left threaded.

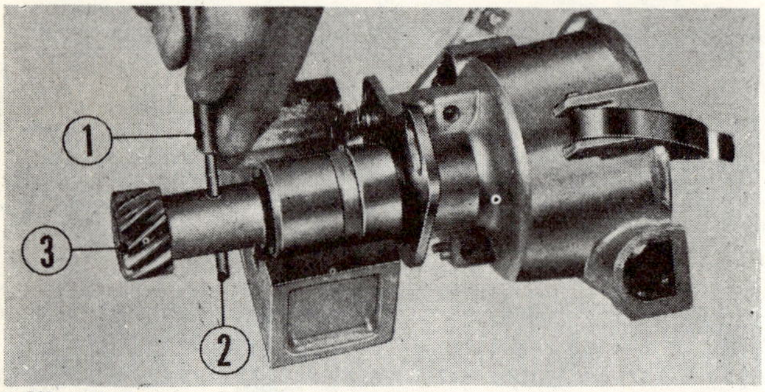

Fig. J-13
① Pin driver ② Set pin ③ Drive pinion

8. Pull out the distributor drive pinion set pin using the special tool and separate the pinion gear from the shaft. Then, pull the shaft out of the distributor body.

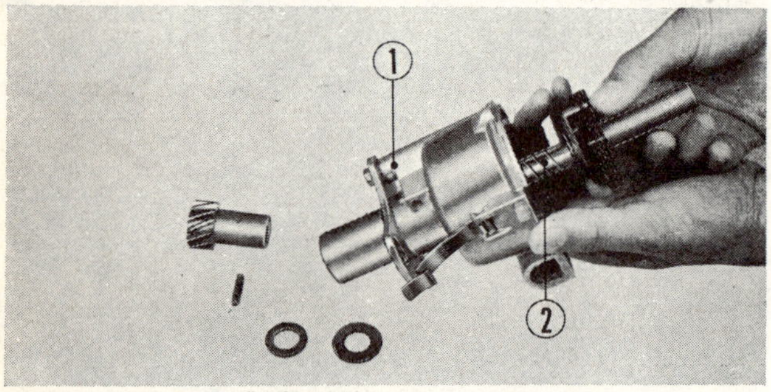

Fig. J-14
① Distributor body ② Shaft

9. Remove the governor weights and springs from the shaft.

b-3. Inspection and adjustment

1. Check the each part for wear or damage.
2. Inspect the rotor and cap.
3. Inspect the contact points.
   - Point surfaces
   - Point gap
   - Point pressure
4. Inspect the ignition timing.
5. Inspect advancers.
   - Vacuum advancer
   - Centrifugal governor

b-4. Assembly

To assemble the distributor, reverse the disassembly procedures. Pay attention to the following points:

1. Apply a thin coat of grease to each sliding surface.
2. Use new set pins.

Installation

To install the distributor, **see page 100**.

Starting system

a. Starting motor

The Honda CIVIC uses a DC series wound type starting motor with an electromagnetic switch. Closing the starter switch will cause the magnetic switch to plunge out the pinion into mesh with the flywheel ring gear for engine cranking.

Construction

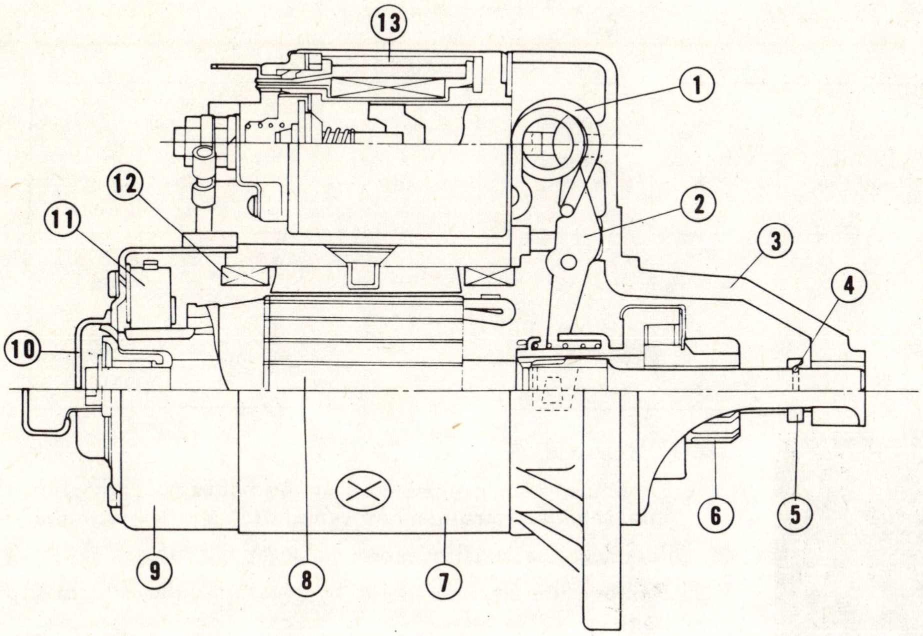

Fig. J–15

① Torsion spring
② Shift lever
③ Gear case
④ Stopper clip
⑤ Pinion stop collar
⑥ Pinion
⑦ Yoke
⑧ Armature
⑨ Rear cover
⑩ Rear dust cover
⑪ Brush
⑫ Field coil
⑬ Magnetic switch

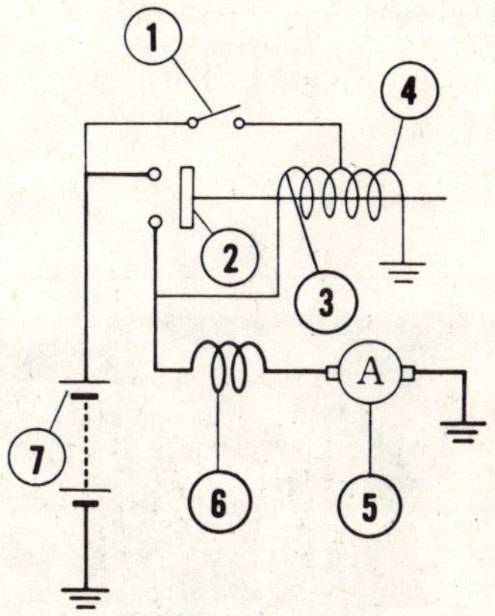

Fig. J–16 Starting motor circuit

① Starter switch
② Magnetic switch contact
③ Magnetic switch coil
④ Magnetic switch holding coil
⑤ Armature coil
⑥ Field coil
⑦ Battery

Specifications and Characteristics

	Standard type	Cold area type
Type	Electromagnetic switch	
Nominal output	0.7 KW	
Nominal voltage	12 V	
Hour rating	30 seconds	
Direction of rotation	Clockwise as viewed from pinion side	
Weight	4.8 kg (9.48 lb)	5.1 kg (11.24 lb)
Yoke outside diameter	80 mm (31.5 in.)	
Overall length	216.5 mm (8.52 in.)	236.5 mm (9.31 in.)

Characteristics

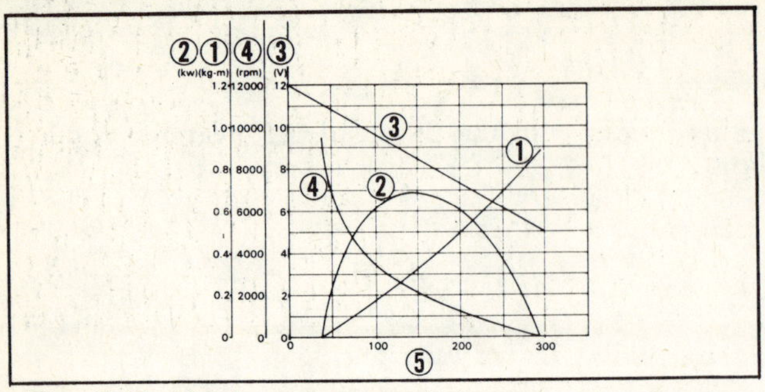

Fig. J-17 Honda-matic transmission type
① Torque ② Output ③ Voltage (V) ④ RPM
⑤ Current (A)

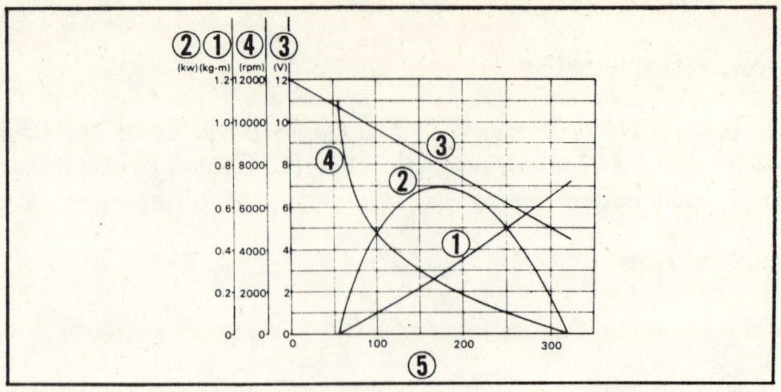

Fig. J-18 Manual transmission type
① Torque ② Output ③ Voltage ④ RPM
⑤ Current (A)

	Pinion plunging-out voltage, V	No-load			When locked		
		Terminal voltage, V	Ampere, A	Speed, rpm	Terminal voltage, V	Ampere, A	Torque, kg-m (lb-ft)
Manual transmission type	8, max.	12	70, max.	7000, min.	4.9	400, max.	0.7 (5.06), min.
Honda-matic transmission type						380, max.	0.75 (5.42), min.

Fig. J-19
① Attaching bolts

a-1. Removal

1. Disconnect the ground cord at the battery negative terminal and the starting motor cable at the positive terminal.
2. Disconnect the starting motor cable at the motor.
3. Remove the starting motor by loosening the attaching bolts.

a-2. Disassembly

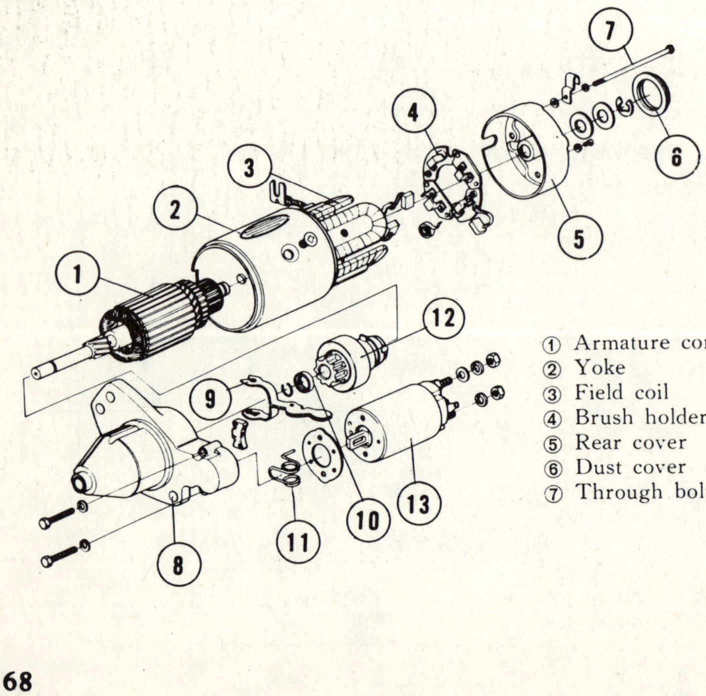

Fig. J-20

① Armature complete
② Yoke
③ Field coil
④ Brush holder
⑤ Rear cover
⑥ Dust cover
⑦ Through bolt
⑧ Gear case
⑨ Shift lever
⑩ Pinion stop collar
⑪ Torsion spring
⑫ Pinion complete
⑬ Magnetic switch

1. Remove the magnetic switch by loosening the attaching bolts.
2. Remove the rear cover dust cover with a screwdriver and remove the E-type clip and thrust washer. Remove the two brush holder plate retaining screws.

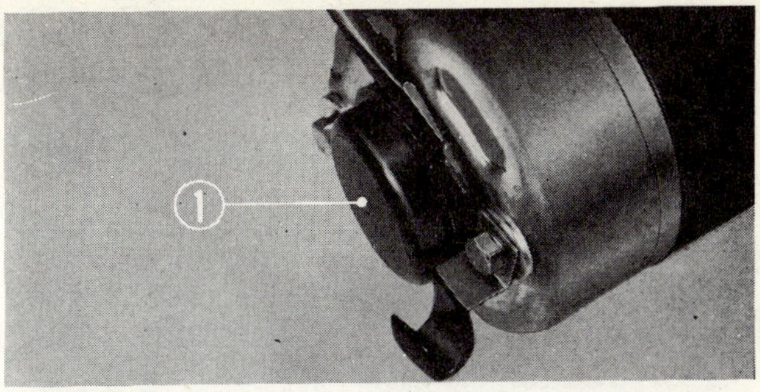

Fig. J-21
① Rear cover dust cover

3. Remove the through bolts. Lightly tap the rear cover with a mallet and remove it.
4. Remove the gear case. Remove the brushes from the commutator and remove the armature toward the gear case.
5. Remove the brushes from the holders and separate the holders from the yoke.

Fig. J-22
① Gear case ② Yoke ③ Brush holder plate
④ Rear cover

6. Remove the pinion gear from the armature.

   Put the armature shaft with the pinion side facing upward and pull the clutch stop collar downward (toward pinion). Then, remove the pinion stopper clip and pull the stopper and gears as a set upward.

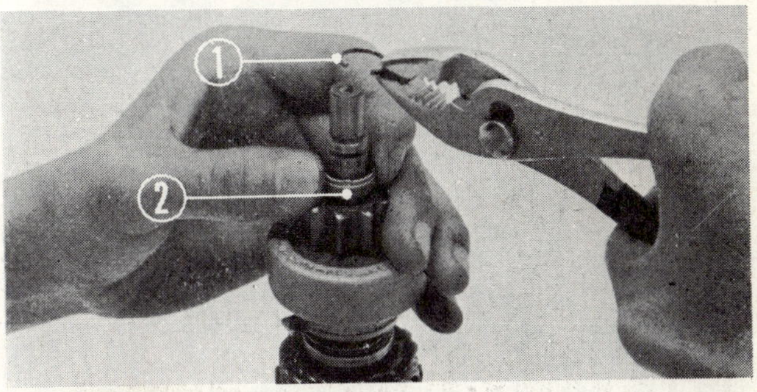

Fig. J-23
① Stopper clip ② Stop collar

a-3 Inspection

Commutator

Check the commutator surface for contamination or wear. If the surface is dirty, recondition it with a #500 or #600 emery paper. If the wear is within the specification, undercut the mica with a hacksaw blade.

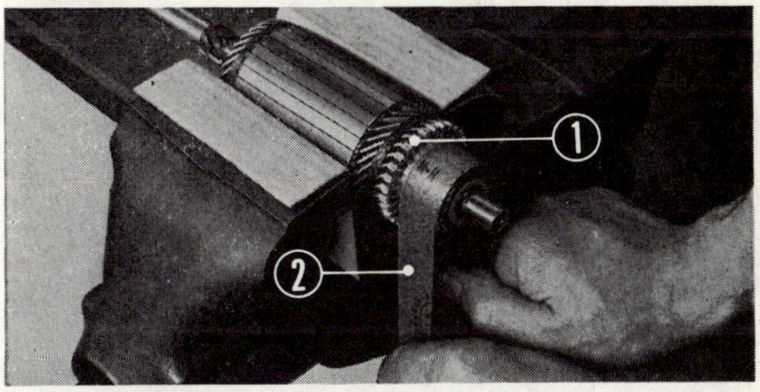

Fig. J-24
① Commutator ② Sandpaper

NOTE:

The commutator surface must be free from scores, scratches or burr.

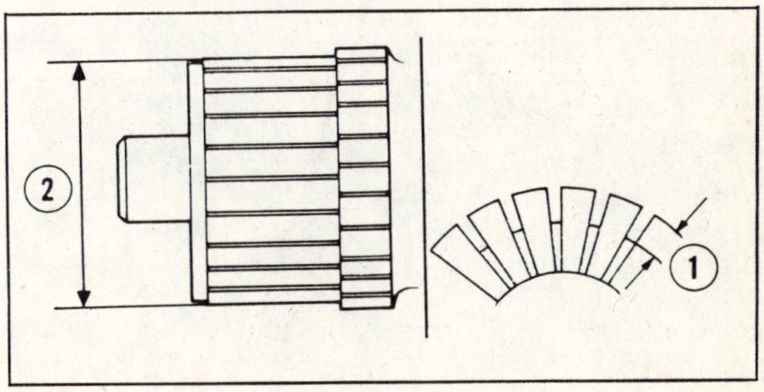

Fig. J-25
① Repair limit: 0.2 mm (0.008 in.)
② Specified outside diameter and service limit
Specified outside diameter: 33 mm (1.29 in.)
Service limit: 32 mm (1.26 in.)

Armature coil

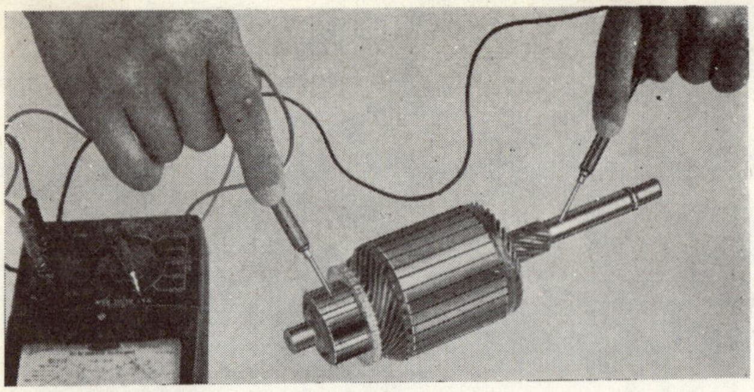

Fig. J-26 Checking armature coil for short circuit

1. Inspect for short circuit between the coil and shaft

   Using a radio tester, check to see if there is any continuity between the commutator and armature shaft. If there is any continuity, replace the armature.

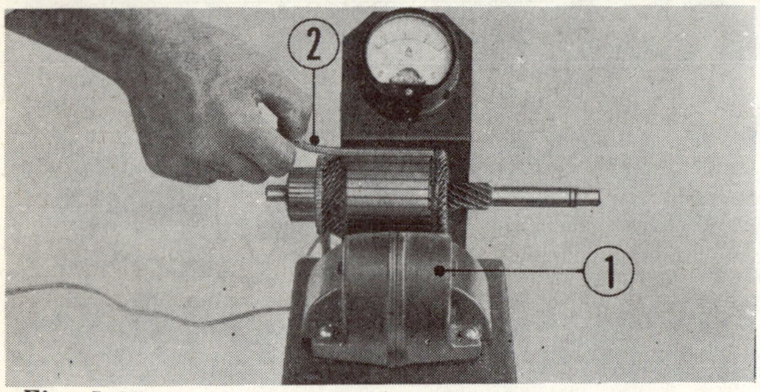

Fig. J-27
① Growler tester ② Steel strip

2. Inspect for short circuit in coils

   Place the armature on a growler tester and rotate it slowly while holding a steel strip over the armature. If the strip vibrates, replace the armature.

Field coil

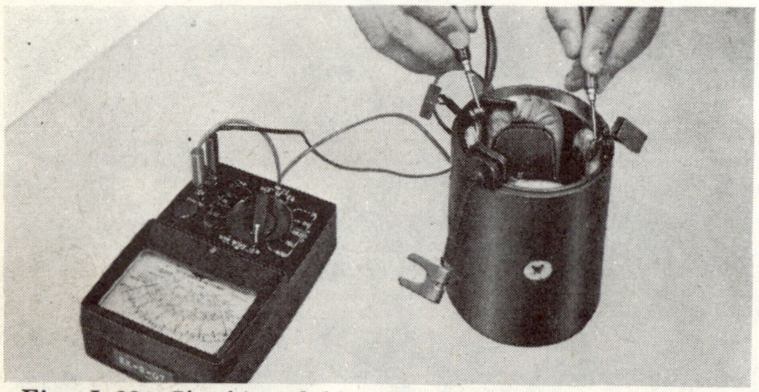

Fig. J-28 Checking field coil for continuity

Using a radio tester, check for open or short circuit in the coil. If the coil is defective, replace the coil.

1. If there is no continuity between the field coil terminals and coil brushes, the coil has an open circuit.
2. If there is any continuity between the one terminal of the coil and yoke, the coil has a short circuit.

Brushes, springs and brush holders

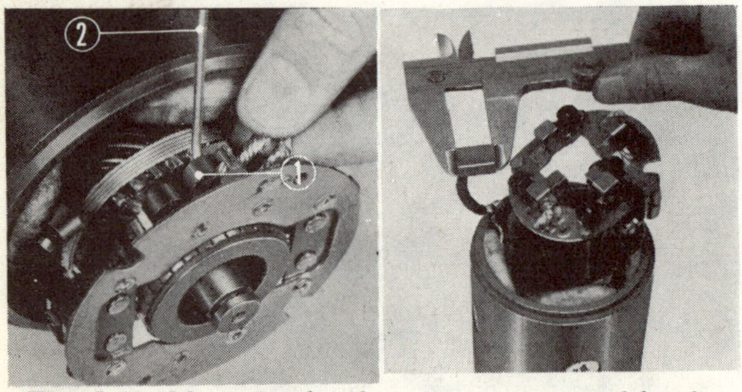

Fig. J-29 Measuring brush spring pressure and brush length
① Spring scale ② Brush spring
Measuring brush length

1. Measure the length of the brush. Replace the brush if it is worn beyond the service limit.

   NOTE:

   When replacing the brush, polish it with a #500 or #600 emery paper.

2. Using a spring scale, measure the pull on the brush spring.

   If the pull is below the service limit, replace the spring.

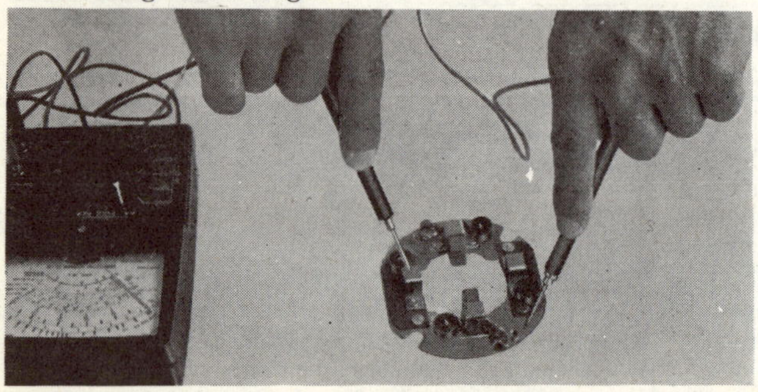

Fig. J-30 Checking brush holder for short circuit

3. Using a radio tester, check to see if there is any continuity between the holder plate and the holder on the field coil. If there is any continuity, replace the holder.

Pinion complete (Overrunning clutch)

1. Move the overrunning clutch along the shaft. If there is any sign of binding, or if the clutch slips when the armature is rotated with the pinion gear held, replace the pinion.
2. Check the pinion gear teeth for wear or damage and replace the pinion if necessary.

   NOTE:

   When the pinion gear teeth are damaged, also check the ring gear teeth for damage.

Fig. J-31
① Overrunning clutch

Magnetic switch

1. Inspect for open circuit in the holding coil

   Using a tester, check for continuity between the magnetic switch terminal "S" and coil case. If there is no continuity, the coil has an open circuit.

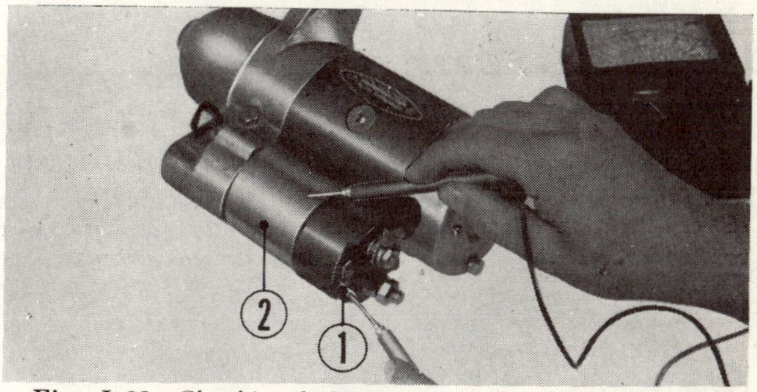

Fig. J-32 Checking holding coil for continuity
① Switch body ② Terminal "S"

2. Inspect for open circuit in the pull-in coil

   Using a tester, check for continuity between the magnetic switch terminal "S" and terminal "M". If there is no continuity, the coil has an open circuit.

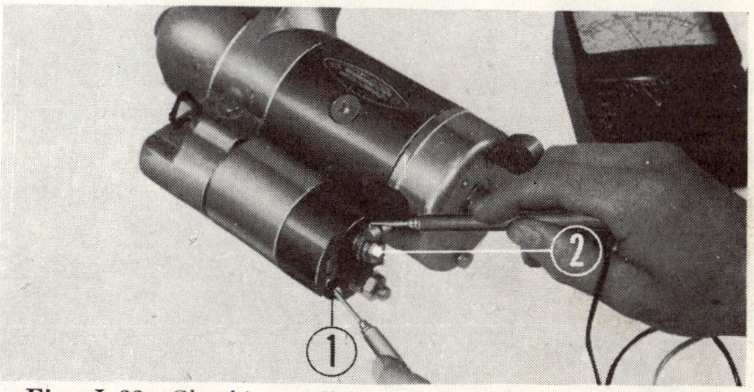

Fig. J-33 Checking pull-in coil for continuity
① Terminal "S" ② Terminal "M"

a-4. Assembly

To assemble the starting motor, reverse the disassembly procedures. Pay attention to the following points:

1. Use new clips.
2. Be careful of the installation direction of the shift lever.

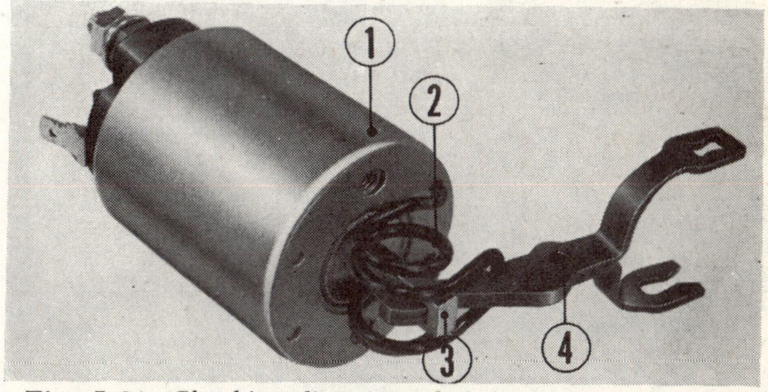

Fig. J-34 Checking direction of shift lever
① Magnetic switch ② Torsion spring ③ Plunger
④ Shift lever

After assembly, the following inspections must be made:

1. Pinion-to-stopper clearance

   When the pinion is plunged out by means of the magnetic switch, check to see if the clearance between the pinion and pinion stopper is within the specification (0.3~1.5 mm or 0.012~0.059 in). If not, adjust the clearance by changing the adjusting plate. The plates are available in two sizes (thicknesses), 0.4 mm (0.016 in.) and 0.8 mm (0.031 in.).

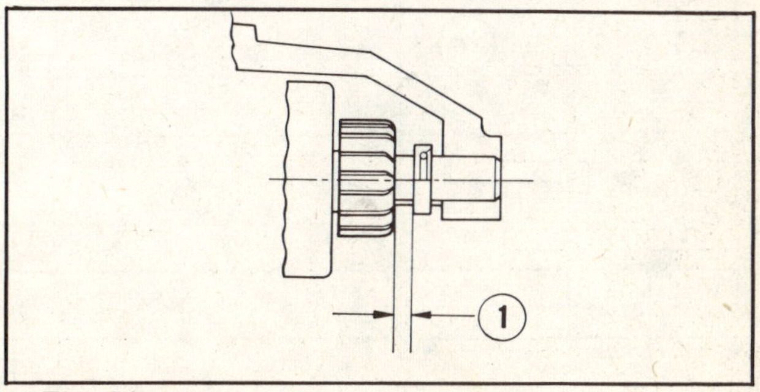

Fig. J-35
① Pinion-to-stopper clearance

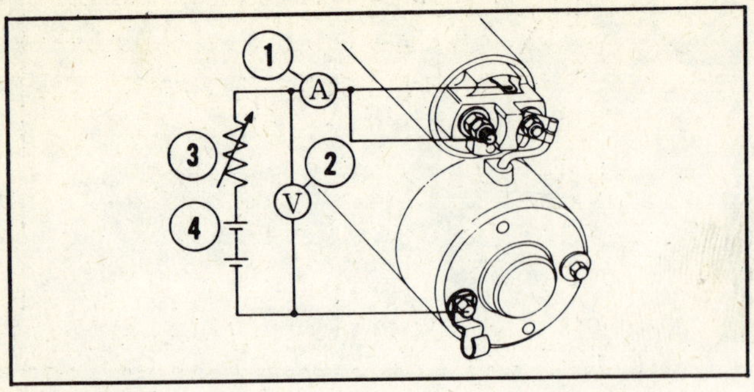

Fig. J–36
① Ammeter ② Voltmeter ③ Variable resistor
④ Battery

2. Performance test
No-load test

Using a fully charged 12-volt battery, connect as shown. The starting motor must turn smoothly at over 7000 rpm and the current must be less than 70A.

J–3. Charging system

The charging system of the Honda CIVIC consists essentially of an AC generator and a regulator. The rotor of the generator rotates within the stator as shown.
As the rotor runs, three-phase alternating current is induced in the stator coils. This alternating current is full wave rectified by six silicon diodes (rectifiers) built in the end frame and is converted into a DC current. The regulator provides a means of limiting the current flowing to the field coil and thereby maintains the generator output constant.

a. AC generator

Construction

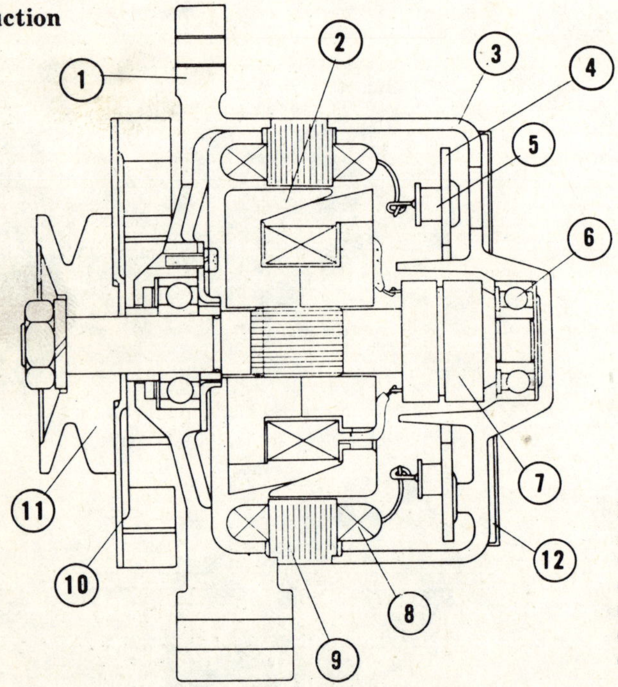

Fig. J–37
① Drive end frame
② Rotor
③ Rectifier end frame
④ Rectifier holder fin
⑤ Rectifier
⑥ Ball bearing
⑦ Slip ring
⑧ Stator coil
⑨ Stator core
⑩ Fan
⑪ Pulley
⑫ Protector cover

Wiring diagram

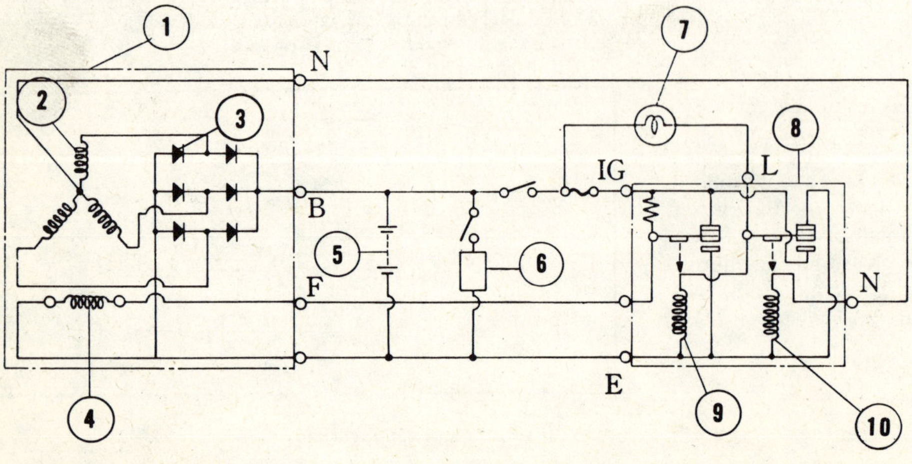

Fig. J–38
① AC generator
② Stator coil
③ Silicon diode
④ Field coil
⑤ Battery
⑥ Load
⑦ Battery charge lamp
⑧ Regulator
⑨ Voltage regulator
⑩ Battery charge lamp relay

Specifications (Car w/o air conditioner)

Nominal voltage	12 V
Nominal output	35 A
Excitation current rating	2.5 A
Polarity (ground)	Negative
Direction of rotation	Counterclockwise as viewed from pulley
Outside diameter of pulley	80.6 mm (3.173 in.)
Outside diameter of stator	118 mm (4.646 in.)

Specifications (Car w/ air conditioner)

Nominal voltage	12 V
Nominal output	40 A
Excitation current rating	2.5 A
Polarity (ground)	Negative
Direction of rotation	Counterclockwise as viewed from pulley
Outside diameter of pulley	77.6 mm (3.055 in.)
Outside diameter of stator	132.5 mm (5.217 in.)

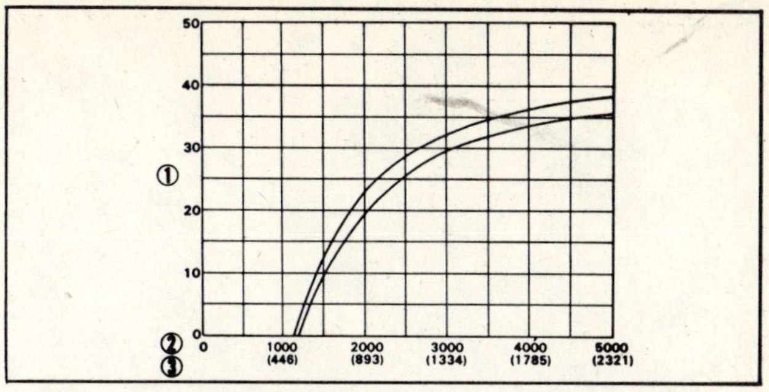

Fig. J-39 Output characteristics (w/o Air control) (14V-constant)
① Output current (A) ② AC generator speed (RPM)
③ (Engine speed, RPM)

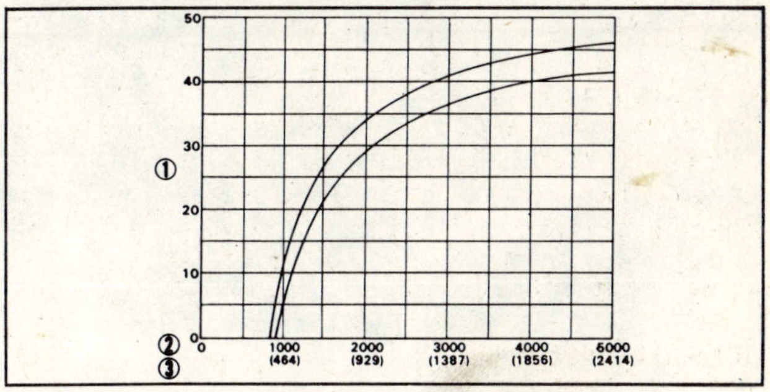

Fig. J-40 Output characteristics (w/ Air control) (14V-constant)
① Output current (A) ② AC generator speed (RPM)
③ (Engine speed, RPM)

Disassembled view

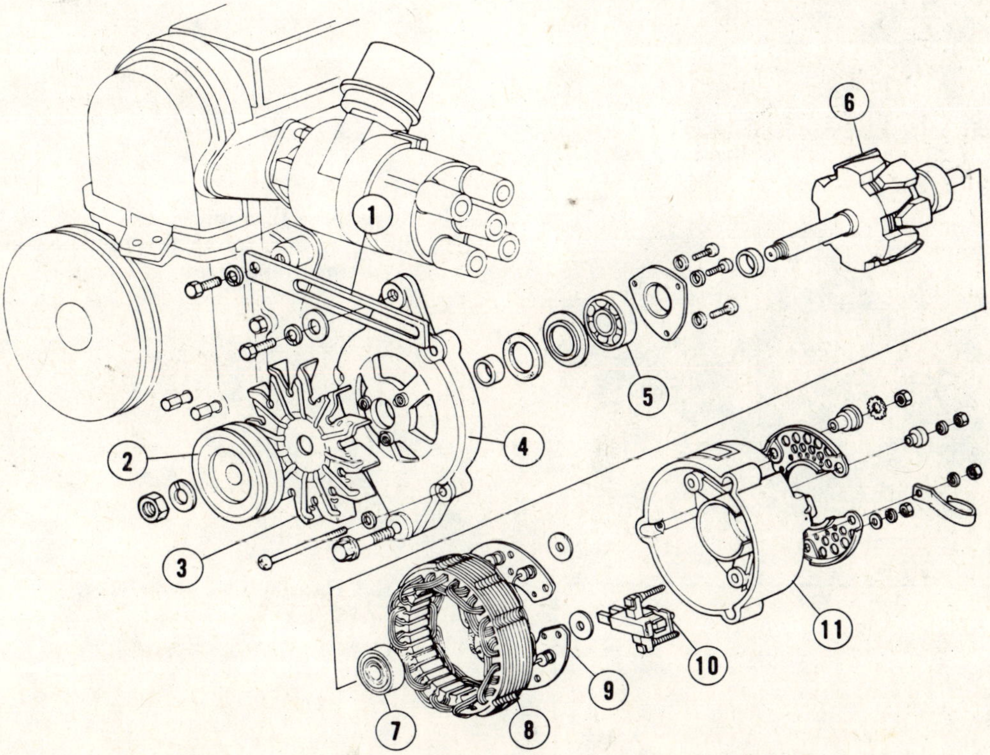

Fig. J-41

① Generator stay
② Pulley
③ Fan
④ Drive end frame
⑤ Ball bearing
⑥ Rotor
⑦ Ball bearing
⑧ Stator
⑨ Rectifier
⑩ Brush holder
⑪ Rectifier end frame

Precautions for handling

1. Make sure that the connections of the positive and negative terminals are not reversed. The missconnection will allow the current to flow in the reversed direction, resulting in a damaged diodes and burnt wire harness.
2. The battery voltage is applied to the terminal B of the generator. Take care not to short that terminal.
3. If the generator is turned at higher speeds with its wiring disconnected, the diodes may be damaged.
4. When charging the battery quickly, be sure to disconnect the generator wiring to prevent punctuation of the diodes.

Fig. J–42
① Attaching bolts

a–1. Removal

1. Disconnect the cord at the battery negative terminal.
2. Remove the generator coupler.
3. Loosen two generator bolts and remove the generator together with the V-belt.

Fig. J–43
① Box wrench ② Allen wrench

a–2. Disassembly

1. Using an Allen wrench (width across flats: 6 mm) and a box wrench, loosen the pulley lock nut as shown and remove the pulley and fan.

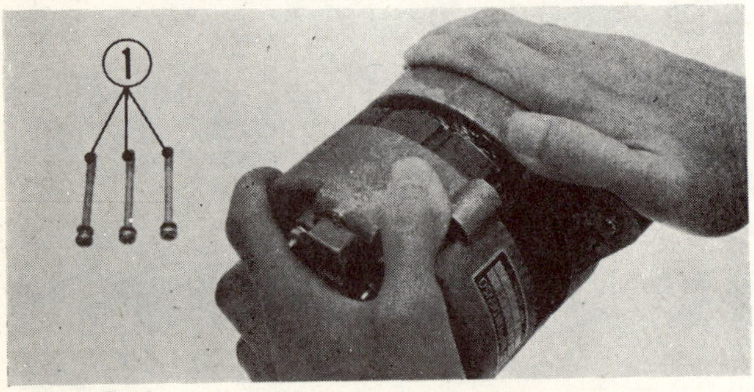

Fig. J–44
① Through bolts

2. Remove three through bolts. Give light blows of a mallet to the drive end frame through bolt attaching boss and separate the frame from the stator.

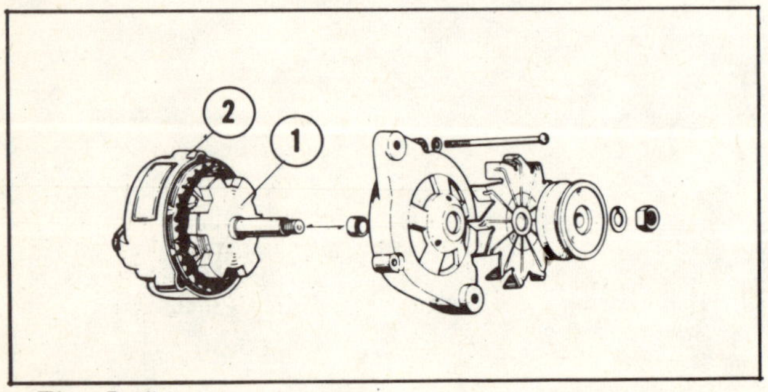

Fig. J–45
① Rotor ② End frame

3. Pull the rotor out of the stator.

4. Remove the nuts and washers from the rectifier end frame and separate the stator and rectifiers from the frame as shown.

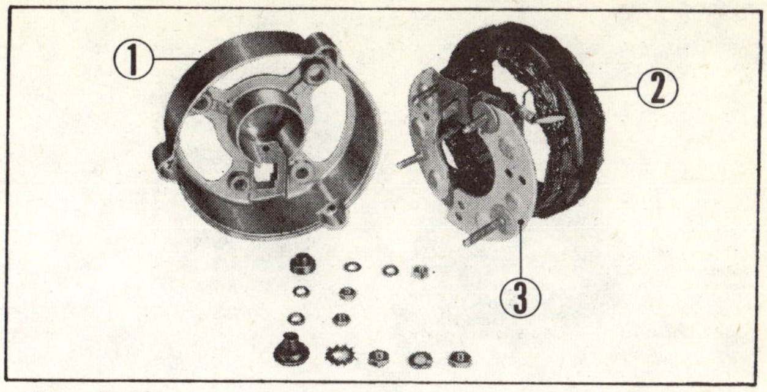

Fig. J-46
① End Frame ② Stator ③ Rectifier

5. Unsolder the brush holder lead and separate the brush holder from the stator.

Fig. J-47
① Brush holder ② Stator coil wire lead

6. Unsolder the stator coil leads and disconnect the rectifier terminals from the stator coils.

NOTES:

1) Diodes cannot stand high temperatures. Therefore unsolder quickly.

2) Hold the soldered points with a nose pliers to prevent heat from transferring to the diodes.

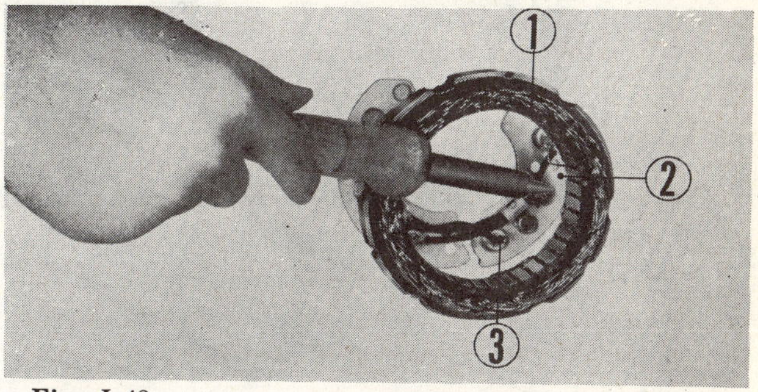

Fig. J-48
① Stator coil ② Rectifier holder fin ③ Rectifier

a-3. Inspection

Rotor coil

1. Check for continuity between two slip rings with a radio tester. If there is no continuity, the coil has an open circuit. If the coil resistance is excessively below the specification, the coil has a short circuit. Coil resistance specification: 4.1～4.3Ω.

NOTE:

If the connection between the rotor coil lead and slip ring is found to be separated, check to see if there is an excessive clearance between the rotor coil and shaft. If necessary, replace the rotor.

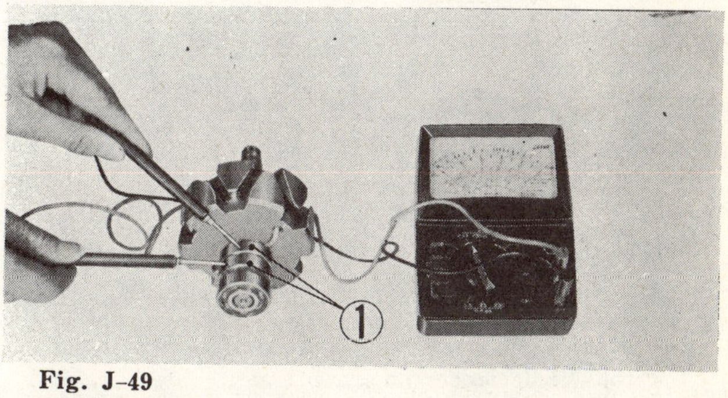

Fig. J-49
① Slip rings

2. Check for continuity between the slip ring and rotor shaft. If there is any continuity, the insulation between the two is poor. Replace the rotor.

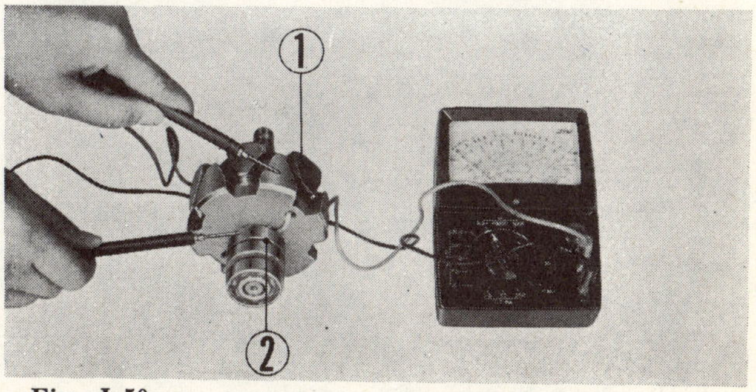

Fig. J-50
① Rotor core ② Slip ring

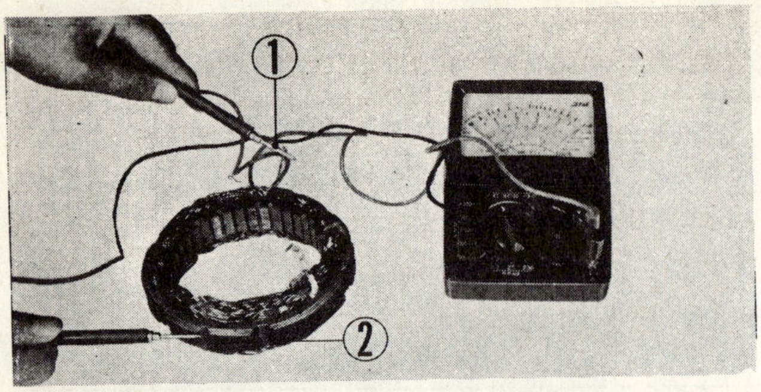

Fig. J-51
① Stator coil terminal ② Stator core

Stator coils

1. Check for continuity between the stator coil terminal and stator core. If there is any continuity, the insulation between the two is poor. Replace the stator.

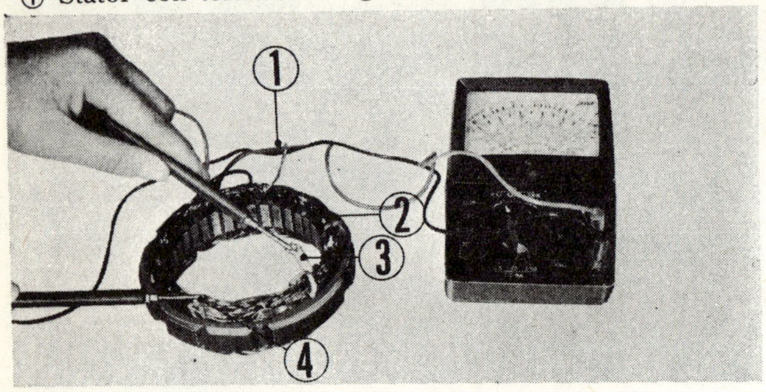

Fig. J-52
① Coil terminal ② Coil terminal ③ Terminal "N"
④ Coil terminal

2. Check for continuity between the stator coil terminals. If there is no continuity, the coil has an open circuit. Replace the stator.

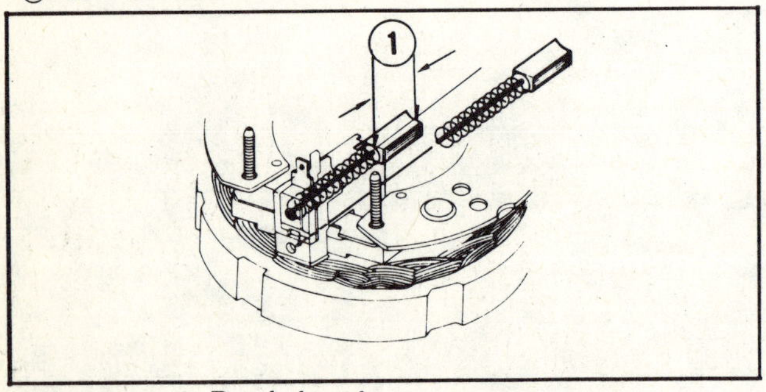

Fig. J-53 ① Brush length

Brushes

1. Inspection

Measure the length of each brush as shown. If it is below the service limit, replace the brush. **See the Maintenance Standard.**

2. Installation

1. First, separate the worn brush from the holder by unsoldering.
2. Remove deposits or flux front the bottom of the holder and clean it.
3. Install a new brush in the holder and solder the wire to the bottom of it.

   Be sure to solder the end of the wire to the holder.

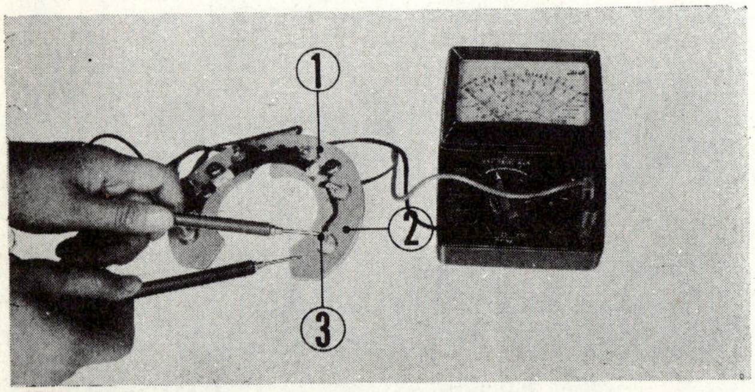

Fig. J-54
① Positive side holder fin ② Plate
③ Rectifier terminal

Rectifiers

1. Positive diodes (with red mark)

   Hold the negative probe of a radio tester against one terminal of the diode and the positive probe against the plate. If there is any continuity, the diode is normal. Reverse the positions of the tester probe and check the diode for continuity. If there is no continuity, the diode is normal.

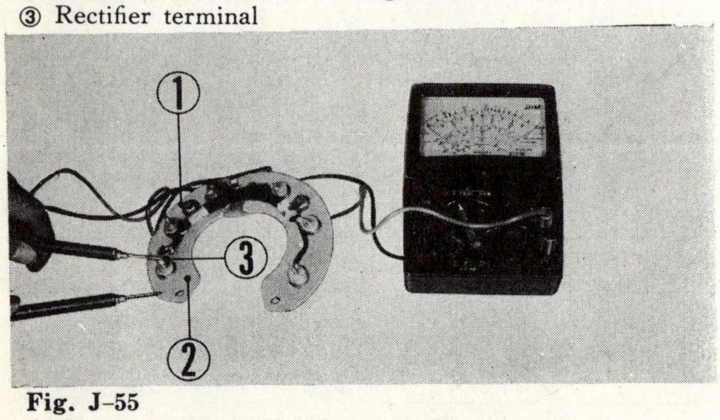

Fig. J-55
① Negative side holder fin ② Plate
③ Rectifier terminal

2. Negative diodes (with black mark)

   Hold the positive of the tester against the terminal of the diode and the negative probe against the plate. If there is any continuity, the diode is normal.

   Reverse the positions of the tester probes and check the diode for continuity. If there is no continuity, the diode is normal.

   NOTE:

   If there is any continuity in both directions or no continuity, the diode is defective.

a-4. Assembly

To assemble the generator, reverse the disassembly procedures. Pay attention to the following points:

1. When soldering the terminals of the diodes, use a 120W (approx.) soldering iron. Soldering must be performed quickly. Take care not to put excessive solder on the terminals.

   NOTE:

   It is advisable to hold the soldered points with nose pliers or the like to dissipate heat which may transfer to the diodes.

2. When installing the rotor to the rectifier end frame, first push each of the brushes into the holder until it reaches the bottom and insert a steel pin in the brush lifting hole in the end frame to keep the brushes away from the slip rings when assembling. After installing, pull out the pin to permit the brushes to seat on the slip rings.

3. The pulley lock nut must be tightened to a specified torque of 3.5 to 5.0 kg-m (25.32 te 36.17 lb-ft).

   After assembling, hand-rotate hte pulley and check to be sure it rotates smoothly.

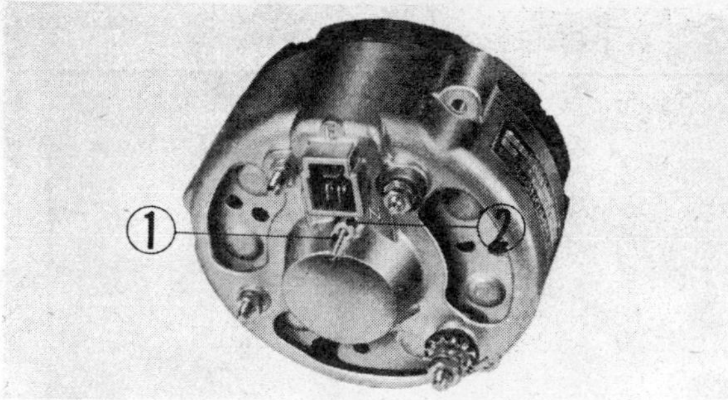

Fig. J-56
① Pin (Finger-move up two brushes and insert pin)
② Brush lifting hole

a-5. Performance tests

No-load test

Connect the generator in the test circuit as shown and measure its speed in rpm. Close the switch "S_1" and open the switch "S_2". Gradually increase the motor speed and read the generator rpm when the generator voltage rises to 14V on the voltmeter. A good generator will run up to 1050 to 1250 rpm.

Output test

Use the same test circuit as for the no-load test. Close the switches "S_1" and "S_2". Gradually increase the motor speed and adjust the variable resistor so that a voltage of 14V is read on the voltmeter. If the generator speed is below 4000 rpm when the output current is 35A, the generator is in good condition.

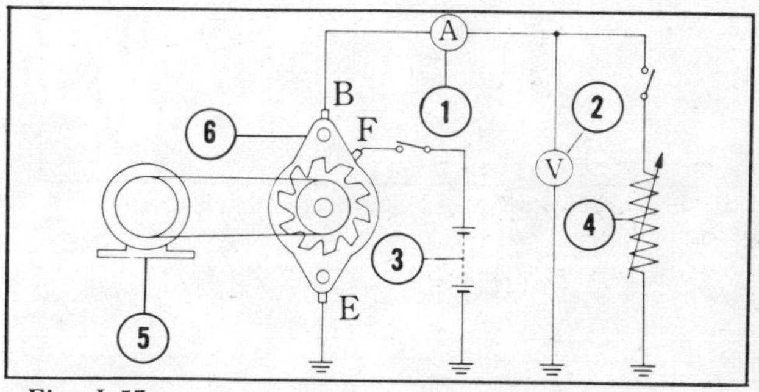

Fig. J-57
① Ammeter ② Voltmeter ③ Battery
④ Variable resistor ⑤ Variable-speed motor
⑥ AC generator

b. Regulator

Wiring diagram

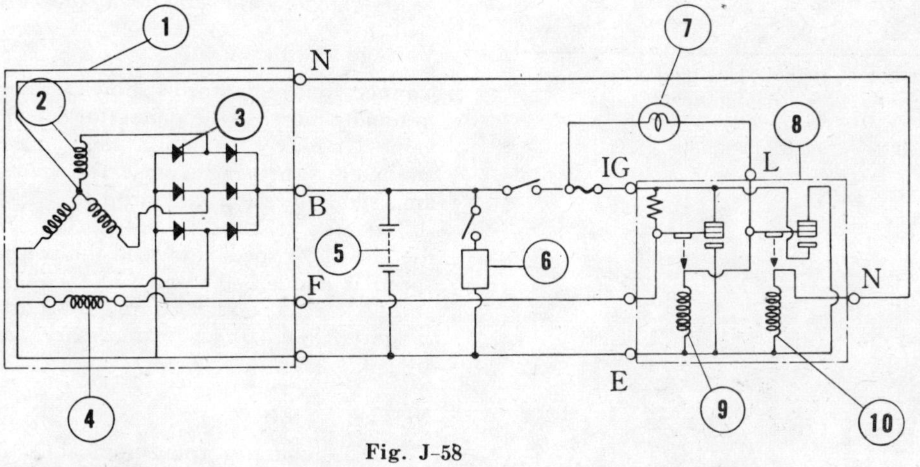

Fig. J-58

① AC generator
② Stator coil
③ Silicon diode
④ Field coil
⑤ Battery
⑥ Load
⑦ Charging lamp
⑧ Regulator
⑨ Voltage regulator
⑩ Voltage relay

Name of the parts

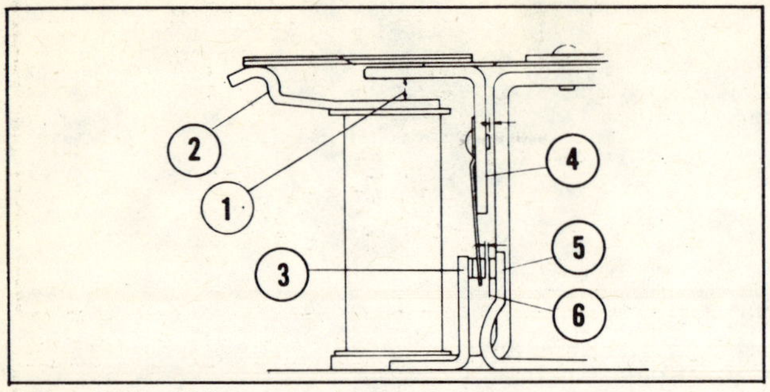

Fig. J-59 Voltage regulator
① Armature gap ④ Angle gap
② Adjusting arm ⑤ Point P
③ Point P_1 ⑥ Point gap

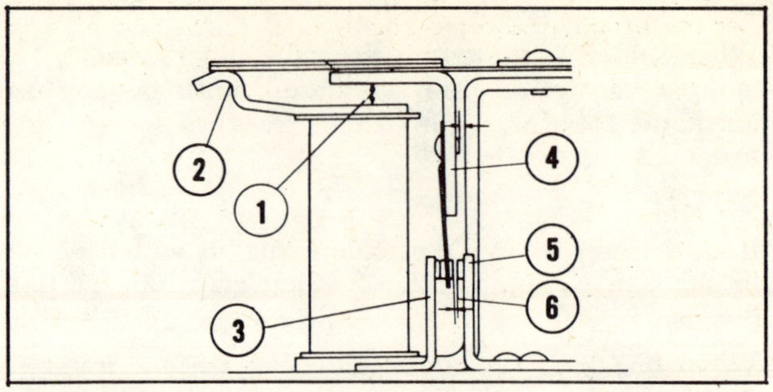

Fig. J-60 Voltage relay
① Armature gap ④ Angle gap
② Adjusting arm ⑤ Point P
③ Point P_1 ⑥ Point gap

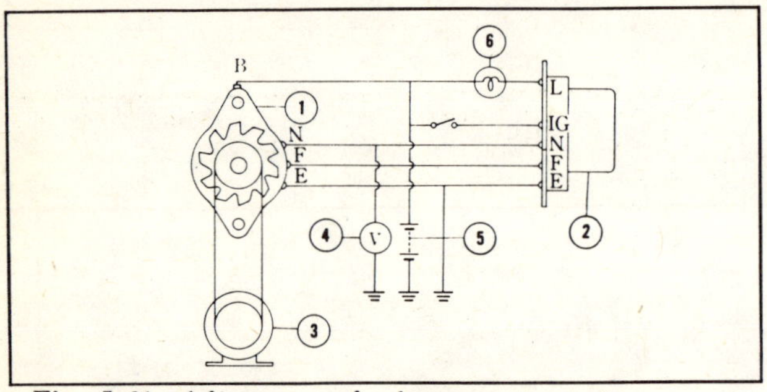

Fig. J-61 Adjustment of relay operating voltage
① Generator ④ DC Voltmeter
② Regulator ⑤ Battery
③ Variable-speed motor ⑥ 12V–3W lamp

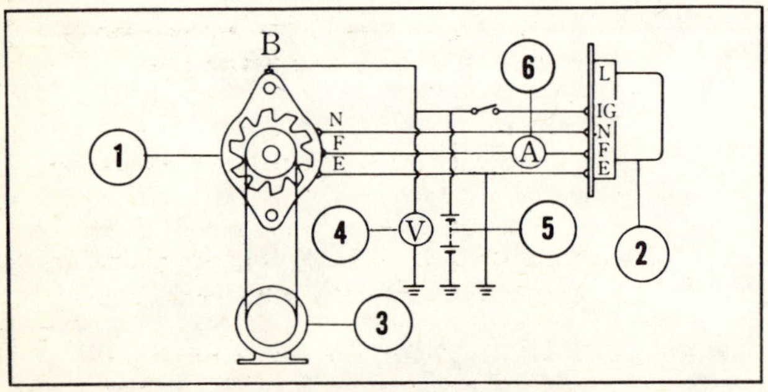

Fig. J-62 Adjustment of regulated voltage
① Generator ④ DC Voltmeter
② Regulator ⑤ Battery
③ Variable-seed motor ⑥ DC ammeter (5A)

b-1. Inspection and adjustment

NOTES:

1) When adjusting, be sure to turn the ignition switch off.
2) Use a fully charged battery.
3) When bending the adjusting arm for adjustment, do not bend it more than necessary.
4) The regulator adjustment must be made when the generator speed is increasing; otherwise the voltage will drop 0.3 to 0.5V due to hysteresis of the voltage regulator.

Voltage relay (**charge lamp relay**)

Connect the relay as shown. Close the switch " S " and gradually increase the generator speed. Read the voltage (relay operating voltage) on the voltmeter just the moment the lamp goes out. The specified voltage is 4.5 to 5.8V. If out of specification, adjust by bending the adjusting arm.

NOTE:

If a variable-speed motor is not available, the generator may be operated by the engine in the car.

Voltage regulator

Connect the regulator as shown. Close the switch " S " and gradually increase the generator speed. . Read the maximum value (If. max.=maximum field current) indicated by the ammeter. Further increase the generator speed, and the ammeter indication will drop to 1/2 of the maximum value (1/2 If. max.). Read the voltage at this time. Again increase the generator speed and read the voltage when the generator speed is about 3000 rpm. The voltage (regulated voltage) at 1/2 If. max. and 3000 rpm must be within the specified range (13.5~14.5V). If necessary, adjust by bending the adjusting arm.

NOTE:

The AC generator rpm should be correctly measured on a test.

Resistance between terminals

If the reading of the voltmeter is not still satisfactory, even after the procedure given above is used, check the resistance between the terminals of the coupler with a tester.

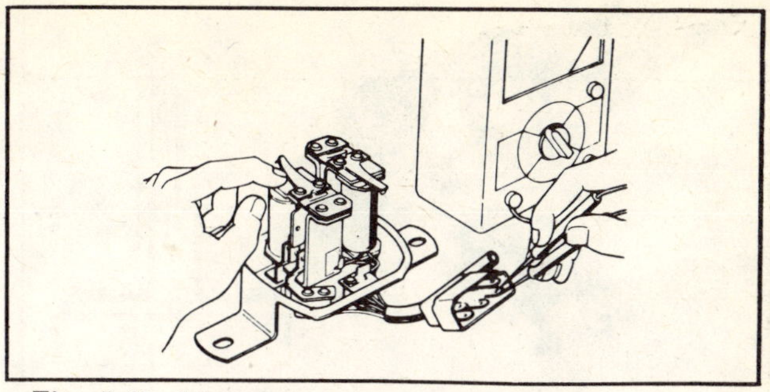

Fig. J-63 Measurement of resistance between terminals

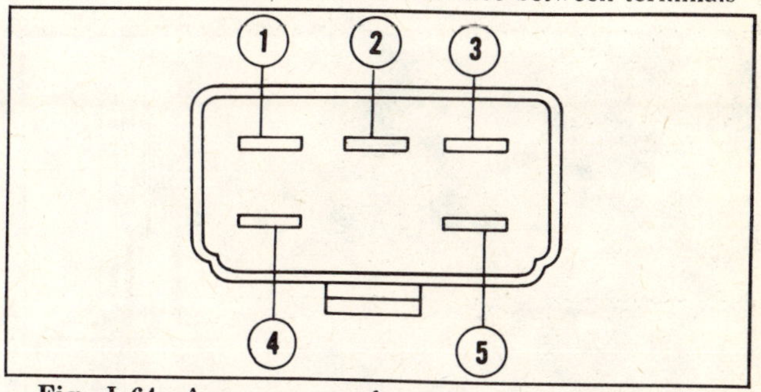

Fig. J-64 Arrangement of regulator terminals
① Terminal E
② Terminal N
③ Terminal L
④ Terminal IG
⑤ Terminal F

NOTES :

1) Use the minimum ohm range of a tester.
2) Be careful not to misconnect the terminals of the regulator cord coupler.
3) If the point surfaces are dirty, wipe them clean with a piece of paper.
4) If the point surfaces are excessively damaged, replace the regulator. Be sure to locate problem causing damage so as to prevent recurrence of the same trouble.

* To check the function of the regulator according to the table below, perticular points should be forced to be contacted by hand to see if the points function properly.

Terminal	Condition of voltage relay	Condition of voltage regulator	Specification (Ω)	Remarks
IG–F	—	Not working	0	If reading is over 0, low-speed contact points of voltage regulator are defective.
	—	working	11, approx.	If reading is ∞, regulator resistor has open circuit.
L–E	Not working	—	0	If reading is over 0, contact point P_1 of voltage relay is defective.
	working	—	100, approx.	If reading is 0, contact point P_1 of voltage relay is melted. If it is ∞, voltage coil has open circuit.
N–E	—	—	23, approx.	If reading is 0, pressure coil has short circuit. If it is ∞, pressure coil has open circuit.

● Mechanical adjustment

If the regulator does not function even after the electrical adjustment as described above has been performed, the mechanical adjustments are necessary.

Fig. J-65 Measuring voltage relay point gap

Voltage relay

1. Measure the point gap with a feeler gauge. It should be 0.4 to 1.2 mm (0.016~0.047 in.).

   Adjust by bending the point holder P_1 (B), if necessary.

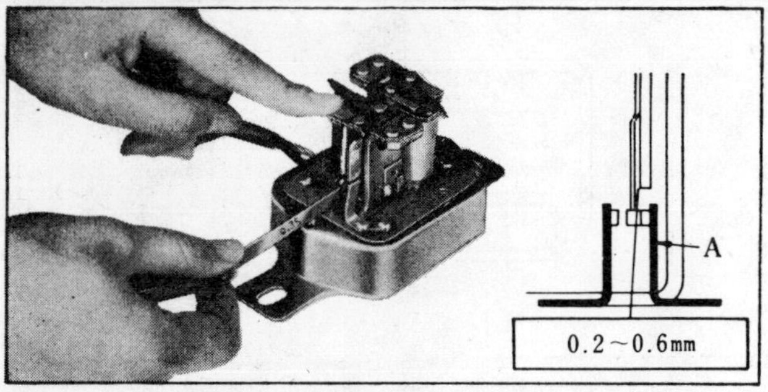

Fig. J-66 Checking voltage relay contact spring for deflection

2. Check the contact springs for deflection. Press the armature against the coil and measure the deflection with a feeler gauge. It should be 0.2 to 0.6 mm (0.008~0.024 in.).

   Then, the armature gap should be 0 and the angle gap be less than 0.5 mm (0.020 in.). Adjust by bending the point holder P_2(A), if necessary.

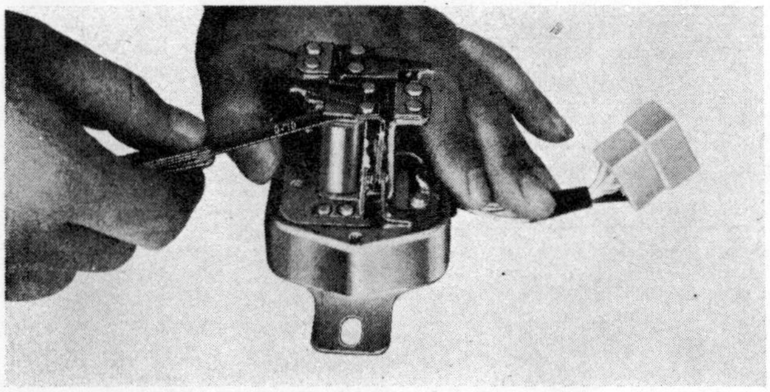

Fig. J-67 Measuring voltage relay armature gap

Voltage regulator

1. Measure the armature gap with a feeler gauge. It should be not less than 0.3 mm (0.012 in.). Adjust by bending the low-speed point holder.

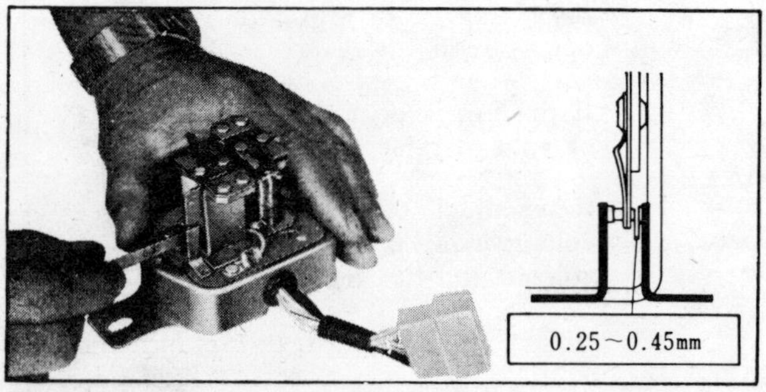

Fig. J-68 Measuring voltage regulator point gap

2. Measure the point gap with a feeler gauge. It should be 0.25 to 0.45 mm (0.010~0.018 in.). Adjust by bending the high-speed point holder.

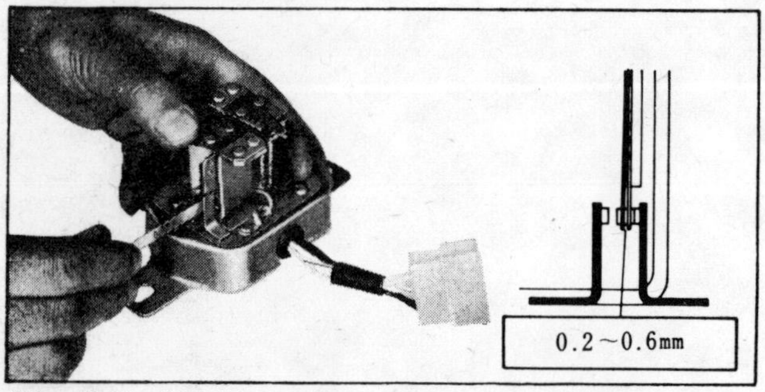

Fig. J-69 Checking voltage relay contact spring for deflection

3. Check the contact springs for deflection. Press the armature against the coil and measure the amount of deflection with a feeler gauge. It should be 0.2 to 0.6 mm (0.008~0.024 in.).

4. Check the angle gap. Press the armature against the coil and measure the gap with a feeler gauge. It should be not less than 0.2 mm (0.008 in.).

Fig. J-70 Measuring voltage regulator angle gap

Cooling system

The Honda CIVIC cooling system is of a forced water circulation type and features a fan motor. As the water tempera ture rises beyond 90°C (194°F), the thermoswitch operates and the fan motor starts to turn.

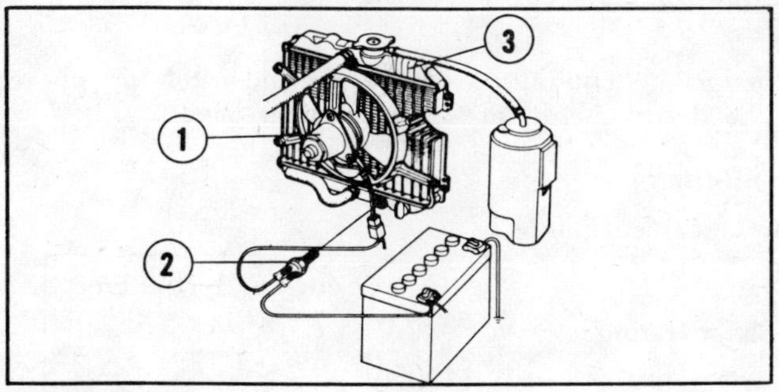

Fig. J-71
① Fan motor ② Thermoswitch ③ Radiator

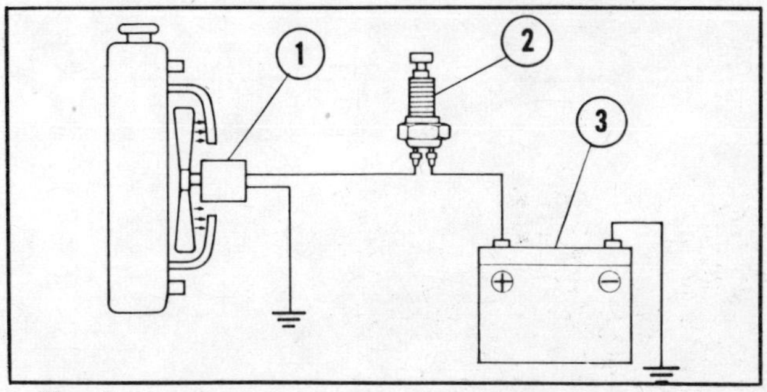

Fig. J-72
① Radiator fan motor ② Thermoswitch ③ Battery

a. Thermoswitch

a-1. Construction

Specifications

Rating	7 A
Operating temperature	
ON	90±1.5°C (191~197°F)
OFF	ON temperature–5±1.5°C (6~12°F)

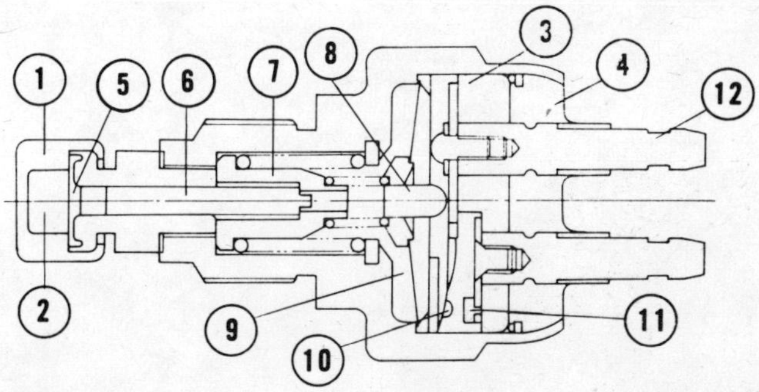

Fig. J-73 Construction of thermoswitch
① Case
② Wax
③ Terminal base
④ Sealant
⑤ Diaphragn
⑥ Piston
⑦ Rod
⑧ Plunger
⑨ Snap plate
⑩ Movable point
⑪ Stationary point
⑫ Terminal

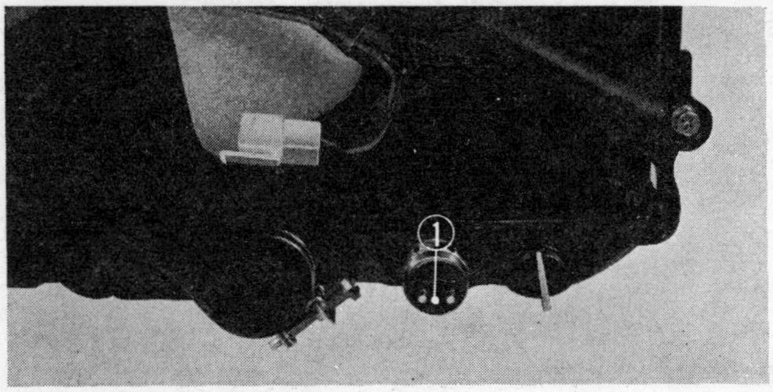

Fig. J-74
① Thermoswitch

The thermoswitch is located at the lower radiator tank. When the water temperature reaches 88.5~91.5°C the wax in the switch expands to close the switch, causing the fan motor to start. As the water temperature drops 3.5~6.5°C below the designated temperature, the wax contracts and the switch opens to stop the motor.

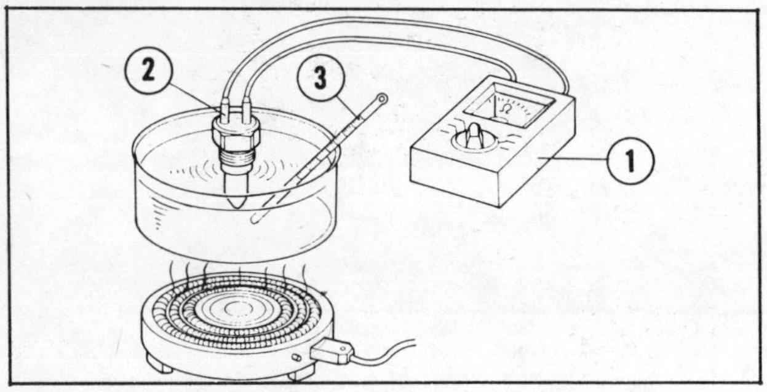

Fig. J-75 Testing thermoswitch for function
① Radio tester ② Thermoswitch ③ Thermometer

a-2. Inspection

As shown below, submerge the switch in heated water in a tub, observing the water temperature with a thermometer. Using a radio tester, check to see if the switch closes when the water temperature reaches 88.5~91.5°C.

Keep stirring the water so that there is no temperature difference anywhere in the tub.

Keep the sensing parts of the thermometer and thermoswitch away from the bottom of the tub.

NOTE:

Do not judge the conditions of the thermoswitch by observing the thermometer on the instrument panel.

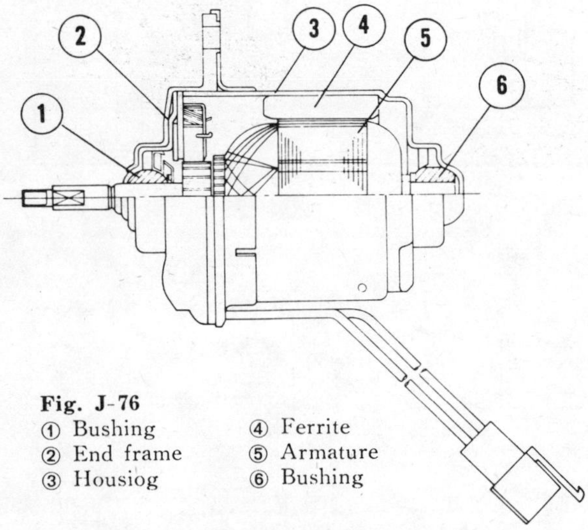

Fig. J-76
① Bushing
② End frame
③ Housiog
④ Ferrite
⑤ Armature
⑥ Bushing

b. Fan motor

Motor specifications

Type	Direct current, ferrite type
Voltage rating	12 V
No-load rpm	2300
No-load current	1.1A, max.
Direction of rotation	Clockwise as viewed from drive side
On-load torque rating	1.15 kg-cm (0.083 lb-ft) at 1900±250 rpm and 3.5±0.4A

b-1. Removal

1. Disconnect the fan motor coupler.
2. Loosen four shroud mounting bolts and remove the shroud together with the motor.
3. Remove the fan and motor from the shroud.

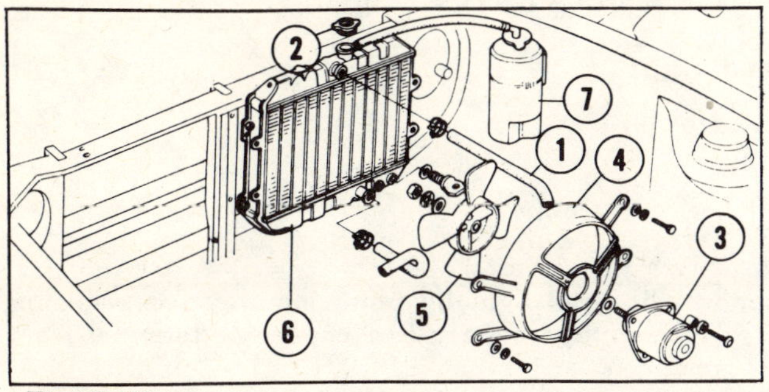

Fig. J-77
① Upper water hose ④ Shroud
② Filler cap ⑤ Cooling fan complete
③ Fan motor ⑥ Radiator

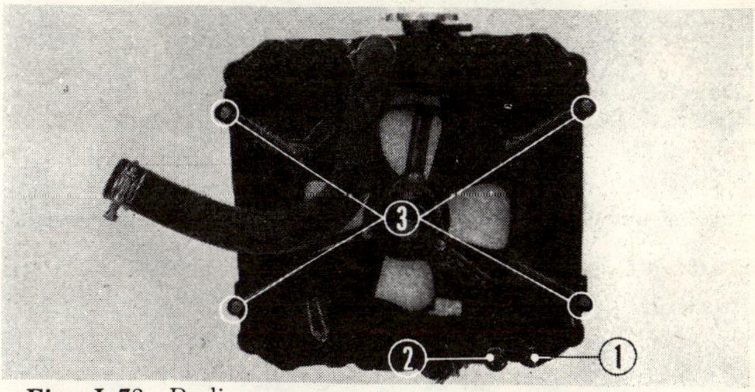

Fig. J-78 Radiator
① Drain cock ② Thermoswitch
③ Shroud attaching bolts

b-2. Inspection

1. Using a radio tester, check for continuity between two terminals of the coupler. If there is no continuity, the coil may have an open circuit. Replace it.
2. If the coil is normal, connect the positive terminal of a 12V battery to the black/yellow terminal of the coupler and the negative terminal to the black terminal, and check to be sure the motor operates smoothly.

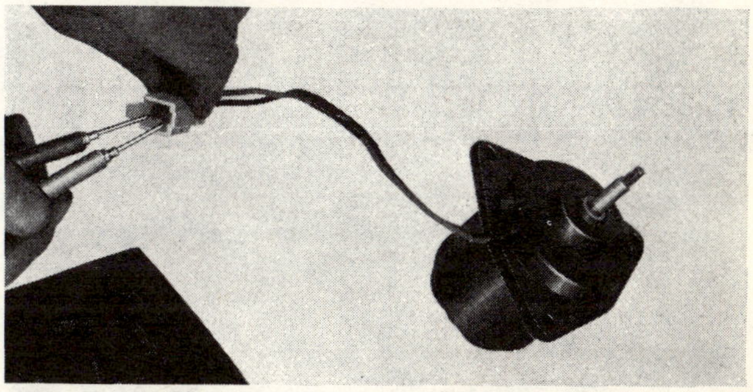

Fig. J-79 Checking fan motor coil for continuity

b-3. Installation

To install the fan motor, reverse the removal procedures.

Body Electrical System

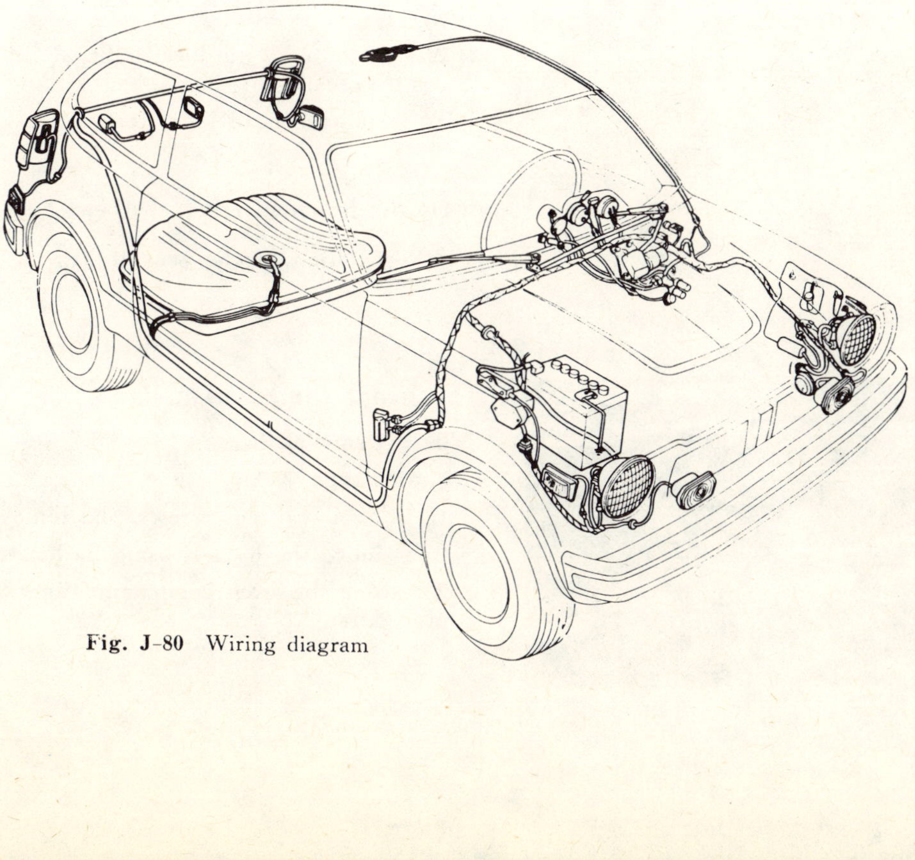

Fig. J-80 Wiring diagram

Fig. J-81

Lights

NOTES :

1) Do not tighten the lens screws excessively; otherwise the lens will be damaged.
2) Securely attach the packing to the lens to prevent entry of water into the light housing.

a. Head light

a-1. Removal

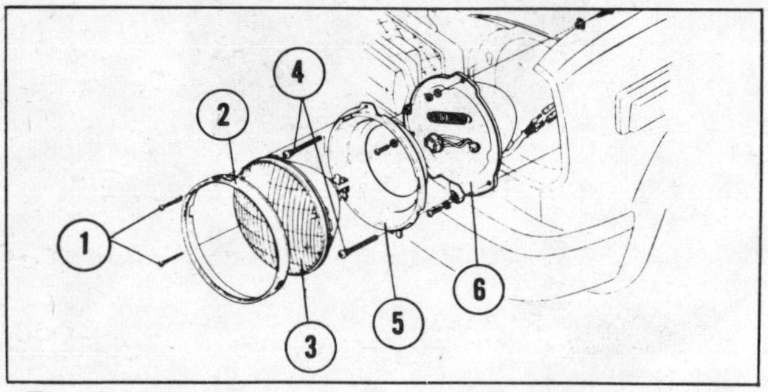

Fig. J-82
① Retaining screw ② Retaining ring ③ Headlight ④ Adjusting screw ⑤ Mounting ring ⑥ Headlight housing

For U.S.A. type

1. Remove the head light by removing the two retaining ring screws, the retainer and connector terminal from the lamp.
2. Remove the mounting ring by removing the spring and two adjusting screws.
3. Remove the head light housing by removing the three screws.

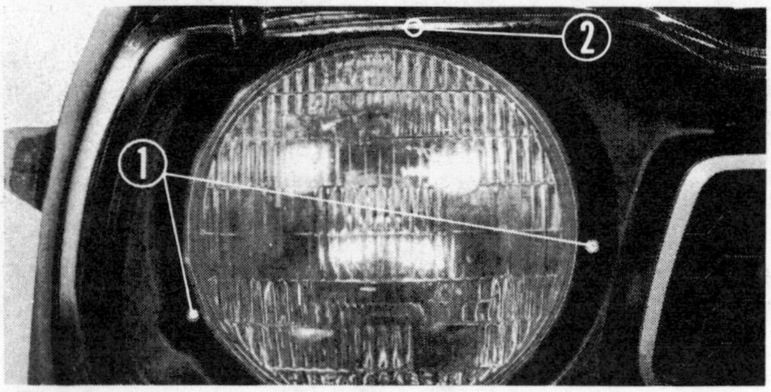

Fig. J-83
① Adjusting screws ② Head light plug

For except U.S.A. type

1. Remove the head light plug.
2. Remove the head light by turning it inward while pressing in.
3. Remove the socket and separate the light from its wire harness.

NOTE :

When removing the head light adjusting screws, take care not to lose their springs.

To install the head light, reverse the removal procedures.

NOTE :

Do not forget to install the head light plug.

* **For aiming of the head light.**

Fig. J-84

b. Front combination lights

b-1. Removal

U.S.A. type

1. Loosen two lens screws, and remove the lens.
2. Remove the light housing by loosening two screws.

To install the front combination light, reverse the removal procedures.

Except U.S.A. type

1. Loosen four lens screws, and remove the lens.

   To install the front combination light, reverse the removal procedures.

2. Remove the light housing by loosening two screws from under the bumper.

Fig. J-85

c. Side turn signal lights (except U.S.A. type)

c-1. Removal

Loosen two lens screws, and the lens can be removed together with the light housing.

To install the side turn signal light, reverse the removal procedure.

Fig. J-86

d. Side marker lamp (For U.S.A. type)

d-1. Removal

Front

1. Loosen the two nuts from inside the front fender and remove the lamp and reflector assembly.

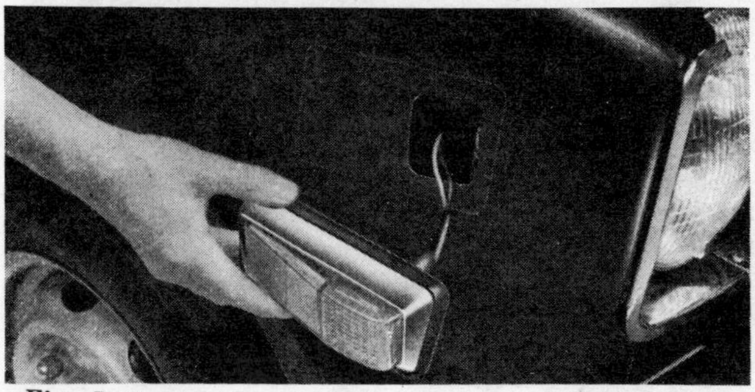

Fig. J-87

2. Loosen the screw on the back and separate the reflector from the lamp assembly. Then two screws appear.

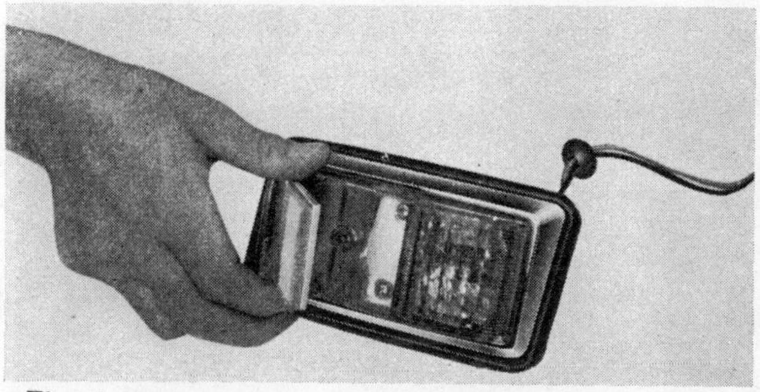

Fig. J-88

3. Loosen the two screws and remove the lens from the lamp.

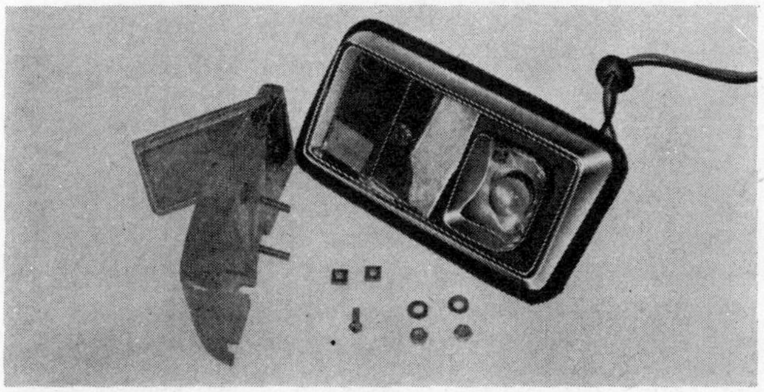

Fig. J-89

Fig. J-90

Rear

For two-door type

1. Open the trunk lid and remove the rear side marker lamp by loosening the two nuts from the inside the trunk.
2. Perform procedures as described in the above 2 and 3.

For three-door type

1. Remove the rear combination lamp. (See bellow f.)
2. Remove the marker lamp by loosening two nuts through the opening.
3. Perform procedures as described in the above 2 and 3.

* To install the side marker lamp, reverse the removal procedures.

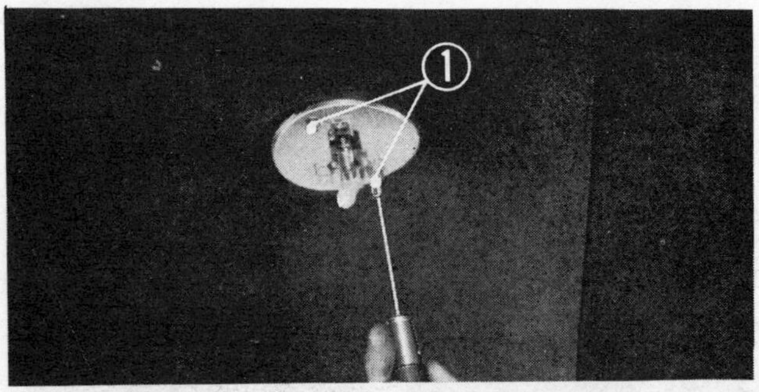

Fig. J-91
① Set nuts

e. Room light

Removal

1. Pull off the lens by hand.
2. Pull the socket base downward by removing two nuts. Then disconnect the connectors.

To install the room light, reverse the removal procedures.

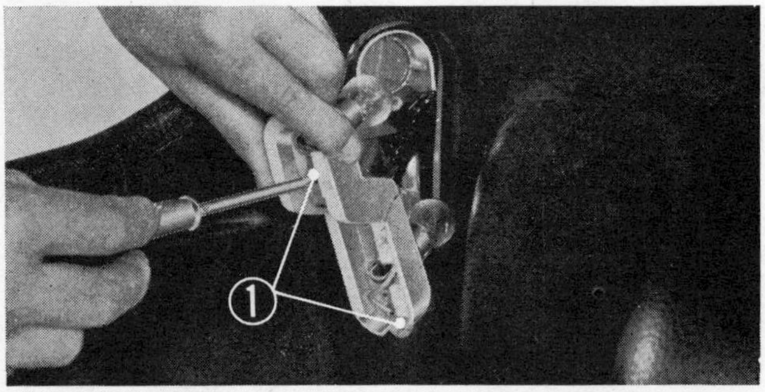

Fig. J-92 Tail light socket
① Socket tapping screws

f. Rear combination lights

Two-door type

f-1. Removal

1. For bulb replacement, open the trunk lid and remove the socket by removing the two tapping screws working from the trunk.

Fig. J-93

2. Remove the light assembly by loosening the upper and lower attaching nuts.

NOTE:

It is impossible to separate the lens from the light body.

Three-door type

Raise the tail gate and remove the rubber plugs. Then, remove the tail light as in the steps for two-door type.

To install the tail light, reverse the removal procedure.

g. Licence number plate lights

g-1. Removal

Remove the light assembly by removing the two nuts from inside the trunk.

Separate the base packing from the housing for bulb replace ment.

To install the licence number plate light, reverse the removal procedure.

NOTE:

Securely install the lens in place.

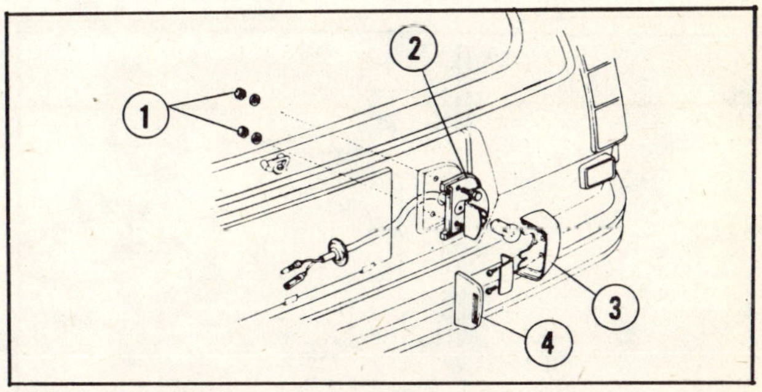

Fig. J-94 Licence number plate light
① Attaching nuts ② Base gasket
③ Light housing ④ Lens

h. Backup lights

h-1. Removal

1. Remove the lens by removing the two retaining ring screws.
2. Remove the light housing by removing the mounting nut.

To install the backup light, reverse the removal procedures.

NOTE:

Take care not to set the lens upside down.

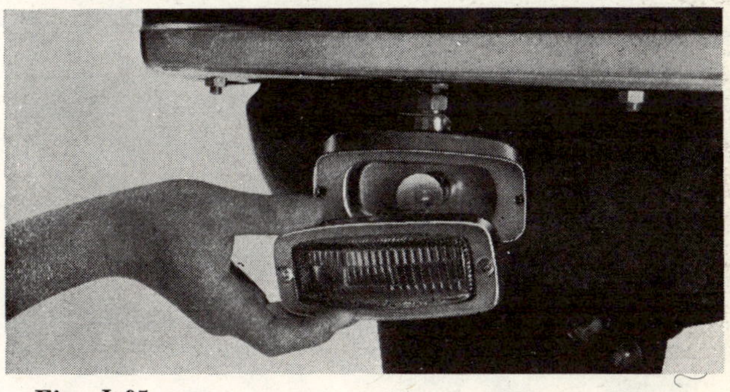

Fig. J-95

Switches

a. Turn signal switch and dimmer switch

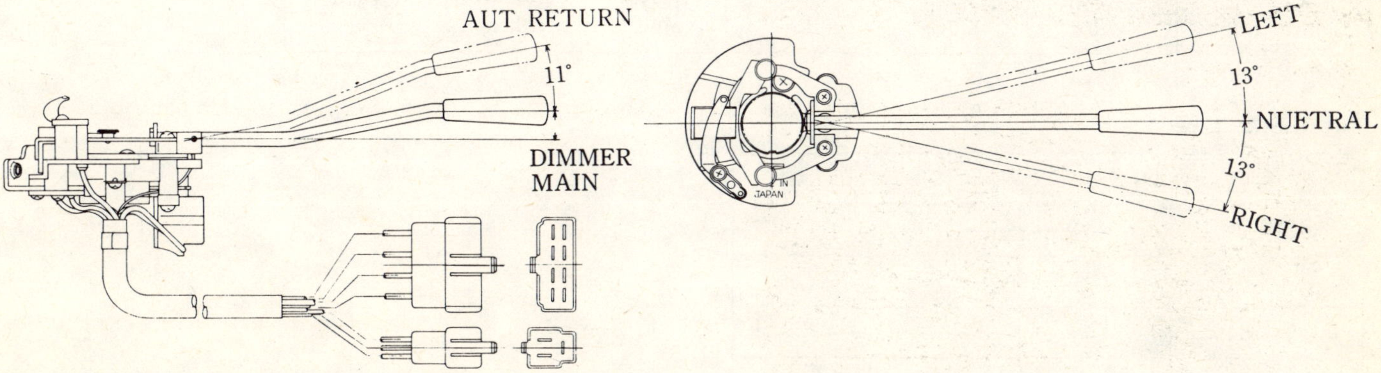

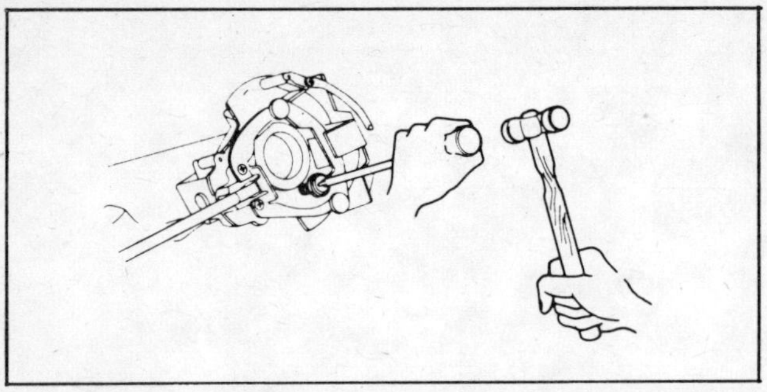

Fig. J-97

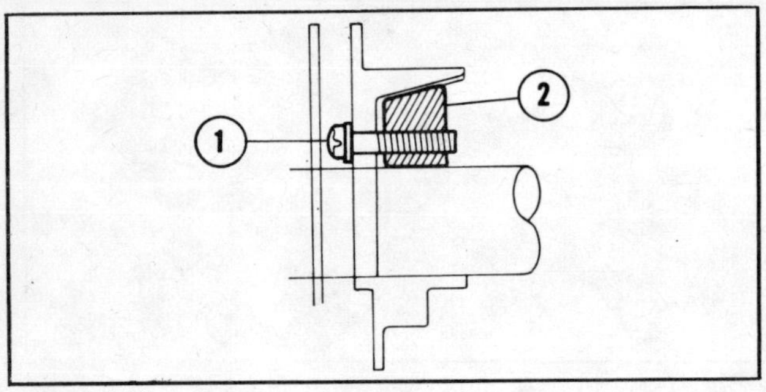

Fig. J-98
① Lock screw ② Lock piece

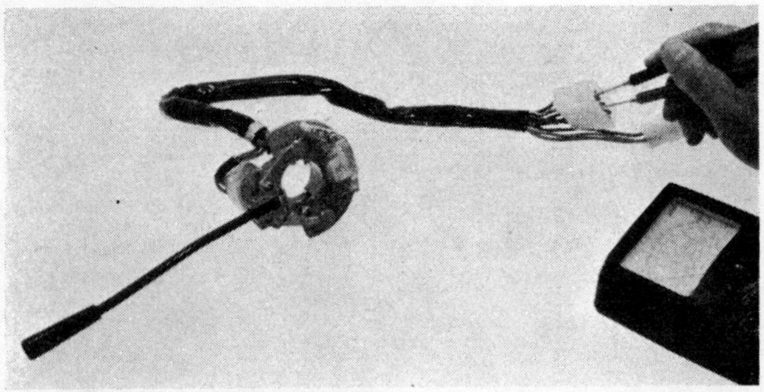

Fig. J-99

a-1. Removal

1. Remove the steering pad.
2. Remove the upper and lower steering column covers by loosening four screws with a screwdriver.
3. Remove the steering wheel by loosening the lock nuts.
4. Loosen the lock screw and lightly tap the screw head with a mallet to permit the lock piece to loosen. Remove the lock piece, taking care not to drop it.
5. Remove the turn signal switch, taking care not to drop the cancel key and washers.
6. Remove the wiper/washer switch by loosening the two screws.

a-2. Inspection

Using a radio tester, check the continuty at each switch lever operating position in accordance with the table below.

For U.S.A. type

TURN SIGNAL SWITCH						
LEFT	O	O		O	O	O
NEUTRAL				O	O	O
RIGHT				O		O
	O		O		O	
Color identification	Green/ white	Green/ blue	Green/ yellow	Green/ blue	Green/ yellow	Green

For U.S.A. type

DIMMER SWITCH			
MAIN	O	O	
⇕	O	O	O
DIMMER	O		O
Color identification	Red/ yellow	Red/ blue	Red/ white

Except U.S.A. type

TURN SIGNAL SWITCH						
LEFT	○	○		○	—	○
NEUTRAL				○	○	○
RIGHT	○	—	○	○	○	
Color identification	Green/ white	Green/ blue	Green/ yellow	Red/ black	Red/ green	Red

Except U.S.A. type

DIMMER SWITCH				OVERTAKING SWITCH		
MAIN	○	○		ON	○	○
DIMMER	○	—	○	OFF		
Color identification	Red/ yellow	Red/ blue	Red/ white		Blue	

a-3. Installation

* To install the switch, reverse the removal procedures. Pay attention to the following points:

1. Fit the protrusion of the turn signal switch in the recess in the steering column and be sure to install the lock piece correctly as shown in Fig. J-98. Then, tighten the lock screw.

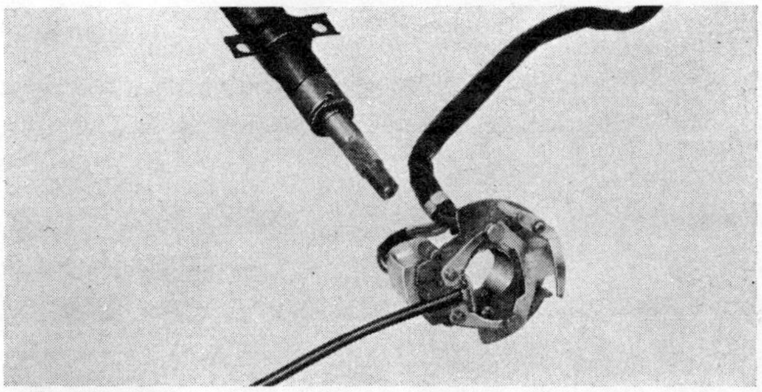

Fig. J-100

2. Fit the two lugs of the turn signal switch cancel key in the recesses in the steering wheel.

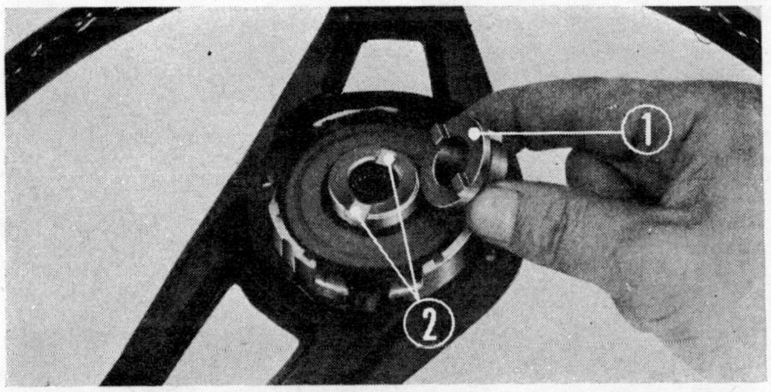

Fig. J-101
① Cancel key ② Grooves

3 Check to be sure that the horn switch spring terminal is contacting the steering wheel contact ring. Apply grease to the ring.

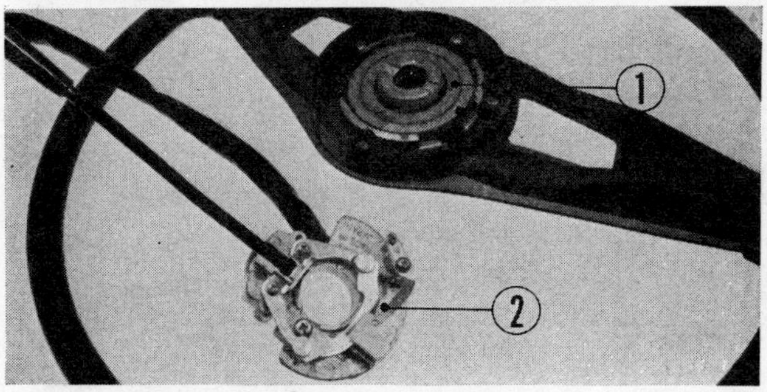

Fig. J-102
① Spring terminal ② Contact ring

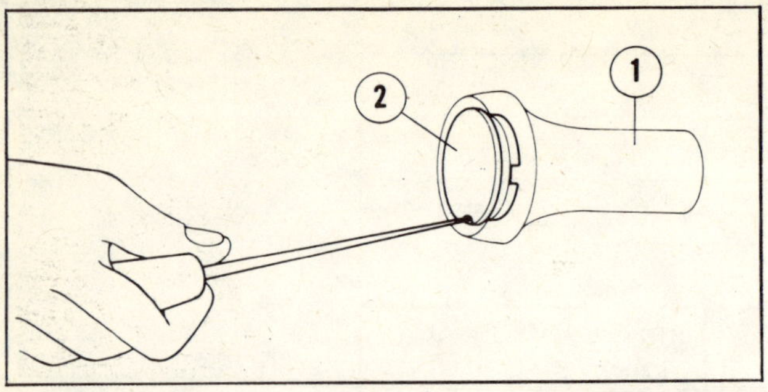

Fig. J-103 Removing mark plate
① Switch knob ② Mark plate

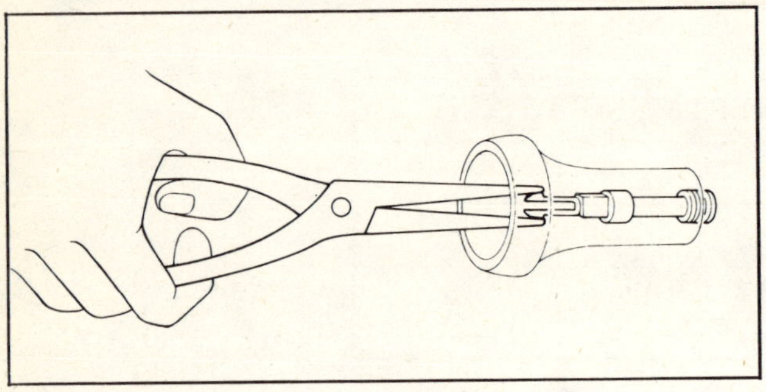

Fig. J-104 Removing switch knob

b. Other switches

b-1. Removal of switch knobs

1. The knobs of other switches are of a press-in type. Insert the pin into the notch in the mark plate on the knob and pry off the plate.

2. Using radio pliers, remove the knob as shown.

NOTES:

1) When removing the switches, do not force to pull out the wires.
2) Carefully remove or install the set ring nuts since they are made of resin.

b-2. Inspection

Check each switch for continuity at each operating position of its knob, in accordance with the continuity table below. If there is no continuity or if there is too much resistance, failure is suspected in the internal parts. Replace the defective switch.

Fig. J-105

1. Ignition switch

OFF	0°				
I	90°	○			○
II	180°	○		○	○
III	210°	○	○	○	
Color identification		White	Black/white	Black/yellow	White/red

2. Lighting switch

U.S.A. type

OFF				
TAIL	○	○		
HEAD	○	○	○	○
Color identifica-tion	Red/ Green	Red/ black	Red	Red/ black

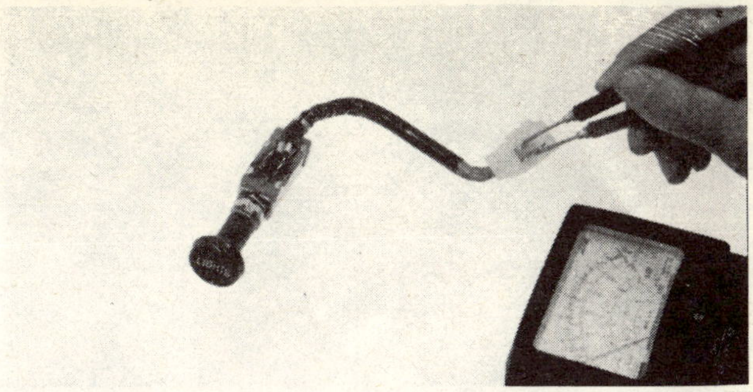

Fig. J-106

Except U.S.A. type

OFF						
TALL	○	○			○	○
HEAD	○	○	○	○	○	○
Color identifica-tion	Red/ Green	Red/ black	Red	Red/ yellow	White	Red/ black

3. Hazard warning switch

OFF						○	○	○	○
HAZARD	○	○	○	○	○				
Color identifica-tion	Green/ red	Green/ yellow	Green/ blue	Green/ yellow	Green/ blue	Green	Green	Green/ white	Green/ white

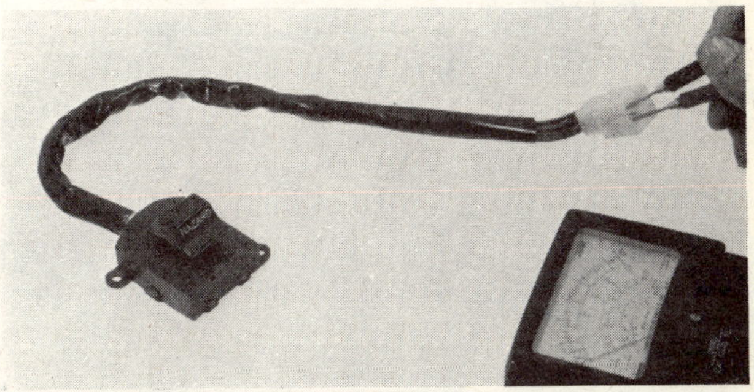

Fig. J-107

4. Wiper/washer switch

OFF	○	○				
WIPER 1		○	○			
2			○	○		
WA-SHER					○	○
Color identi-fication	Blue/ white	Blue	Black	Blue/ yellow	Green/ black	Green/ black

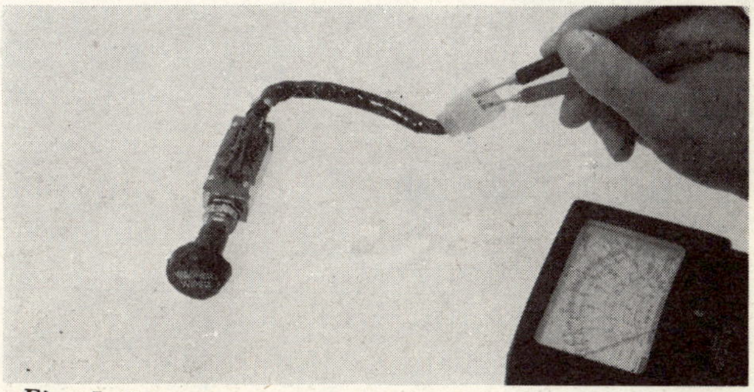

Fig. J-108

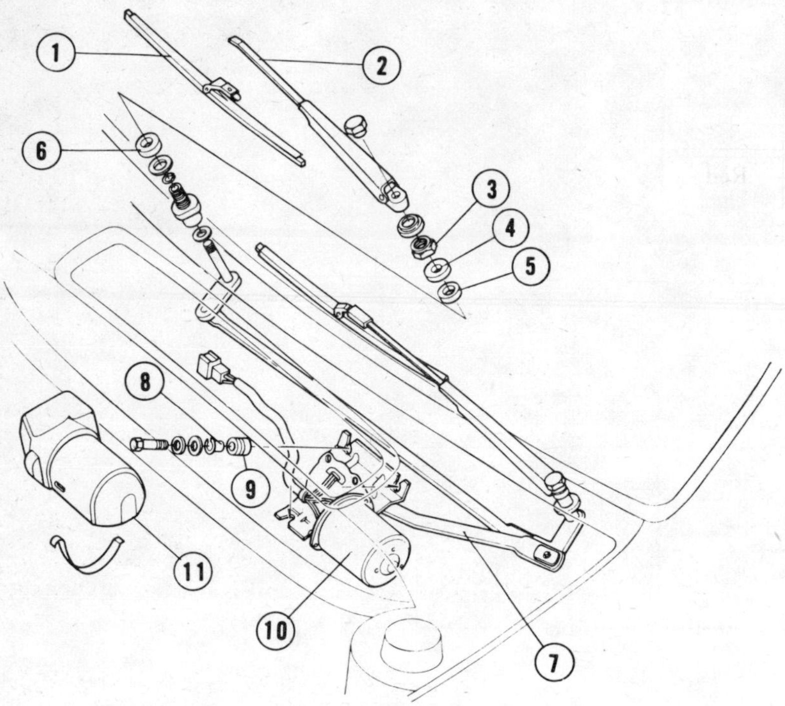

Fig. J-109

① Wiper blade
② Wiper arm
③ Pivot nut
④ Cup washer
⑤ Cushion rubber
⑥ Cushion rubber
⑦ Link complete
⑧ Mount collar
⑨ Mount rubber
⑩ Wiper motor
⑪ Water seal cover

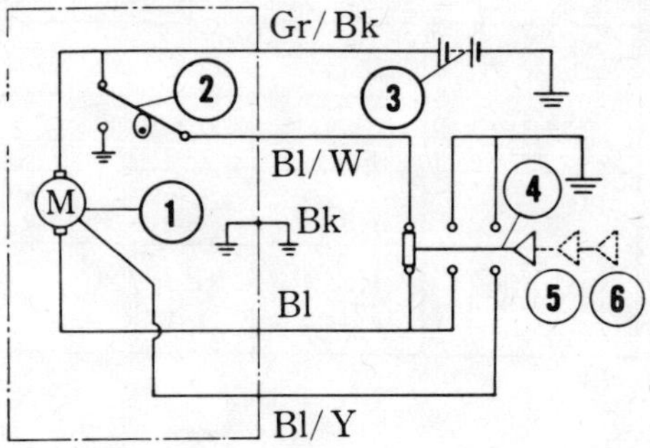

Fig. J-110

① Wiper motor
② Automatic stop switch
③ Battery
④ Wiper switch

Fig. J-111

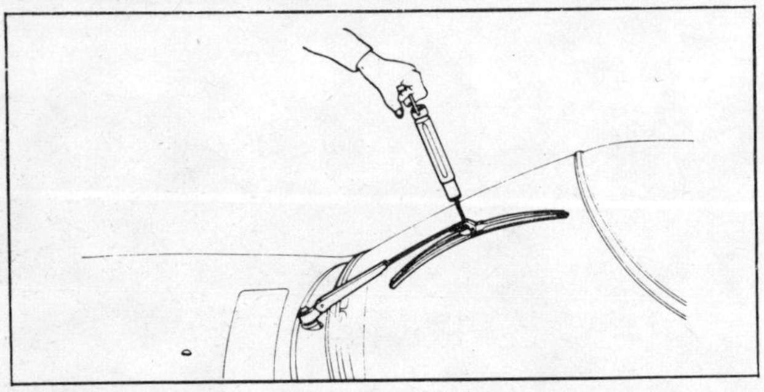

Fig. J-112 Measuring wiper blade pressure

The wiper motor of the Honda CIVIC is a compound wound motor. This type of motor develops large torque and the variation in speed as the load varies is slight. An automatic stop device is built in this motor to stop the wipers to the predetermined position.

NOTE:

When checking the wipers for operation, cover the surface of the windshield with washer fluid or water.

a. Inspect the wipers for operation

1. For installation of the wiper arms; see Fig. J-109.
2. If the wiper operation is erratic, measure the wiper blade pressure with a spring scale. Stop the wiper arms in the position where they are perpendicular to the weather strip. Hook the scale to the arm tip as shown. Observe the reading when both ends of the blade are raised off.
3. Check the blade rubber for deterioration or other damage, and replace if necessary.

Unit: g (oz)

	Specification	Service limit
Blade pressure	470±30 (17.64~15.42)	Replace arm when pressure is below 400 (14.11)

b. Removal

1. Remove the special nut by loosening it with a wrench.
2. Remove the right and left pivot nuts and push down the pivots.

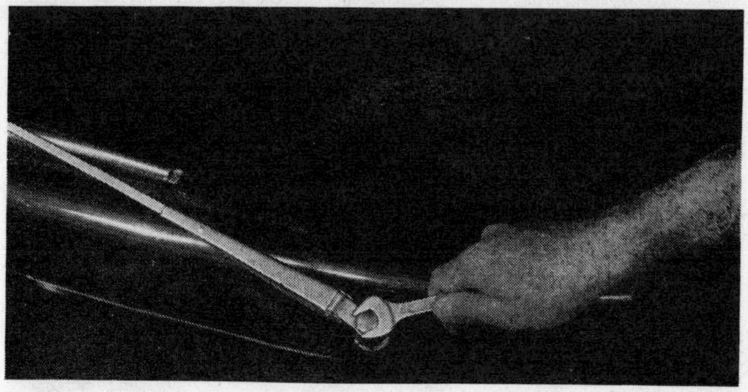

Fig. J-113

3. Remove the three wiper motor mounting bolts and disconnect the cord coupler, and remove the wiper assembly.

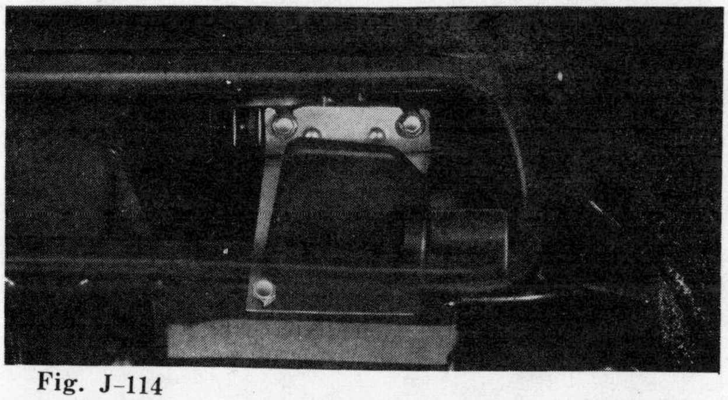

Fig. J-114

4. Pull out the motor arm cotter pin and separate the linkage from the motor.

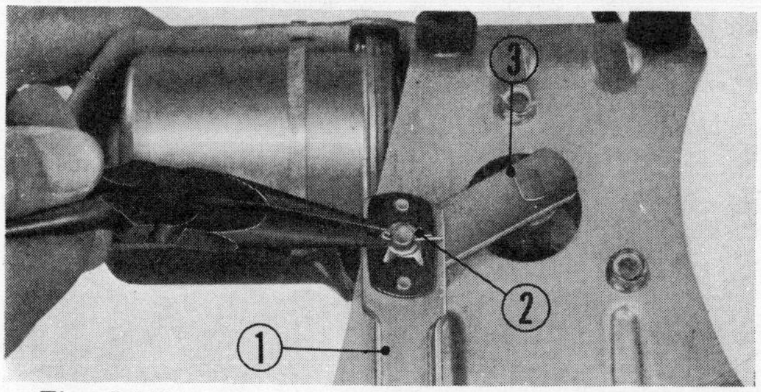

Fig. J-115
① Link complete ② Cotter pin
③ Wiper motor arm

c. Inspection

c-1. Wiper arm linkage

1. Check each joint of the linkage for wear or rattle. If the linkage does not operate smoothly, replace it.

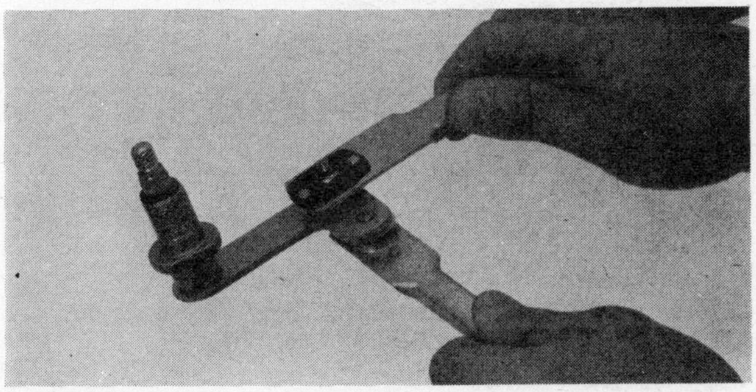

Fig. J-116

2. Check the wiper pivot for wear or rattle and replace it if necessary.

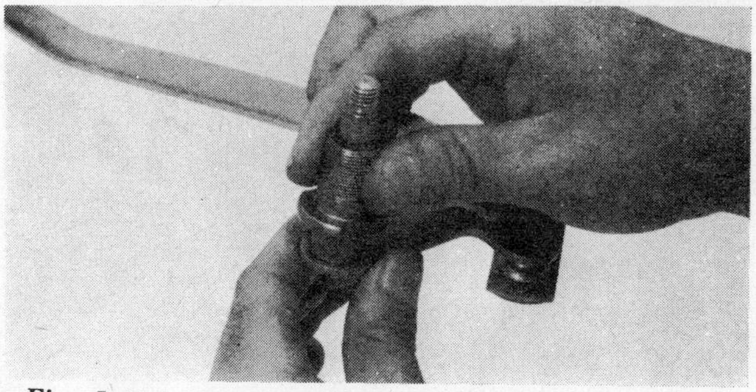

Fig. J-117

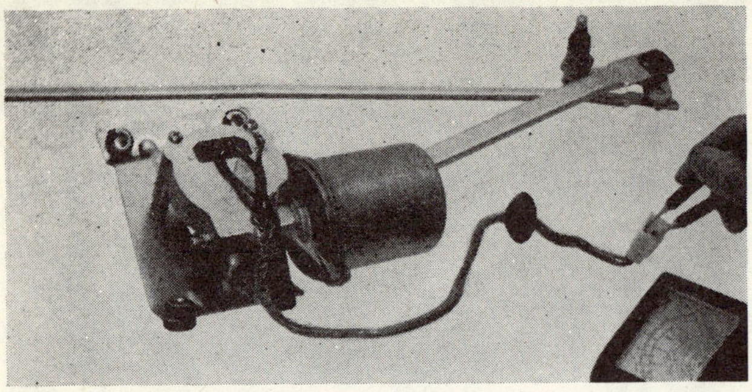

Fig. J-118

c-2. Wiper arm motor

1. Inspect the motor coil

   Using a radio tester, check the coil for open circuit.

   If there is any continuity between the green/black terminal and blue terminal of the coupler and the green/black terminal and blue/white terminal, the coil is normal.

2. Inspect the motor

   Connect a 12V battery as shown in table below. If the motor turns smoothly without any sign of binding, it is in good condition.

Low speed side	Positive terminal to green/black terminal	Negative terminal to blue terminal
High speed side	Positive terminal to green/black terminal	Negative terminal to blue/white terminal

To install the wipers, reverse the removal procedures. Pay attention to the following points:

1. Install the motor in **"automatic stop"** position.
2. Take care not to install the linkage inversely.
3. After making sure all parts are installed in correct position, check the wipers for operation by sprinkling washer fluid or water over the windshield.

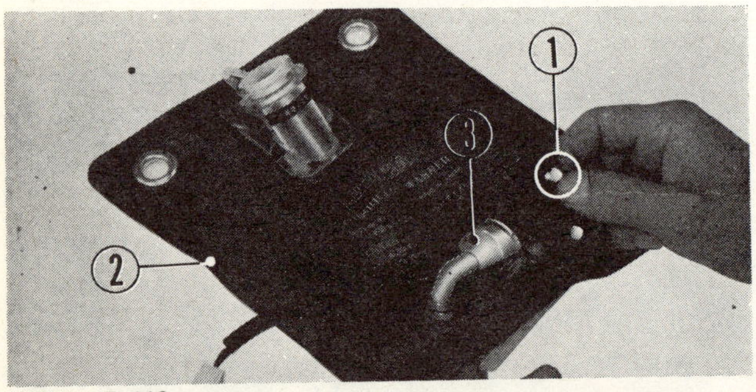

Fig. J-119
① Clip ② Washer bag ③ Washer motor

c-3. Washer motor

Removal

1. Disconnect the washer tube at the washer fluid container.
2. Disconnect the couplre at the motor.
3. Remove the three resin clips from under the washer fluid container, taking care not to damage them. Then, remove the motor.

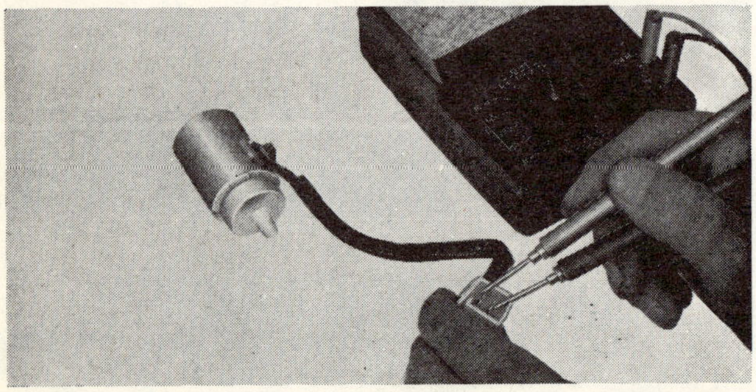

Fig. J-120 Checking washer motor coil for continuity

Inspection

1. Connect the positive terminal of a 12V battery to the green/black terminal of the motor coupler and the negative terminal to the black terminal. Check the motor for operation.
2. If the motor operates in the above test, check to be sure that water is forced out of the washer pipe. If not, the pump is defective. Replace the motor because the pump is incorporated in the motor housing.

To install the washer motor, reverse the removal procedures.

Thermometer and fuel meter

a. Thermometer

The thermometer consists essentially of a thermister type thermounit and a bimetal type meter. A regulator is built in the meter to prevent indication error due to fluctuation of battery voltage.

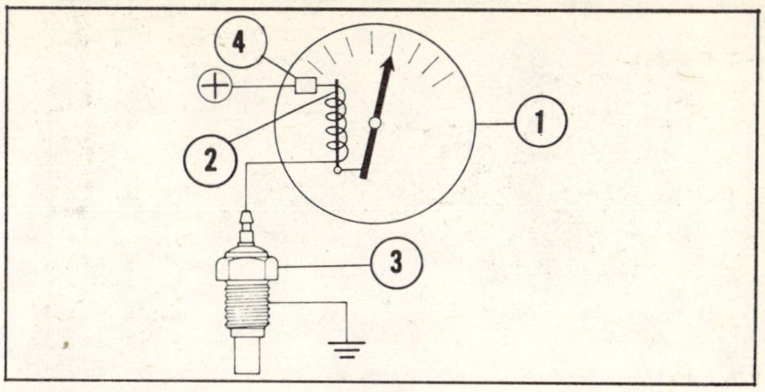

Fig. J-121
① Thermometer ② Bimetal
③ Thermosensor ④ Regulator

Temperature	50°C (122°F)	80°C (176°F)	100°C (212°F)	120°C (248°F)
Unit resistance	154Ω	52±4Ω	27.5±1.5Ω	16Ω

b. Fuel meter

The fuel meter consists essentially of a slide resistance type tank unit and a bimetal type meter. A regulator is built in the meter to prevent indication error due to fluctuation of battery voltage.

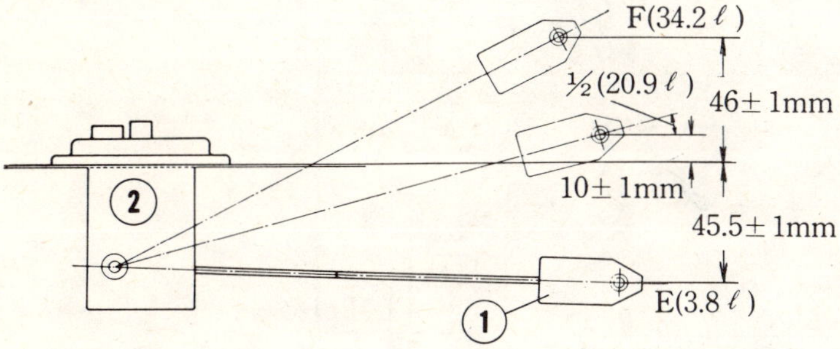

Fig. J-122 Fuel meter tank unit
① Float ② Unit

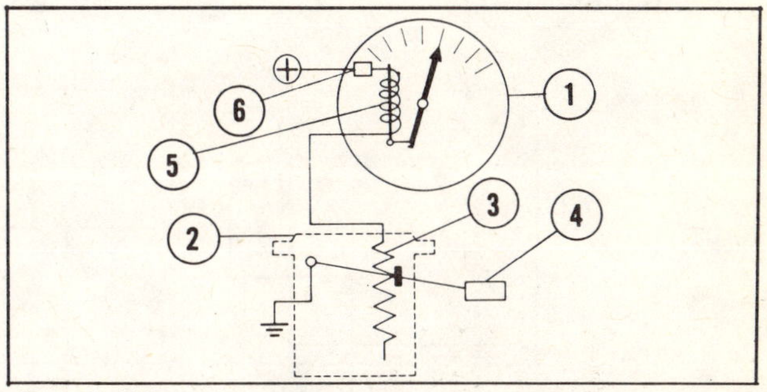

Fig. J-123
① Fuel meter ② Meter unit ③ Resistor
④ Float ⑤ Bimetal ⑥ Regulator

Float level	E (3.8 ℓ or 1.00 gal)	½ (20.9 ℓ or 5.52 gal)	F (34.2 ℓ or 9.03 gal)
Unit resistance	105±1Ω	32.5±7Ω	4±1Ω

c. Inspection

1. If the meter does not work and stays on the zero, disconnect the unit to meter lead and ground the meter side lead in a moment. If the pointer oscillates, the unit or circuit is defective. If the pointer will not oscillate, the meter is defective.
2. The meter pointer registers a certain value regardless of the temperature and float level. Disconnect the unit leads. Connect a 16Ω resistor (for thermometer) or 4Ω resistor (for fuel meter) to the wire harness-side lead and ground the other lead to the body.

   If the meter pointer oscillates excessively, the unit is defective. If the pointer will not oscillate, the meter is defective.

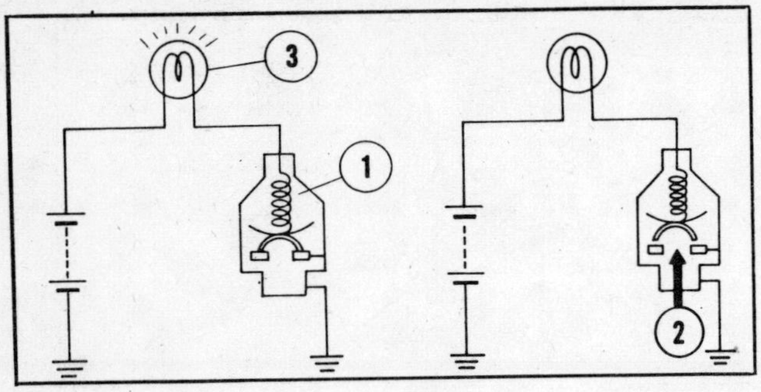

Fig. J-124
① Oil pressure switch ② Hydraulic pressure
③ Oil pressure alarm lamp

Oil Pressure Warning Lamp

The oil pressure warning lamp indicates the condition of engine oil recirculating within the engine during operation. When the pressure reaches the specified value, the pressure switch points will open and the warning lamp will go out.

Warning lamp turn off pressure: 0.2~0.4 kg/cm^2 (2.8~5.7 lb/in^2)

Battery Charge Warning Lamp

The battery charge warning lamp indicates whether or not the AC generator is operating properly. When the generator output voltage reaches the specified value, the voltage relay of the regulator operates, causing the warning lamp to go out.

Warning lamp turn off voltage: 4.5~5.8V (neutral terminal)

Wiring diagram

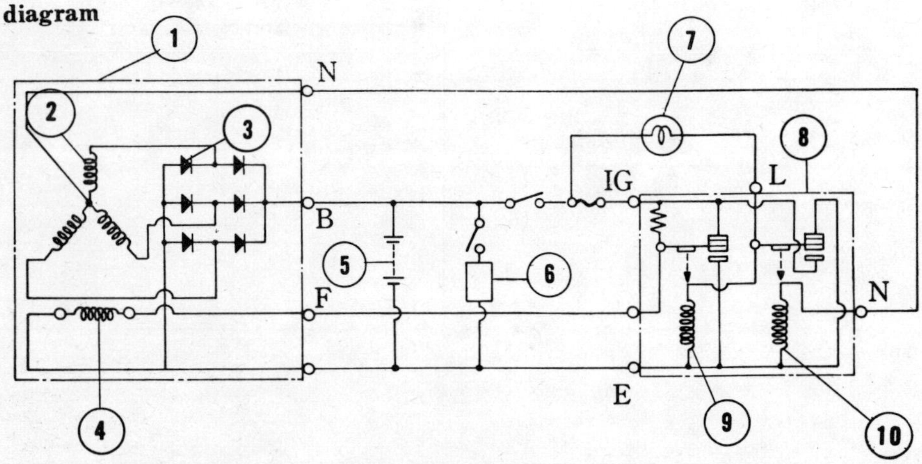

Fig. J-125

① AC generator
② Stator coil
③ Silicon diode
④ Field coil
⑤ Battery
⑥ Load
⑦ Charging lamp
⑧ Regulator
⑨ Voltage regulator
⑩ Voltage relay

DRIVE SHAFTS

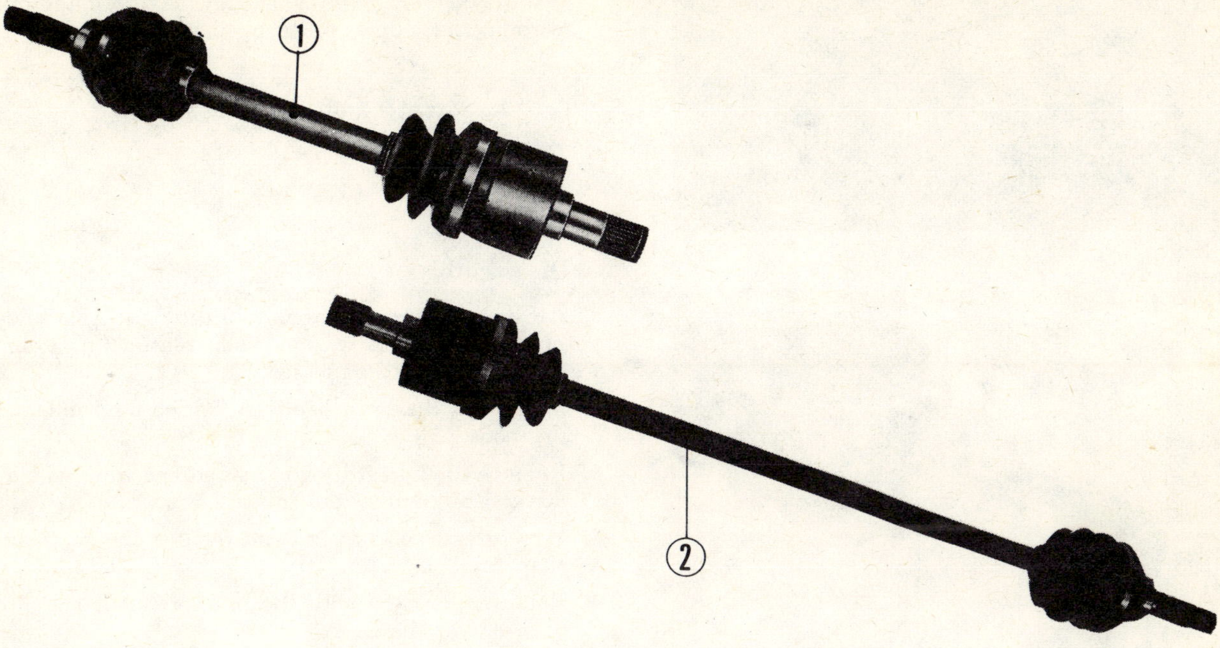

Fig. K–1 ① Right drive shaft ② Left drive shaft

a. Removal

1. Loosen the wheel clip nuts with a wheel wrench.
2. Remove the wheel center cap.
3. Pull out 4 mm (0.157 in.) cotter pin and loosen the spindle nut.

   NOTE:

   To pull out the pin, first straigten it. Then use a standard-type screwdriver at the pin tip and drive it out with a hammer. As the pin comes out of the hole, it interferes with the wheel disc. Bend the pin to remove it with pliers.

4. Jack up the front of the car and support it with safety stands. (See page A–13.)
5. Remove the wheel clip nuts and the the wheel.
6. Remove the spindle nut.
7. Give blows of a mallet to the brake drum and remove it with hands. (except disc brake type)
8. Remove the lower arm ball joints from the knuckles.

   NOTE:

   It is advisable to remove the tie rod ball joints from the knuckles first to do the subsequent job better.

9. Remove the drive shafts.
   - Hold the knuckle with hand and move it toward backward. Then remove the drive shaft from the knuckle.
   - Separate the inboard joint side of the drive shaft from the differential case.

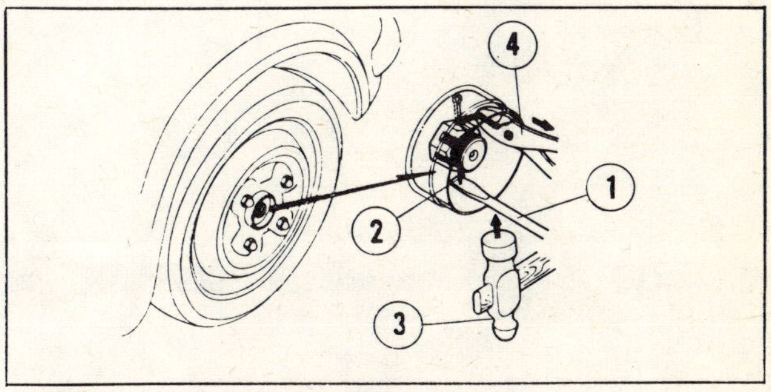

Fig. K–2
① Standard type screwdriver ② Cotton pin
③ Hammer ④ Pliers

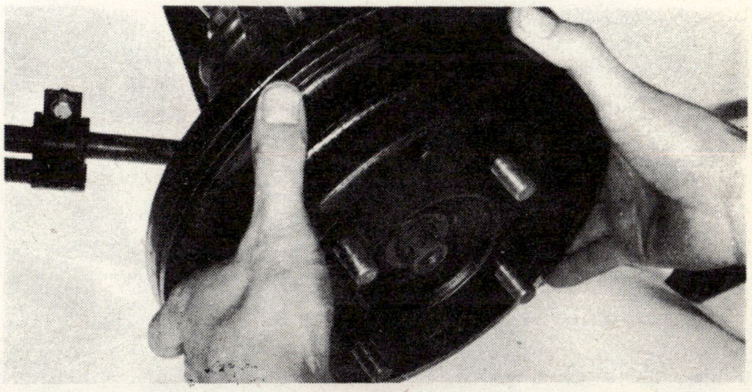

Fig. K–3 Removing front drum

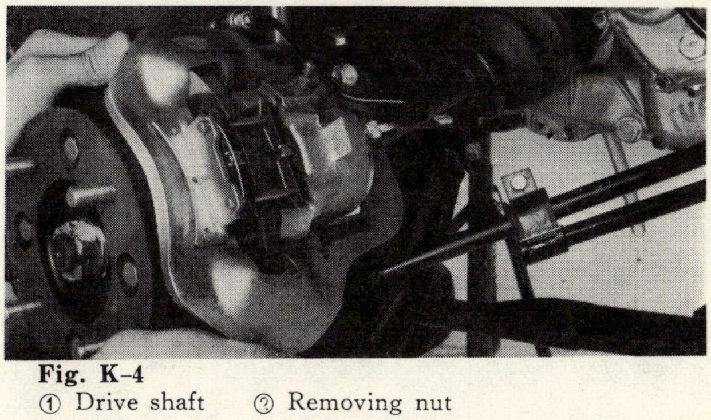

Fig. K–4
① Drive shaft ② Removing nut

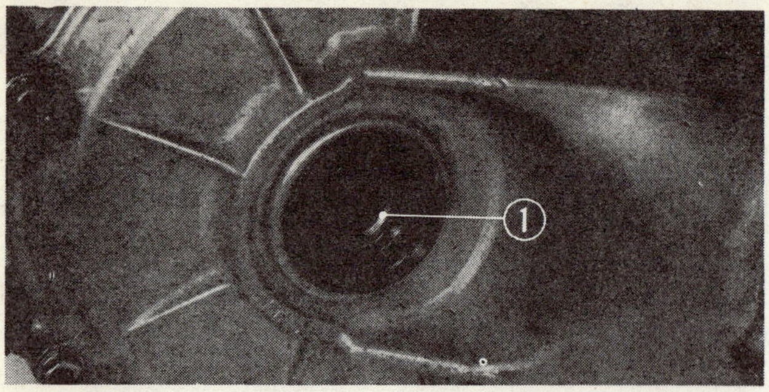

Fig. K–5
① Insert joint here

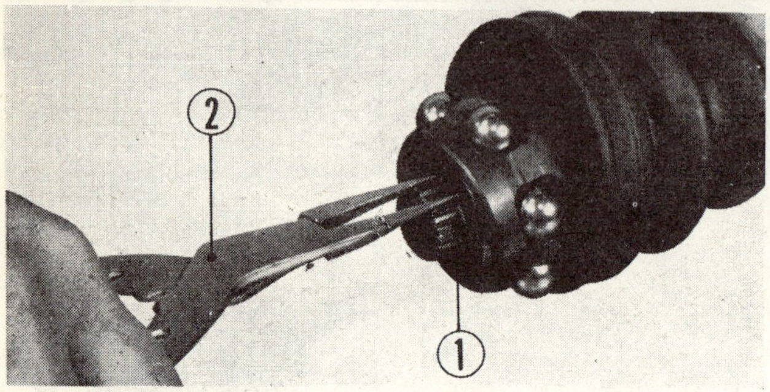

Fig. K–6
① Snap ring ② Snap ring pliers

NOTE:

When the joint side of the shaft is separated, oil may come out. Plug the joint hole in the differential case with lintfree clothes to keep oil in and foreign materials out.

b. Disassembly

The drive shafts are lifetime-lubricated, so that they must not be disassembled without valid reason (for example, as when the boots are damaged.).

1. Straighten the boot bands A, loosen the band, and move the boot A.
2. Remove the inboard joint shaft stopper ring with a screw-driver and remove the inboard joint.
3. Wipe grease off the joint and remove the snap ring with snap ring pliers.
4. Remove the cage with the inner retainer and press the balls out of the cage.
5. Loosen the boot bands B and remove the boot B.

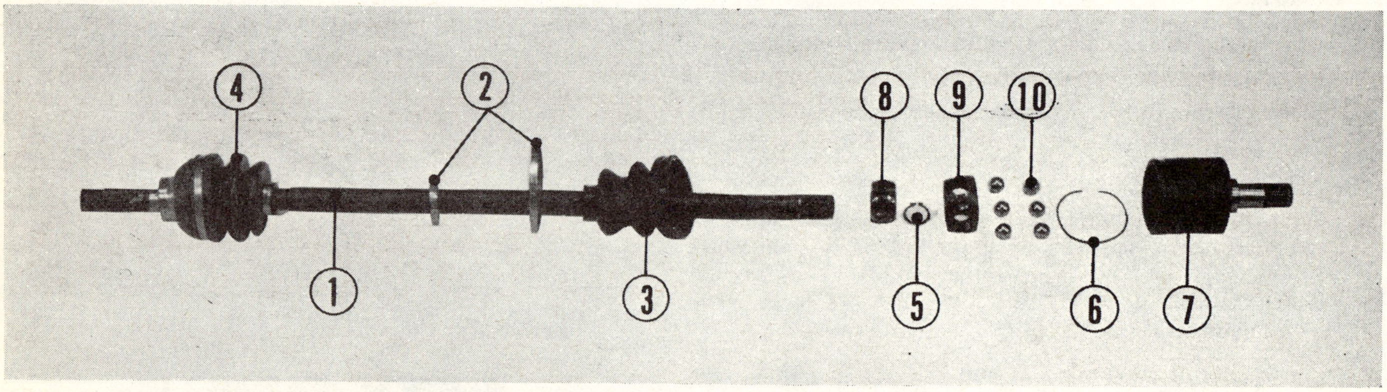

Fig. K–7

① Drive shaft
② Boot bands
③ Boot B
④ Boot A
⑤ Snap ring
⑥ Stopper ring
⑦ Inboard joint
⑧ Inner retainer
⑨ Cage
⑩ Ball

c. Inspection

Wash the drive shaft components removed and check for conditions.

See the Maintenance Standard.

Inspection item	Remedy
Twist, runout or damage of drive shaft	Replace shaft assembly.
Sticking, rust, rattle of inboard joint side of shaft	Replace shaft assembly.
Sticking, rust, rattle of outboard side of shaft	Replace shaft assembly.
Deterioration, cracks, wear or other damage of boots	Replace boot.

d. Assembly

To assemble the drive shaft, reverse the disassembly procedures. Pay attention to the following points:

When installing the boots A and B, take care not to damage them.

Lubricate the specified points with new grease, exercising care to keep foreign materials out.

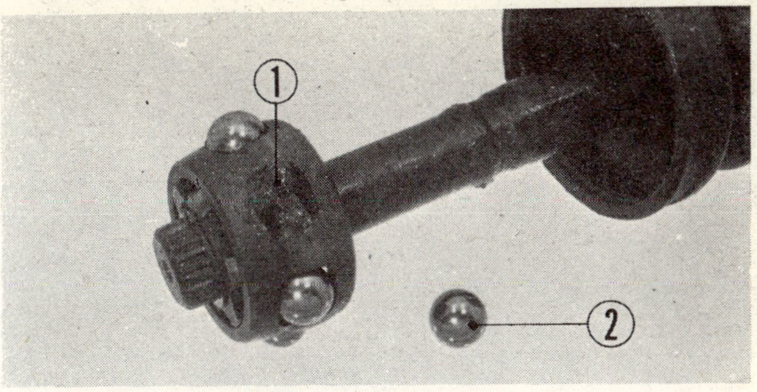

Fig. K–8
① Apply grease here ② Ball

1. Install the inner retainer in the cage. Then insert the cage into the drive shaft and secure the shaft with a new snap ring.
2. When installing the balls to the cage, apply a coat of grease to the cage groove, the balls to the groove, tapping it lightly.
3. Insert the drive shaft with the cage, inner retainer and balls installed in the inboard joint as required and secure the shaft with a snap ring.

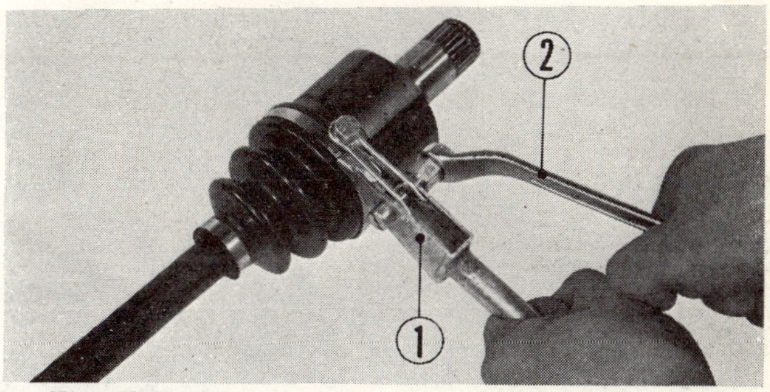

Fig. K–9
① Bellows band fastener ② Box wrench

4. Install the boots in place. Clamp each boot band with the clamping tool and upset it at two places with the punch furnished. Cut the overlapped end of the band remaining 10 mm (0.394 in.) length. Finally fold back this length of the band over the upsetted points.

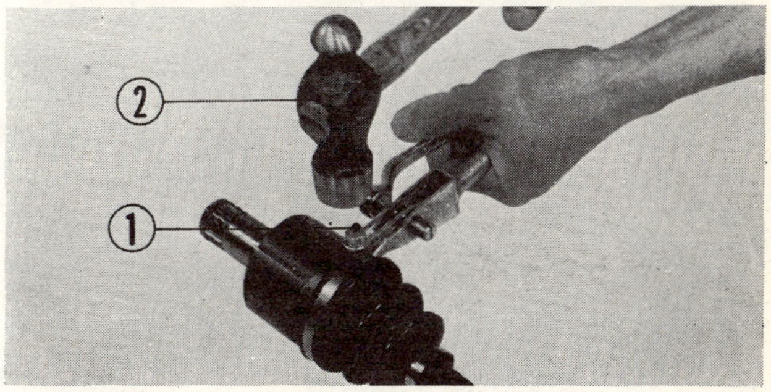

Fig. K–10
① Upsetting with a chisel ② Hammer

Use only

GENUINE

HONDA SPARE PARTS

When Servicing the

CIVIC

Steering Wheel

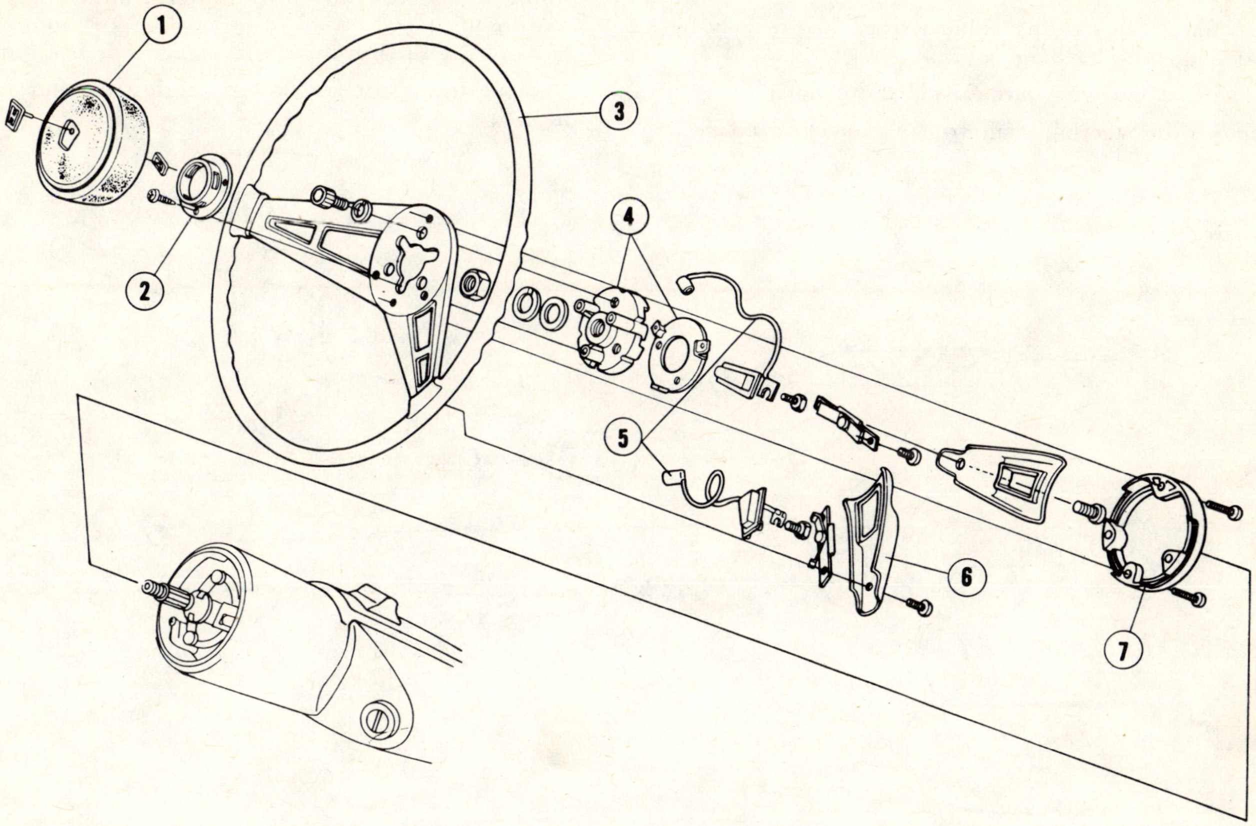

Fig. L–1

① Steering pad
② Pad holder
③ Steering wheel
④ Hub core complete
⑤ Horn cords
⑥ Horn cover
⑦ Steering pad lower retainer

a. Removal

The illustration shows the component parts of the steering wheel (USA type).

1. Remove the steering pad by lifting it off.
2. Remove the steering wheel by loosening the attaching nuts. Then the hub core assembly can be removed together with the steering wheel.

NOTE:

If it is difficult to remove the hub core assembly from the steering shaft, lightly tap the shaft head with a mallet.

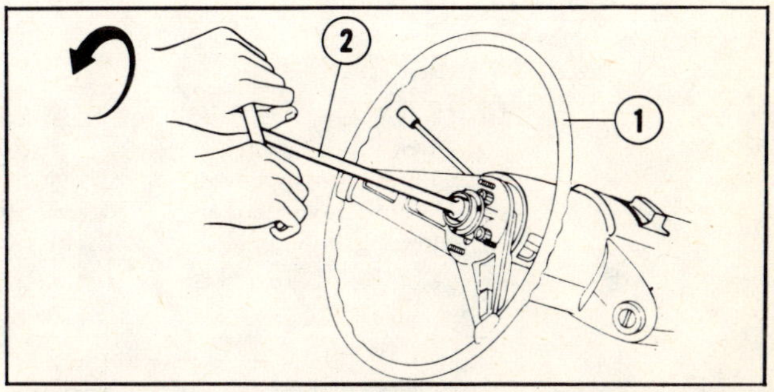

Fig. L–2
① Steering wheel ② T-type socket wrench

3. Separate the steering wheel from the hub core assembly by removing the spoke set bolts.
4. Disconnect the horn cord at the hub core.
5. Remove the horn cord cover and remove the horn springs and horn holder.

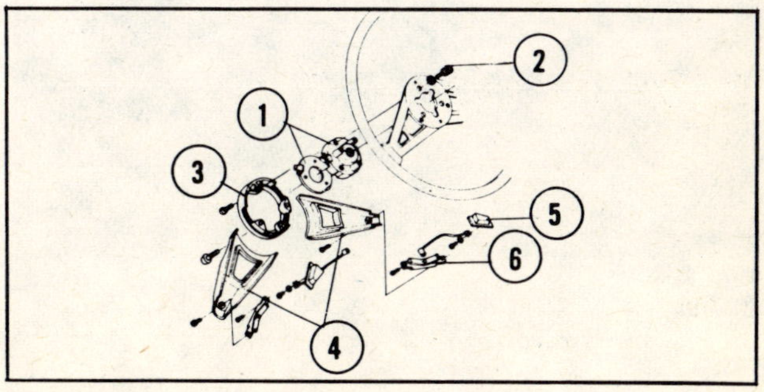

Fig. L–3
① Hub core complete ② Spoke set bolt
③ Steering pad lower retainer ④ Horn cord covers
⑤ Horn button ⑥ Horn spring assembly

Steering Shaft

a. Removal and disassembly

1. Separate the steering joint at the serration by loosening the yoke bolt.
2. Disconnect the steering column from the car body by loosening the attaching bolts.
3. Disconnect the wire harness and the coupler.
4. Remove the steering shaft from the steering column.
5. Remove the upper and lower column covers.
6. Loosen the screw on the turn signal switch key nut and lightly tap its head to permit the key nut to turn freely. Then remove the turn signal switch assembly.
7. Remove the upper and lower steering column bushings.

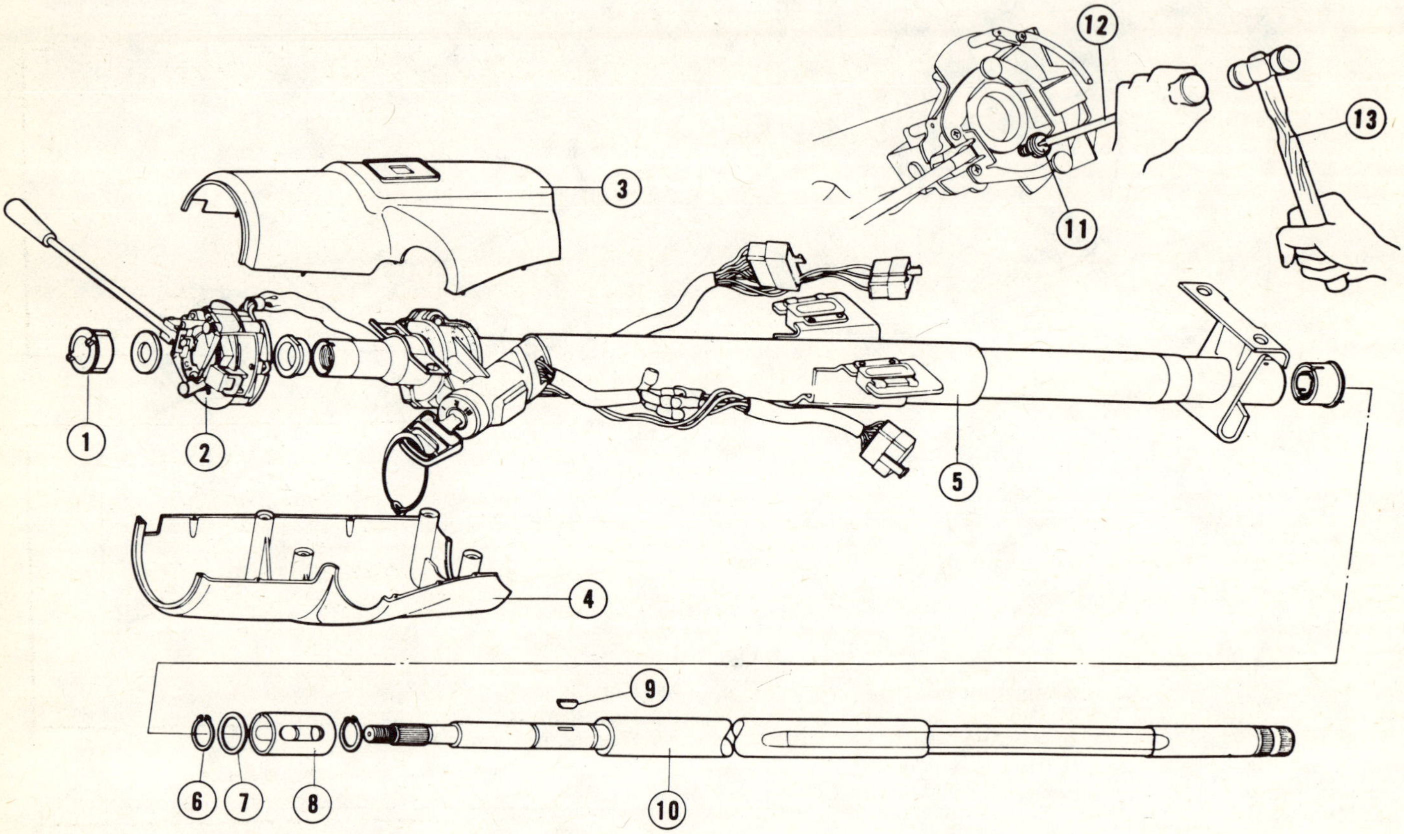

Fig. L-4

① Cancel cam
② Turn signal switch
③ Column upper cover
④ Column lower cover
⑤ Steering column
⑥ Snag ring
⑦ Spacer
⑧ Steering lock collar
⑨ Steering key
⑩ Steering shaft
⑪ Attaching screw
⑫ Screwdriver
⑬ Hammer

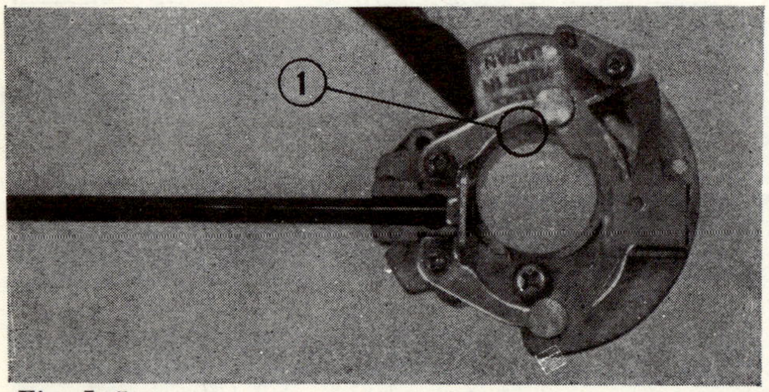

Fig. L-5
① Lug

b. Assembly and installation

To assemble the steering shaft, reverse the disassembly procedures. Pay attention to the following points:

1. When installing the turn signal switch assembly to the steering column, engage the locating tab of the column to the notch in the switch.
2. Be careful of the direction for installation of the key nut which serves as a wedge.

Steering Gear Box

a. Removal

1. Loosen the right front wheel clip nuts.
2. Jack up the front of the car and support it with safety stands.
3. Remove the right front wheel and lift the engine just far enough for the chain to take loads.
4. Remove the tie rod ends from the knuckles using a tie rod end ball joint puller.

NOTES:

1) Be careful not to damage the tie rod threads. Be sure to use a tie rod end ball joint puller.
2) Take care not to damage the boots.

5. Disconnect the exhaust pipe at the exhaust manifold.

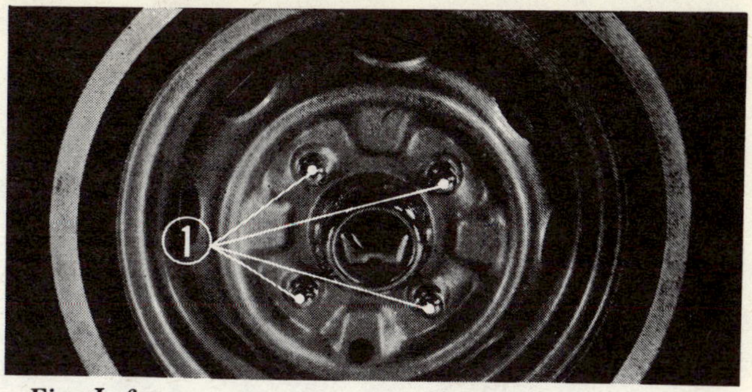

Fig. L-6
① Wheel clip nuts

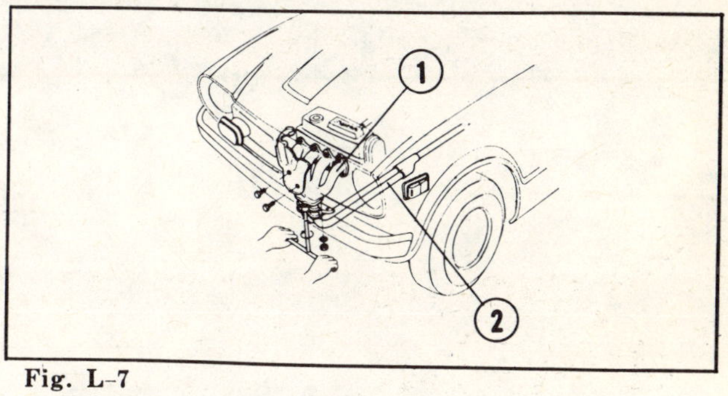

Fig. L-7
① Exhaust manifold ② Exhaust pipe

6. Disconnect the gearshift extension and rod at the engine.
7. Remove the center beam.

NOTE:

When the center beam is removed, the engine may be lowered slightly. Again lift the engine with a chain block.

8. Remove the steering joint and pinion by loosening their attaching bolts.
9. Keep the steering wheel turned to right all the way.

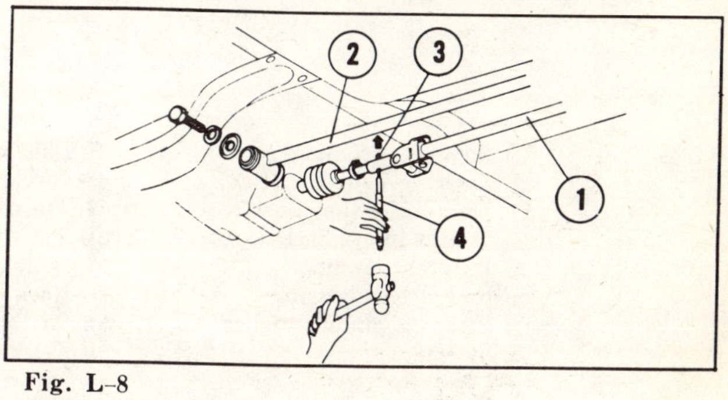

Fig. L-8
① Gearshift rod ② Gearshift extension
③ Spring pin ④ Spring pin driver

10. Remove the steering box bolts with a 12 mm box wrench.
11. Remove the left tie rod end from the rack end. Bend the rack end upward, lower the steering gear box and move it toward right. Then remove the gear box.

NOTE:

To remove the gear box, take care not to damage the dust seal.

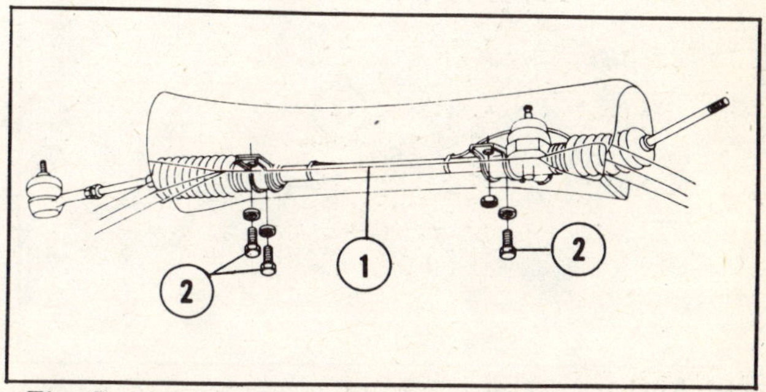

Fig. L-9
① Gear box ② Attaching bolts

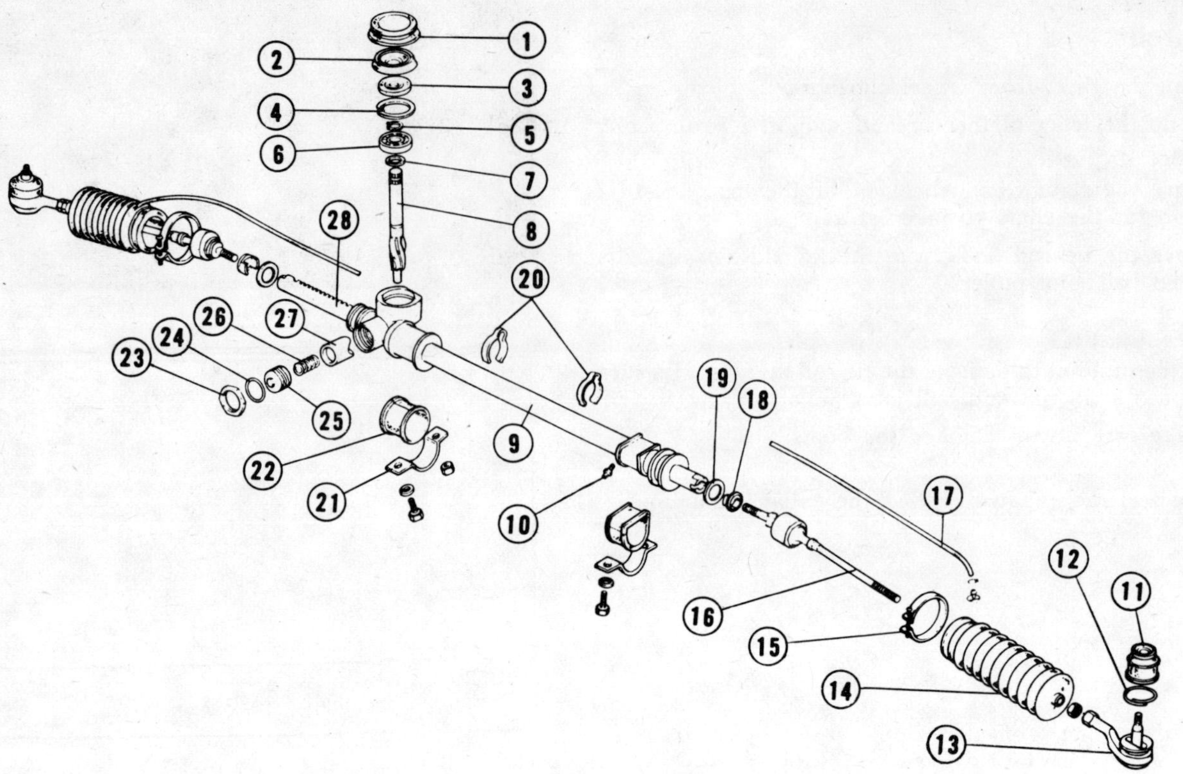

Fig. L-10

① Steering grommet
② Steering grommet B
③ Steering pinion dust seal
④ 45 mm internal circlip
⑤ External circlip
⑥ Radial ball bearing
⑦ Steering pinion washer
⑧ Steering pinion
⑨ Gear box
⑩ Grease fitting
⑪ Ball joint seal
⑫ Circlip
⑬ Tie rod end
⑭ Tie rod dust seal
⑮ Bellows band
⑯ Rack end
⑰ Air tube
⑱ Tie rod lock washer
⑲ Tie rod stopper washer
⑳ Tube clips
㉑ Gear box bracket
㉒ Gear box mount cushion
㉓ Rack screw lock nut
㉔ Rack guide O-ring
㉕ Rack guide screw
㉖ Rack guide pressure spring
㉗ Steering rack guide
㉘ Steering rack

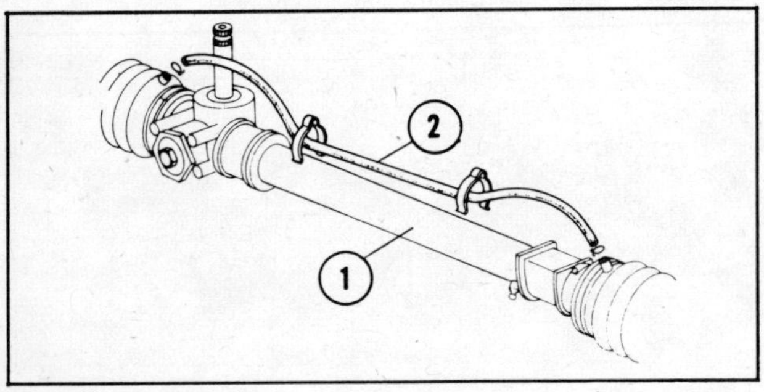

Fig. L-11
① Steering gear box ② Air tube

b. Disassembly

1. Disconnect the air tubes at the tie rod dust seal joints.
2. Move the left tie rod dust seal by loosening the bellows band.

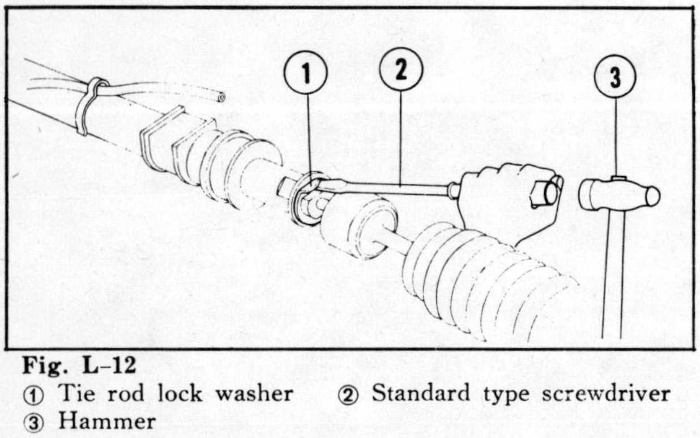

Fig. L-12
① Tie rod lock washer ② Standard type screwdriver
③ Hammer

3. Straighten the lugs of the tie rod lock washers with a standard-type screwdriver, and remove the rack ends from the steering rack by turning them counterclockwise with a wrench.

4. Remove the rack screw lock nut with a lock nut wrench.

   Remove the rack guide screw and then the rack guide pressure spring and steering rack guide.

   NOTE:

   If it is difficult to remove the steering rack guide, insert snap ring pliers into the guide hole and pull out the guide.

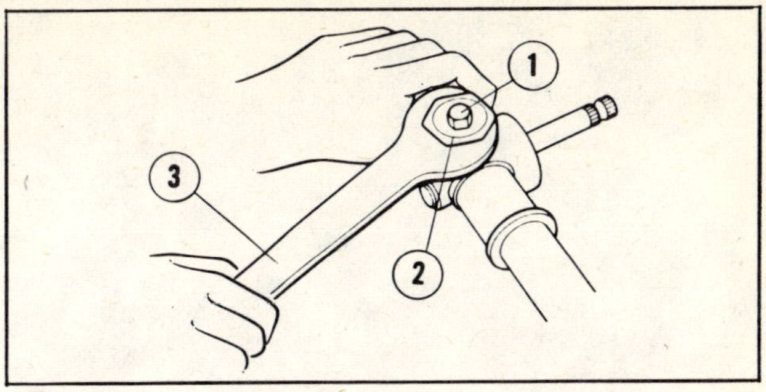

Fig. L-13
① Rack guide screw ② Rack guide screw lock nut
③ Special tool

5. Remove the steering pinion dust seal with a screwdriver.

   NOTE:

   To remove, be carefut not to damage the interior of the gear box.

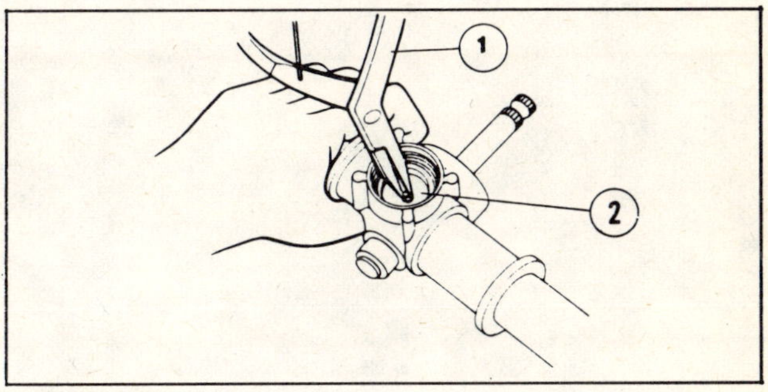

Fig. L-14
① Snap ring pliers ② Snap ring

6. Remove 35 mm (1.38 in.) internal snap ring with snap ring pliers.

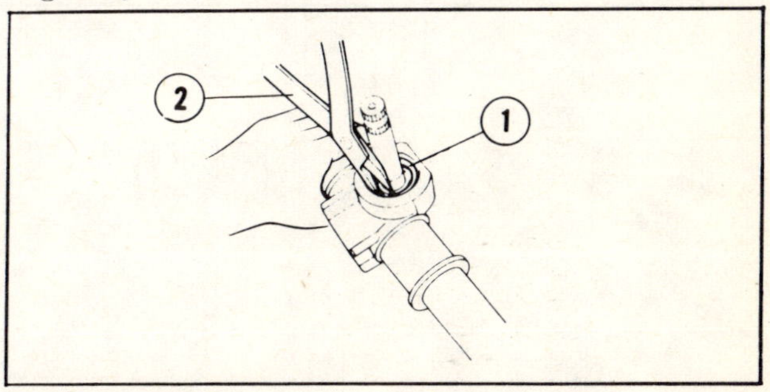

Fig. L-15
① Steering rack guide ② Snap ring pliers

7. Remove the steering pinion from the gear box and then 15 mm (0.59 in.) external snap ring.
8. Remove the ball bearing and washer from the steering pinion.

   NOTE:

   To remove the bearing, hold it with hand and lightly tap the pinion head with a mallet.

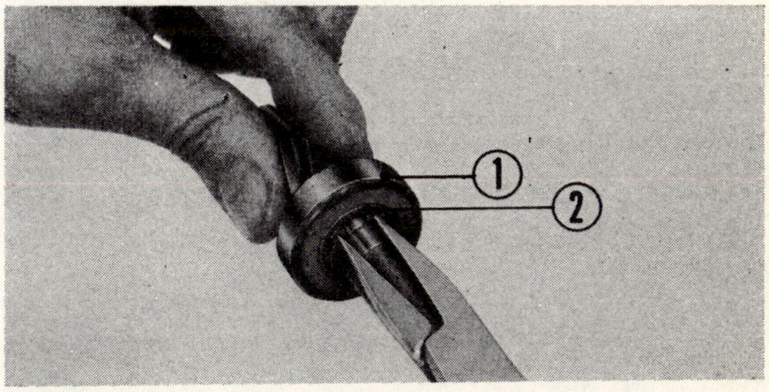

Fig. L-16
① Needle roller bearing ② Snap ring

c. Inspection

1. Check the teeth of the steering rack and pinion for wear. Replace the rack or pinion whichever is excessively worn.

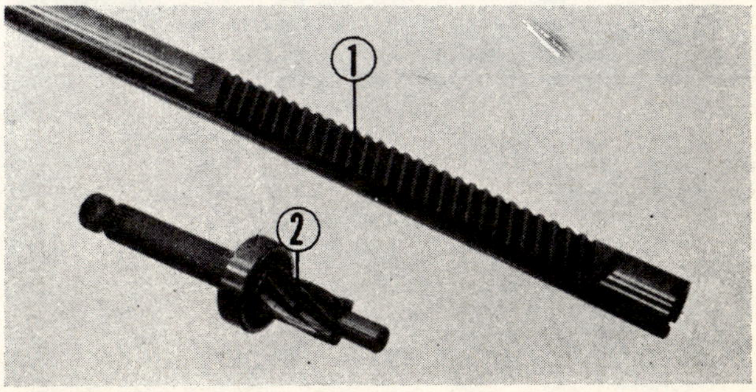

Fig. L-17
① Steering rack ② Steering pinion

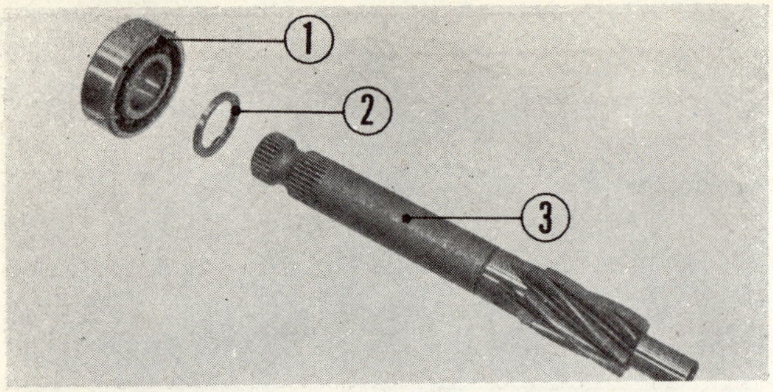

Fig. L-18
① Ball bearing ② Steering pinion washer
③ Steering pinion

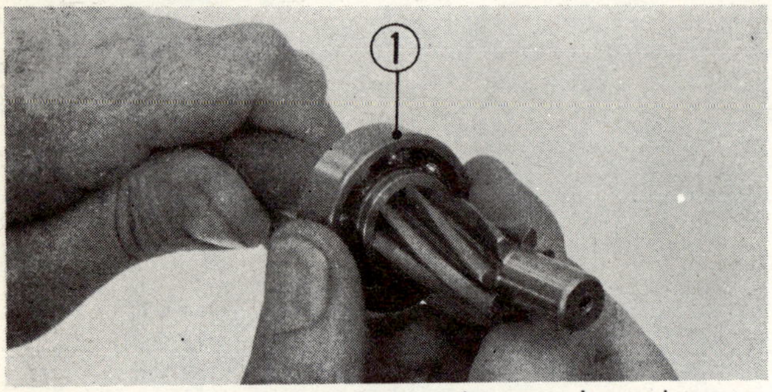

Fig. L-19 Checking ball bearing for smooth rotation
① Ball bearing

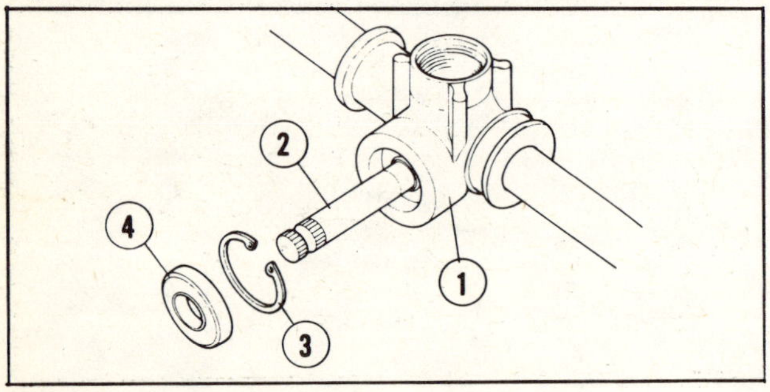

Fig. L-20
① Steering gear box ② Steering pinion
③ Snap ring ④ Steering grommet B

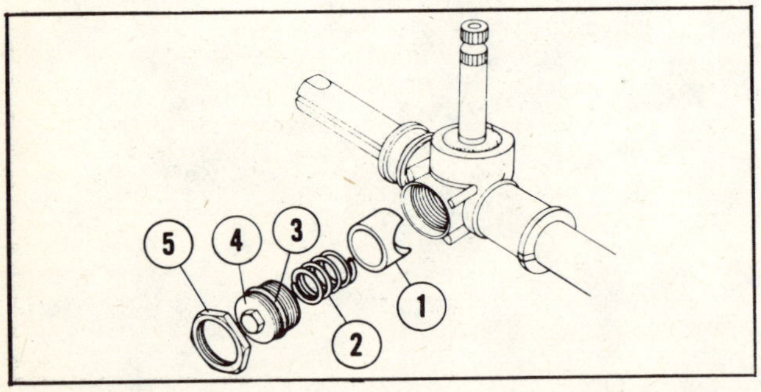

Fig. L-21
① Steering rack guide ② Rack guide pressure spring
③ Rack guide O-ring ④ Rack guide screw
⑤ Rack screw lock nut

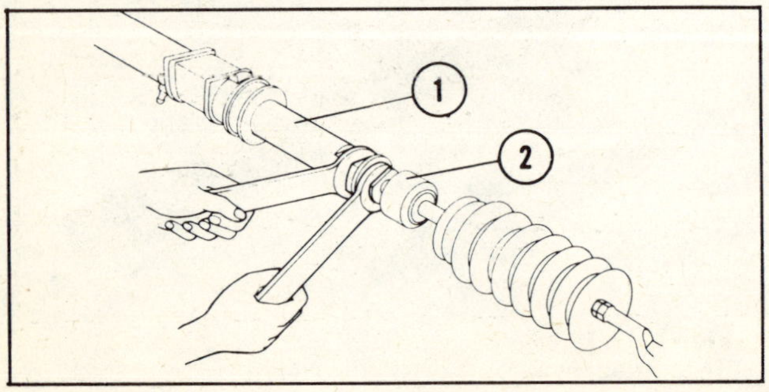

2. Check the steering pinion, ball bearing and washer for wear or damage, and replace them if necessary.

   NOTE:

   To check the bearing, hand-rotate it while pressing it against the pinion and check for abnormal noise or smooth rotation.

3. Measure the free length of the rack guide pressure spring.
   See the Maintenance Standard.

   If the spring is weakened, the meshing pressure of the pinion and rack may drop, resulting in too large a backlash and excessive play of the steering wheel.

d. Assembly

Before assembling the steering gear box, apply a coat of new grease to each sliding part. To assemble, reverse the disassembly procedures. Pay attention to the following points:

1. First install the steering rack to the gear box.
2. Install the steering pinion in place and mesh it with the steering rack.
3. Install the internal snap ring and then a new steering pinion dust seal in place.
4. Install the steering rack guide to the gear box so that it fits to the back of the rack properly.
5. Install the rack guide pressure spring. Put a new O-ring on the rack guide screw and tighten the screw in the gear box to the specified torque.

   Turn the steering rack guide screw until it stops and back off the steering rack 45° from the locked position, and tighten the rack screw lock nut to the specified torque. See Fig. L-21 for correct installation.

6. Install the rack ends to the steering rack. Install the tie rod stopper washers and new lock washers in place and tighten the rack ends to the specified torque.

Fig. L-22
① Steering rack ② Rack end

NOTE:

Upon assembly of the rack end and steering rack, fit the locating lug of the tie rod lock washer in the slot on the end of the steering rack and bend the outer edge of the washer so that the washer is held against the flat part of the end of the tie rod rack end.

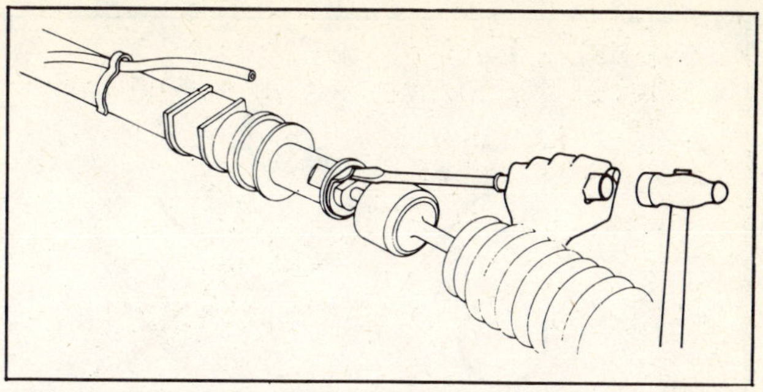

Fig. L-23 Bending locking lug of tie rod lock washer

e. Replacement of tie rod end ball joint seals

1. Remove the tie rod end from the rack end by loosening the lock nut.
2. Remove the snap ring with a screwdriver.
3. Remove the ball joint seal and wipe grease off the seal and ball joint.

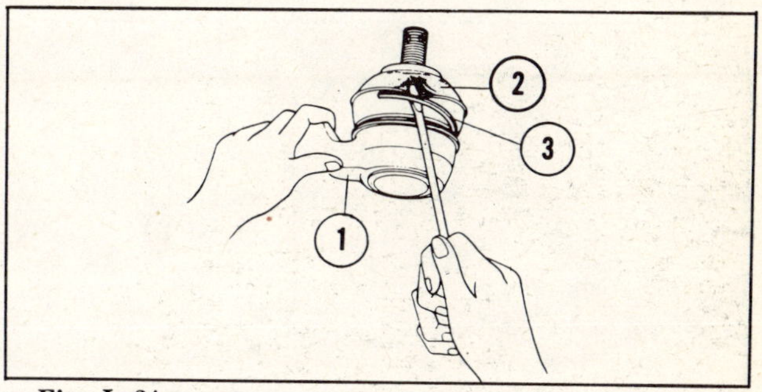

Fig. L-24
① Tie rod end ② Ball joint seal ③ Snap ring

4. Fill a new ball joint seal with grease.

   The recommended amount of grease to be applied is about 7 to 7.5 g (0.247 to 0.265 oz).

   When lubricating, take care not to allow grease to come in contact with the seal lip or with the snap ring groove in the tie rod ball joint.

   Recommended grease:

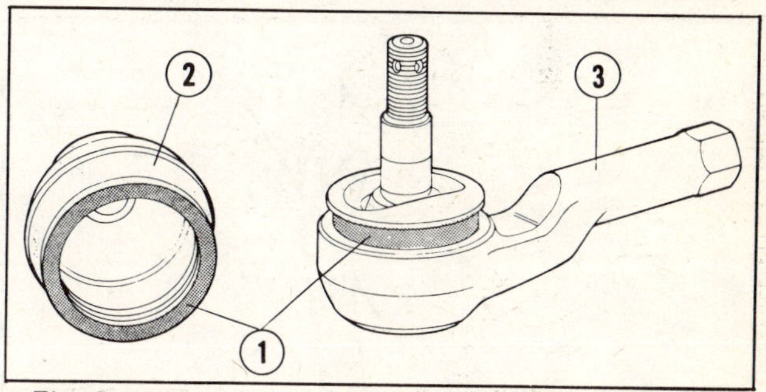

Fig. L-25
① These parts must be free of grease
② Ball joint seal ③ Tie rod end

NOTE:

Remember, too much grease or air within seal will reduce sealing efficiency or to permit seal to come in contact with the knuckle during operation.

Too little grease will also result in reduced sealing efficiency.

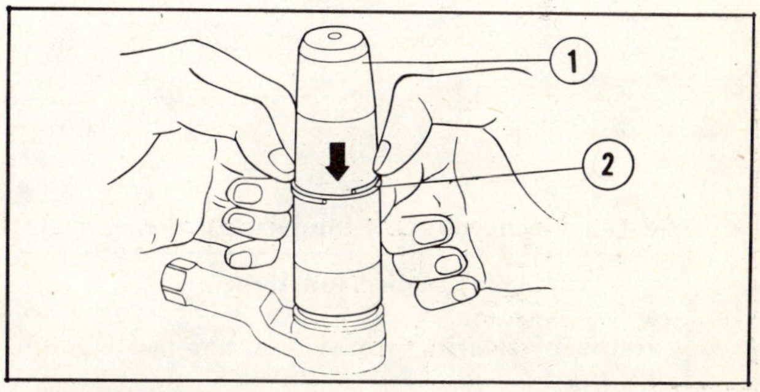

Fig. L-26
① Ball joint snap ring guide ② Snap ring

5. Put the seal on the ball joint. Using a ball joint clip guide, install the snap ring on the seal while pressing it down.

NOTES:

1) Use a new snap ring.
2) When installing the snap ring, take case not to damage the body.
3) When removing and installing the lower arm ball joint seal, refer to the steps for the tie rod end ball joint seals.

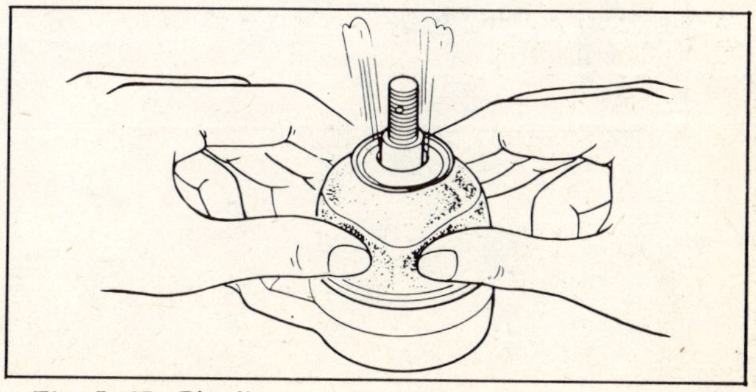

Fig. L-27 Bleeding

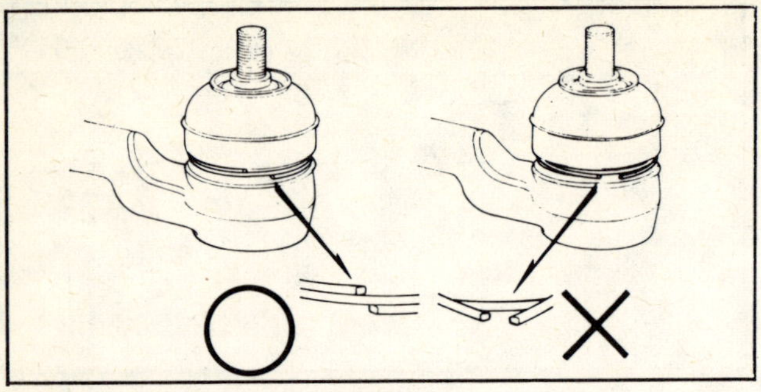

Fig. L-28 Snap ring installed

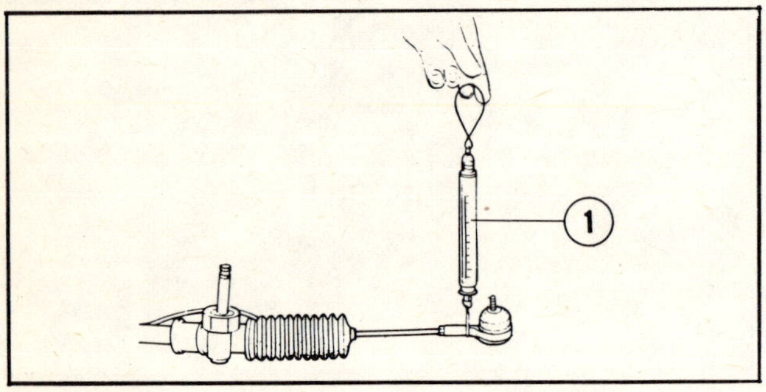

Fig. L-29
① Spring scale

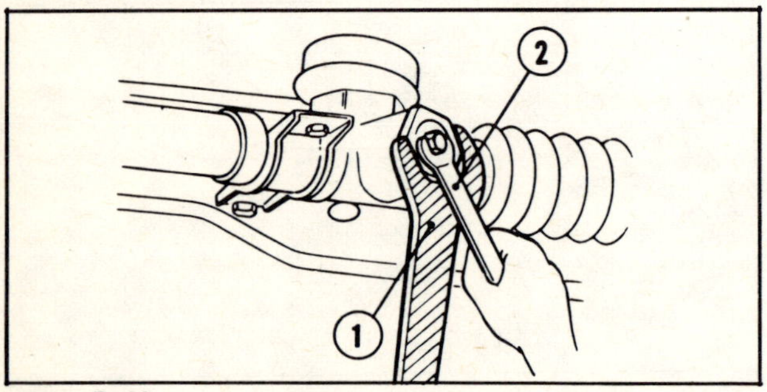

Fig. L-30
① Special tool ② 14 mm wrench

6. Install the snap ring so that the gap is toward the forward direction of car.
7. Make sure the snap ring is properly installed in place; otherwise the ball joint seal may be damaged.
8. Install the tie rod end to the rack end, and tighten the lock nut lightly.

f. Inspection of rack end ball joints

1. Check the rack end ball joint for operation by moving it fore and aft. To check for operating force, use a spring scale. Check to see if the force at which the joint starts moving is within the specification.

   When holding the rack with hand horizontally, the tie rod must not drop by its own weight. For the tightening torque, see the Torque Specifications.
   ~~manual~~.

g. Adjustment of rack guide screw

1. Place the front wheels in the straight-ahead position. Then jack up the front of the car and support it with safety stands.
2. Loosen the rack screw lock nut.
3. Tighten the rack guide screw. Back off the rack 45° from the locked position and tighten the lock nut.
4. After tightening, check for hard steering or for rattle of steering wheel by driving the car.

After disassembling and assembling the steering system, the inspection and adjustment must be made on the following items:

See the Maintenance Standard at the end of this manual.

Inspection item	Remedy
Rattle of steering wheel or other bad steering feeling	Adjust.
Wheel alignment	Adjust.
Steering angles	Adjust.
Toe-in	Adjust.
Side-slip	Adjust.

BRAKE SYSTEM

Precautions for service work

1. Too much emphasis cannot be placed on the importance of keeping brake parts in the best possible condition at all times for your safe driving.
2. Immediately replace any part with a new one upon discovering it is damaged—even if damage is slight. Remember, "A slightly damaged part will be a seriously damaged part after a few minutes of driving."
3. Protect painted surfaces of fenders with fender covers or any other proper covering from attacking of brake fluid when the fluid is spilled on fenders.
4. Before servicing, clean dust and dirt off the brake parts. Wash all removed parts with or in absolute alcohol (for fuel). When installing, wash the parts in fresh brake fluid.

Master Cylinder (Tandem Type)

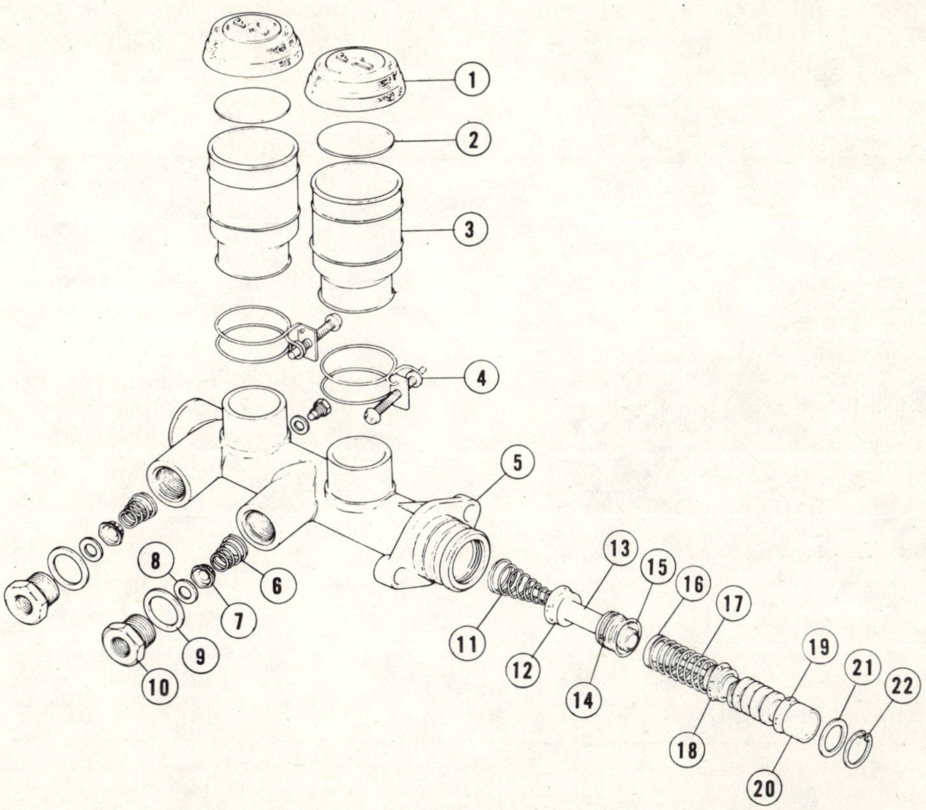

Fig. M-1

① Cap
② Float
③ Reservior tank
④ Reservior tank crip
⑤ Master cylinder body
⑥ Check valve spring
⑦ Check valve
⑧ Check valve washer
⑨ Packing
⑩ Union cap
⑪ Primary spring
⑫ Piston cup
⑬ Primary piston
⑭ Wiper ring
⑮ Piston cup
⑯ Secondary stopper
⑰ Secondary spring
⑱ Piston cup
⑲ Secondary cup
⑳ Secondary piston
㉑ Stopper plate
㉒ Snap ring

a. Removal

• When removing the master cylinder, cover the painted surfaces with a waste cloth to prevent brake fluid from coming in contact with them.

1. Disconnect the brake pipes at the master cylinder.
2. Remove the master cylinder from the master power.

Fig. M-2

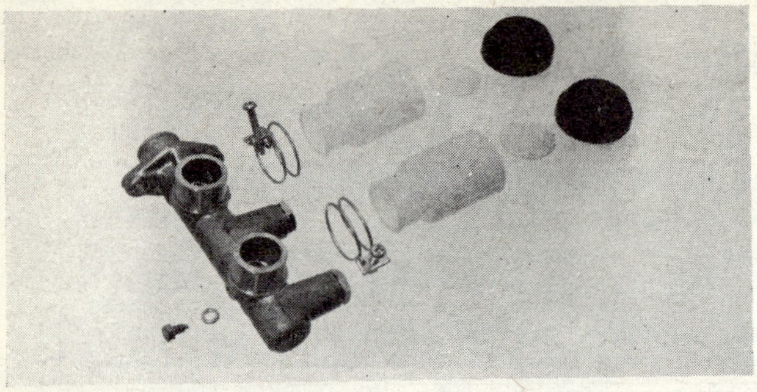

Fig. M-3

b. Disassembly

1. Remove the oil reservoir caps, and remove the floats.
2. Remove the oil reservoirs by loosening the clamps.
3. Remove the stop bolts.

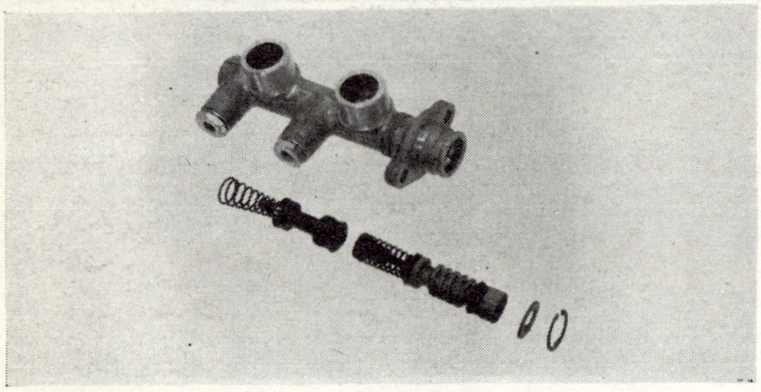

Fig. M-4

4. Remove the master cylinder clip, and remove the return stop plate, secondary piston and primary piston in this order.

   NOTE:

   The primary piston can be removed by removing the stop bolt. To remove it easily, blow the compressed air from the outlet side.

Fig. M-5

5. Loosen the union cups, and remove the gaskets, check valve washers, check valves and springs in this order.

   NOTE:

   When removing the union cup, hold the master cylinder in a vise and use a waste cloth to prevent damage to the oil reservoir attaching surface of the cylinder.

c. Inspection

See the Maintenance Standard.

Inspection item	Remedy
Clearance between master cylinder and piston	Replace.
Clogging of oil holes in pistons and cylinder	Replace.
Damage of cylinder oil reservoir attaching surface	Replace.
Deformation or other damage of check valves	Replace.
Wear, deformation or other damage of piston cups	Replace.

NOTE:

Damage of the piston cups and check valves is apt to be looked over in inspection. The rubber parts are subject to deterioration. Replace the above parts as an assembly at the time of disassembly.

d. Assembly

To assemble, reverse the disassembly procedures. Pay attention to the following points:

1. Before assembling, wash all parts in clean brake fluid. Keep them free from dust, dirt or mineral oil.
2. Insert the primary piston and secondary piston in this order into the master cylinder. To prevent the piston cups from being folded, apply a coat of brake fluid to the inside surface of the cylinder and to the outside surface of the pistons and insert the pistons while turning them slowly.
3. Be careful of the installation direction of each part.
4. Install the master cylinder clip securely.
5. Tighten the union caps and stop bolts to the specified torques.

   Torque specifications:

   Union caps 5.8~6.3 kg-m (42~46 lb-ft)
   Stop bolts 0.9~1.1 kg-m (7~8 lb-ft)

   • When holding the master cylinder in a vise, take care not to damage the cylinder.
6. After assembling, **bleed the brake lines.** **See page B–11.** Then, depress the brake pedal several times and check to see if there is any sign of leak from the brake pipes.

Fig. M-6

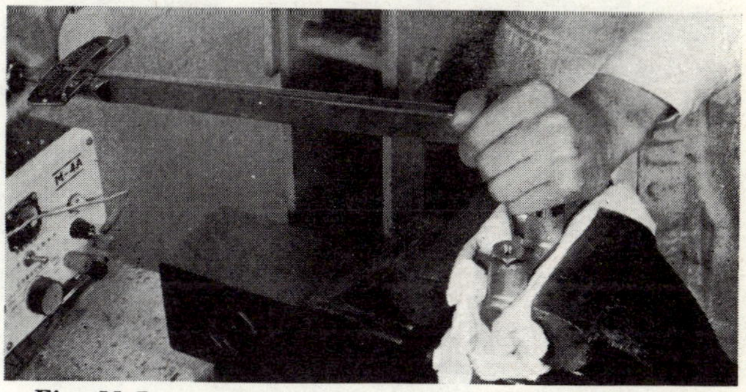

Fig. M-7

Master Cylinder (Single Type)

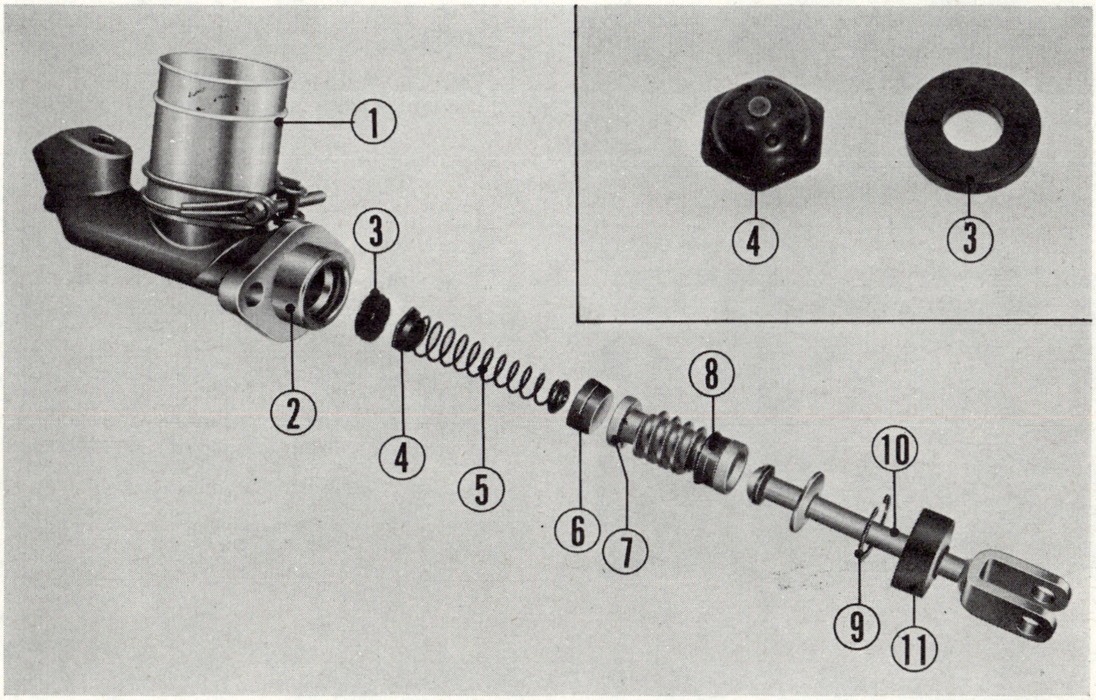

Fig. M-8

① Resevoir tank
② Master cylinder
③ Check valve washer
④ Check valve
⑤ Spring
⑥ Primary cup
⑦ Piston complete
⑧ Secondary cup
⑨ Snap ring
⑩ Push rod complete
⑪ Master cylinder boot

a. Removal and disassembly

Fig. M-9
① Clip pin ② Pedal pin ③ Push rod

1. Remove the clip pin. Remove the pedal pin connecting the brake pedal and push rod.

Fig. M-10
① Flare nut wrench ③ Master cylinder
② Brake pipe ④ Attaching nut

2. Using a flare nut wrench, disconnect the brake pipes at the master cylinder.

NOTES:

1) Use of any tool other than a flare nut wrench may damage the flare nuts.

2) Take care not to spill brake fluid.

3. Remove the master cylinder assembly from the body by loosening its attaching nuts.

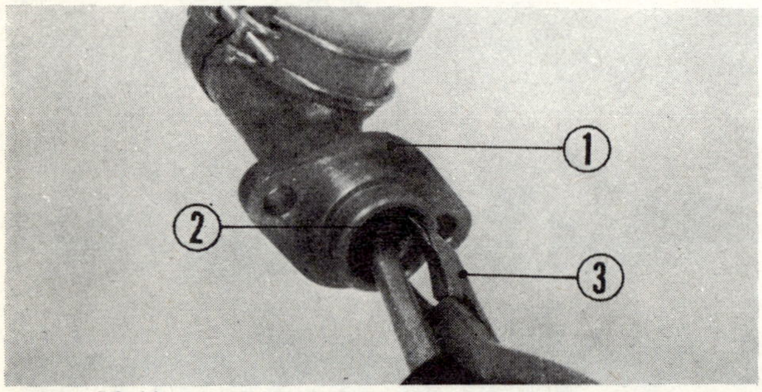

Fig. M-11
① Master cylinder ② Snap ring ③ Snap ring pliers

4. Remove the master cylinder boot. Using snap ring pliers, remove the snap ring. Disconnect the push rod.
5. By disconnecting the push rod, the primary pispon, piston cups, return spring and check valve set can be removed.

NOTE:

When it is difficult to remove the check valve, direct a blast of compressed air to the valve through the pipe connection. Do not use wire.

Fig. M-12
① Piston ② Check valve ③ Check valve washer
④ Secondary cup ⑤ Primary cup

6. Wash the parts removed in absolute alcohol (for fuel).
7. Check the piston cups and check valve for damage, and replace them if necessary.

NOTES:

1) Perform the disassembly and washing of the master cylinder component parts in an enamelled container.

2) After washing, the parts must be free from oil or dust.

b. Inspection

See the Maintenance Standard at the end of this manual.

Inspection item	Remedy
Wear or damage of sliding surfaces of cylinder and piston	Replace.
Restriction of oil holes in cylinder and piston	Repair or replace.
Wear or damage of rubber cups	Replace.
Deformation or damage of return spring	Replace.

c. Assembly and installation

To assemble and install the brake master cylinder, reverse the removal and disassembly procedures. Pay attention to the following points:

1. Before installing, wash the parts in fresh brake fluid.

   NOTE:

   Never use mineral oil such as gasoline for washing brake parts.

2. Apply a sufficient coat of brake fluid to the piston cups and to the cylinder bore.

   Insert the check valve set, piston cups, piston and push rod into the cylinder, and secure them with a new snap ring.

   NOTES:

   1) Do not damage the piston cups.
   2) Replace any damaged part even if slight.
   3) Use a new snap ring.

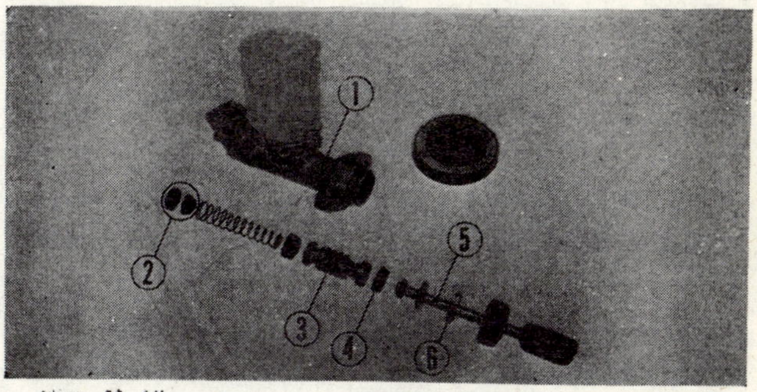

Fig. M-13
① Master cylinder ② Check valve set ③ Piston
④ Piston cup ⑤ Push rod ⑥ Snap ring

3. Install the master cylinder assembly and connect the brake pipe. Tighten them to the specified torque.

   NOTES:

   1) Do not mix brake fluid of different brands; a chemical reaction may occur, resulting in poor braking effect.
   2) Never use any oil other than brake fluid.
   3) Install the oil reservoir filler cap with the mark "F" facing forward. This will prevent rain water from entering the oil reservoir through the breather hole in the cap.
   4) If the brake fluid is poured into the oil reservoir quickly, air may be trapped in the brake lines. Pour the fluid slowly.

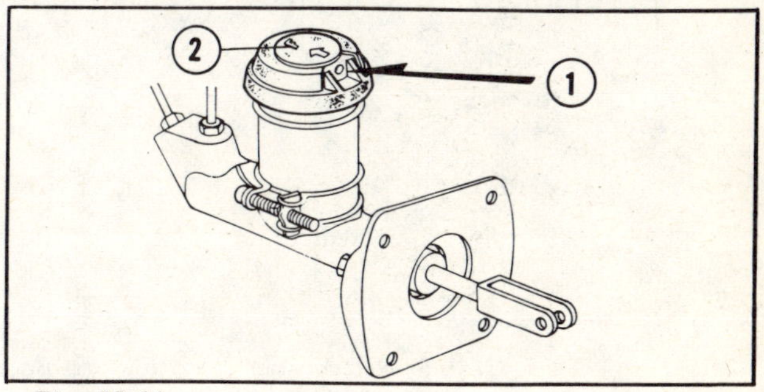

Fig. M-14
① Breather ② Mark "F"

4. Bleed the brake lines, beginning on a wheel cylinder which is farthest from the master cylinder. Continue bleeding until air bubbles are no longer noted in the fluid coming out of the bleeder hole.

● **Bleed the brake lines.**

5. Check for fluid leak and braking effect.

   NOTE:

   Carefully check the brake pipe joint for leak. The braking effect must be checked by using a brake tester or by driving the car. Again check for leak.

Master Power

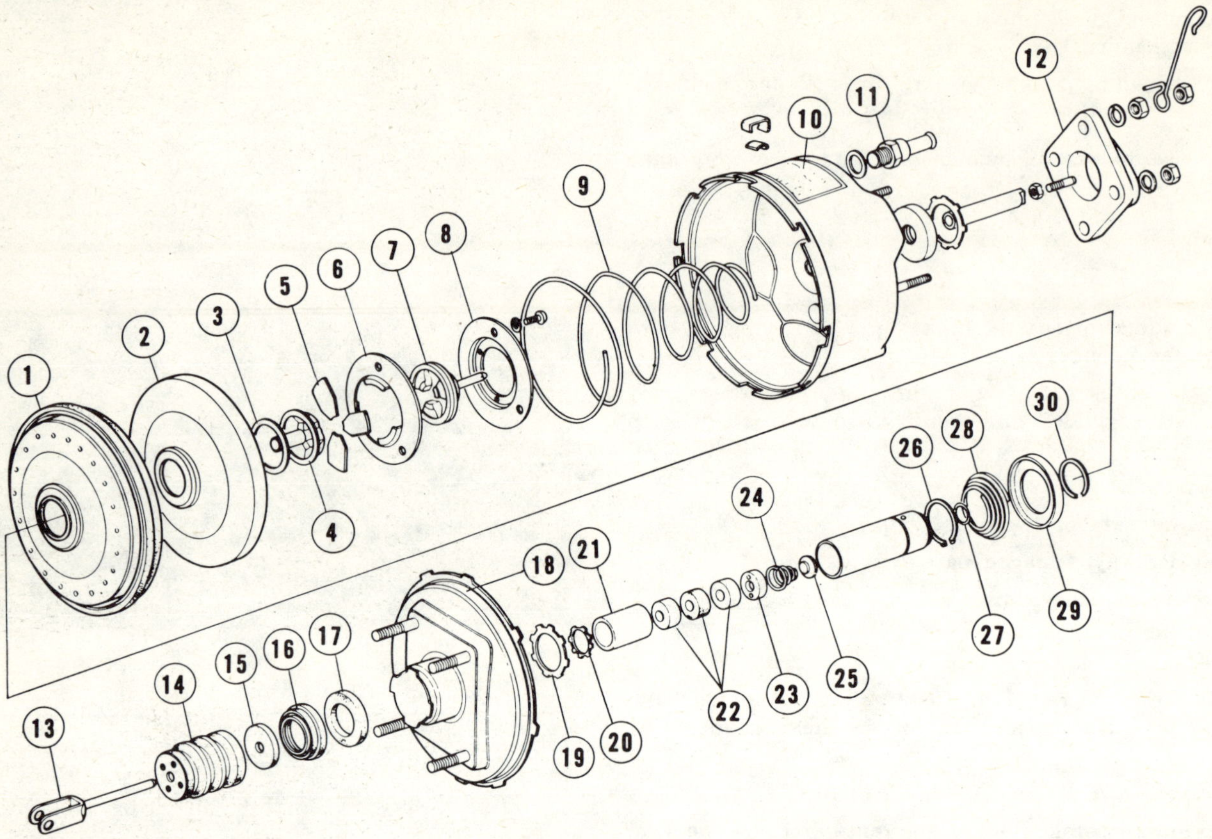

Fig. M-15 ① Diaphragm ② Booster plate ③ Wave spring ④ Center seat ⑤ Reaction plate ⑥ Set cover ⑦ Reaction ring ⑧ Set cover ⑨ Booster spring ⑩ Booster body ⑪ Check valve ⑫ Set block ⑬ Push rod ⑭ Boot ⑮ Filter ⑯ Piston seal ⑰ Bushing ⑱ Booster housing ⑲ Bushing stopper ⑳ Stop washer ㉑ Collar ㉒ Elements ㉓ Push plate ㉔ Spring ㉕ Poppet valve ㉖ Piston stopper ㉗ Center seal ㉘ Valve spring ㉙ Diaphragm holder ㉚ Spring stopper

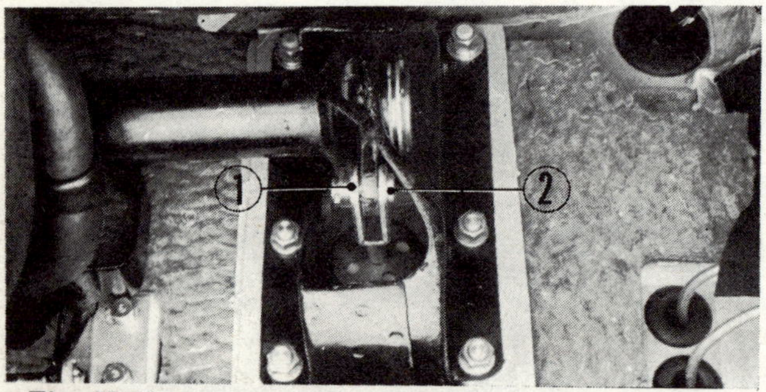

Fig. M-16
① Push rod ② Pedal pin

Fig. M-17

a. Removal

1. Disconnect the vacuum hose at the master power.
2. Disconnect the brake pipes at the master cylinder.

   NOTE:

   Using a waste cloth, prevent brake fluid from coming in contact with other parts.
3. Pull out the push rod by removing the pedal pin from inside the car.
4. Remove the master power together with the master cylinder by removing the master nus.

b. Disassembly

Precautions for disassembly

- Wipe dust and dirt off the master power. Take care not to allow dirt or water to get inside the master power.
- Use the repair kit. Replace rubber parts with new ones.
- Keep the disassembled parts free from dirt and dust.

b-1. Booster body

1. Remove the master cylinder from the master power.

Fig. M-18

2. Remove the lock spring securing the booster body to the housing, and remove the lock plate.

NOTE:

Before disassembling, make a chalk mark across the booster body and housing so that they can be assembled in their original position.

3. Remove the adjusting bolt from the booster body.

Fig. M-19

4. Hold the parallel portion of the booster body set block in a vise.
5. Install the overhaul handle to the booster housing and attach the nuts on the studs. Turn the handle counter-clockwise until the fasteners securing the booster body to the housing is disengaged.
   Then remove the booster body.

NOTE:

The booster housing is loaded by the return springs, and it may jump when the fasteners are disengaged. Turn the handle while holding it down with pressure.

Fig. M-20

6. Remove the seal clip and rod seal from the booster body.
7. Remove the check valve.

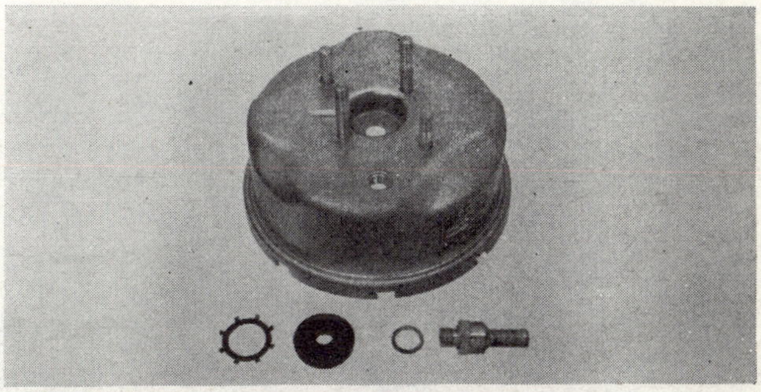

Fig. M-21

b-2. Booster housing

1. Remove the push rod and the rod boot.
2. Remove the booster housing from the valve piston.

Fig. M-22

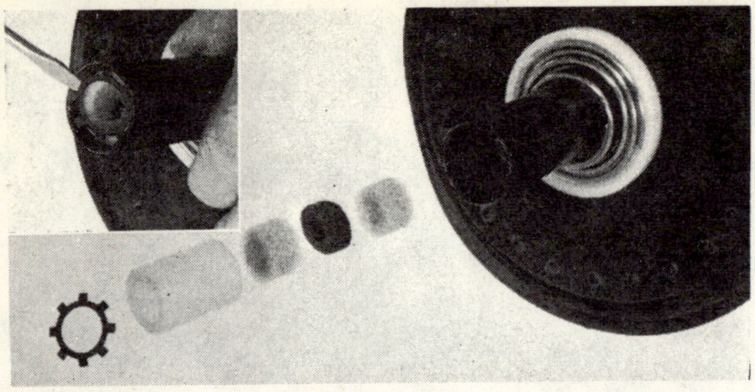

Fig. M-23

3. Remove the stop washer, filter collars and filter from the valve piston.

   NOTE:

   Do not pry off the washer by using a tool like a screwdriver as shown.

Fig. M-24

4. Remove the push plate, poppet spring, poppet valve in this order from the valve piston.

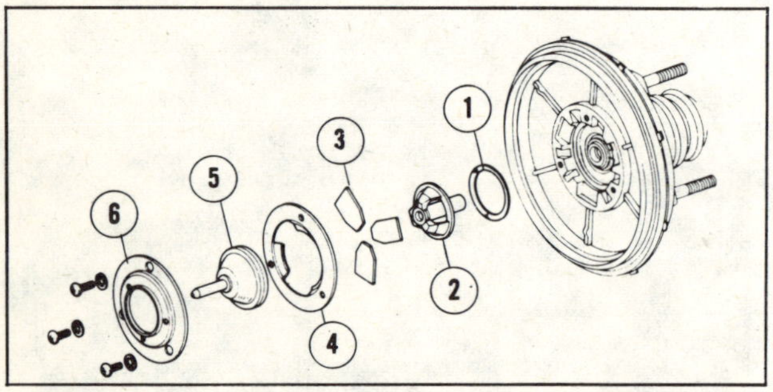

Fig. M-25
① Wave spring ③ Reaction plate ⑤ Reaction ring
② Center seat ④ Spring seat ⑥ Set cover

5. Remove the spring seat, reaction ring, center seat, wave spring, set cover and reaction plate in this order from the booster plate by loosening the spring seat screws.

Fig. M-26

6. Remove the piston stopper and center seal from the valve piston.

   NOTES:

   1) Do not drop the valve piston or apply shock to it. Take care not to damage the sliding surfaces of the piston.
   2) Do not allow oil or grease to come in contact with the diaphragm.

7. Remove the bushing stopper, bushing and piston seal in this order from the booster housing.

c. Inspection

Before inspection, thoroughly wash the disassembled parts in absolute alcohol (for fuel) and dry them.

NOTE:

Never wash brake parts in mineral oil such as gasoline.

Inspection item	Remedy
Deformatioh or other damage of booster body and housing	Replace.
Bend or other damage of push rod	Replace.
Scratches (on sliding surface) or other damage of booster plate	Replace.
Wear of reaction plate, reaction ring and center seat	Replace.
Bend or other damage of adjusting bolt and output rod	Replace.
Deformation or other damage of valve piston	Replace.
Deformation or other damage of springs	Replace.

Repair kit

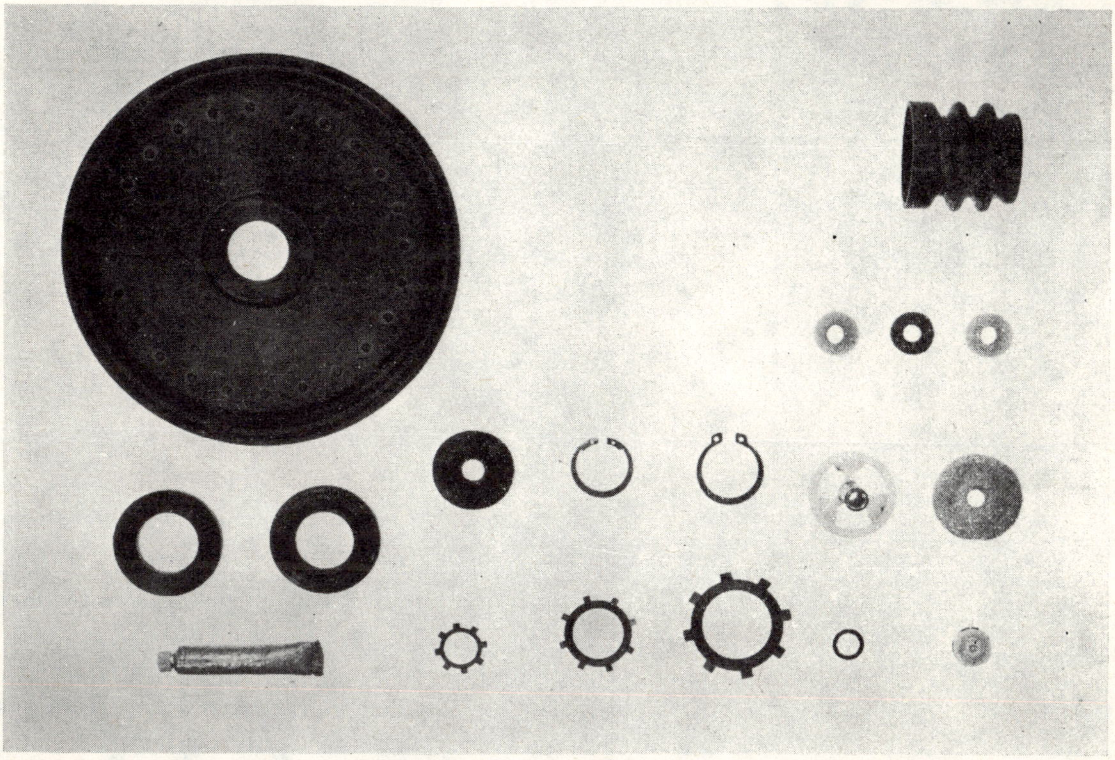

Fig. M-27

d. Assembly

To assemble, reverse the disassembly procedures. Pay attention to the following points:

Precautions for service work

1. Too much emphasis cannot be placed on the importance of keeping master power parts (rubber parts and clips) in the best possible condition at all times for your safe driving.
2. Immediately replace any damaged part with a new one upon discovery of it—even if damage is slight. Remember, "A slightly damaged part will be a seriously damaged part after a few minutes of driving."
3. Be sure to replace all rubber parts with new ones at reassembling. Use the repair kit which contains 17 parts.
4. Take care not to allow foreign materials to enter the master power parts during service work.

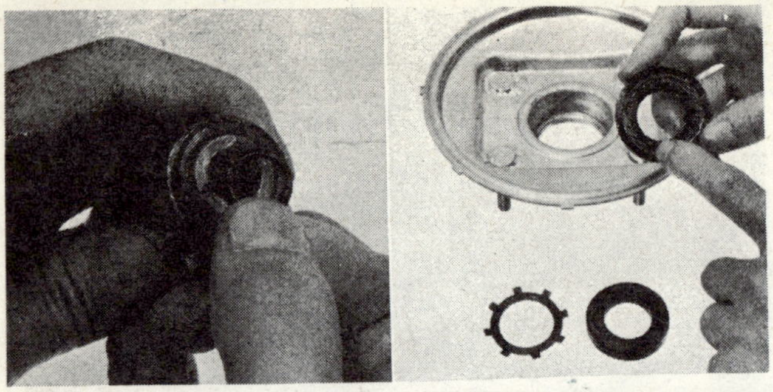

Fig. M-28

1. When installing the piston seal to the booster housing, apply a sufficient coat of silicon grease to the back and lip of the seal.
2. Push the bushing stopper until it lightly bears on the bushing. Be sure to seat it evenly in the recess.
3. Apply a coat of silicon grease to the inside of the valve piston and to the center seal, and install the center seal.

Fig. M-29

4. When installing the diaphragm to the booster plate, apply a coat of silicon grease to it at the points shown. Also apply a coat of silicon grease in the holes in the booster plate.

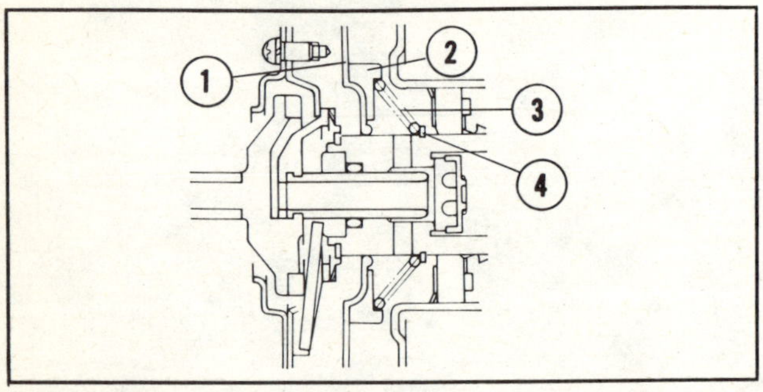

Fig. M-30
① Diaphragm ② Holder ③ Spring ④ Stopper

5. Install the valve spring, diaphragm holder, diaphragm, booster plate and piston stopper in this order to the valve piston. Install the piston stopper while pushing down the booster plate so as to compress the valve piston. At this time, make sure that the stopper is properly seated in the groove in the piston.

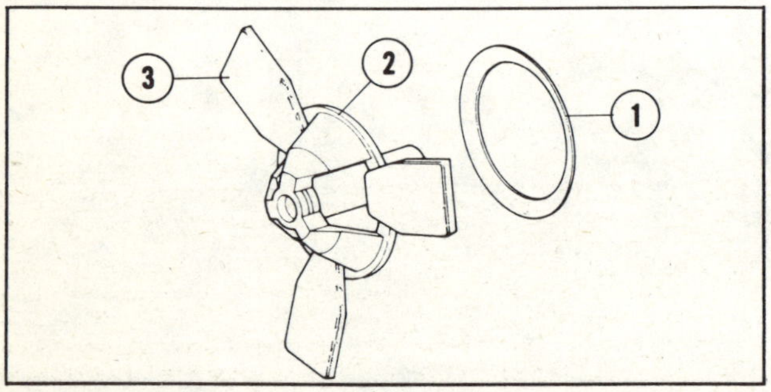

Fig. M-31
① Wave spring ② Center seat ③ Reactioh plate

6. Install the wave spring, center seat, reaction plate, set cover, reaction ring and spring seat in this order to the booster plate, taking care of the following points:
   - Be careful of the installation direction of the wave spring and reaction plate. The reaction plate should be installed with the machined side (R side) facing the front side.
   - Apply a thin coat of silicon grease to the center seat except the top round portion.
   - After installing, make sure that the reaction plate moves smoothly.

Fig. M-32

7. Install the poppet valve, poppet spring, push plate, filter cover, filter and stop washer in this order to the valve piston.
8. Apply a sufficient coat of silicon grease to the periphery and lip of the body seal and install it to the booster body.

   Secure the seal with the clip and then install the set block.

9. Apply a coat of silicon grease to the outside surface of the valve piston, and install the piston assembly in the booster housing.
10. Put the booster spring in the booster body and install the booster housing to the body. To install, align the markings which were made at the time of disassembly. Using the overhaul handle, turn the booster body clockwise until the lug of the housing is properly fit in the groove in the body stopper.
11. After tightening the booster body, secure it with the lock plate and lock spring.
12. Apply a sufficient coat of silicon grease to the outside part of the valve piston which is protruded beyond the booster housing. Install the boots to the housing and then insert the push rod.
13. Insert the output rod and adjusting bolt in the booster body. Before installing the master power and master cylinder, measure the clearance between the master cylinder piston and adjusting bolt using an adjusting bolt gauge as shown. If the clearance is out of specification, adjust the length of the adjusting bolt.

Specification: 0~0.6 mm (0.0236 in.)

On the inspections and testing on the performance of the master power after installation,

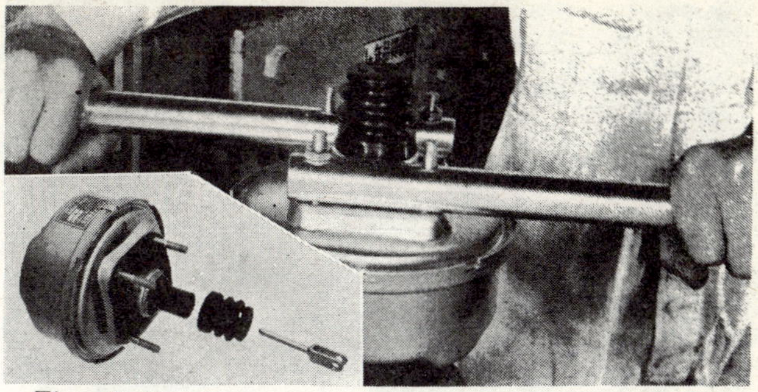

Fig. M-33

Brake Hoses and Pipes

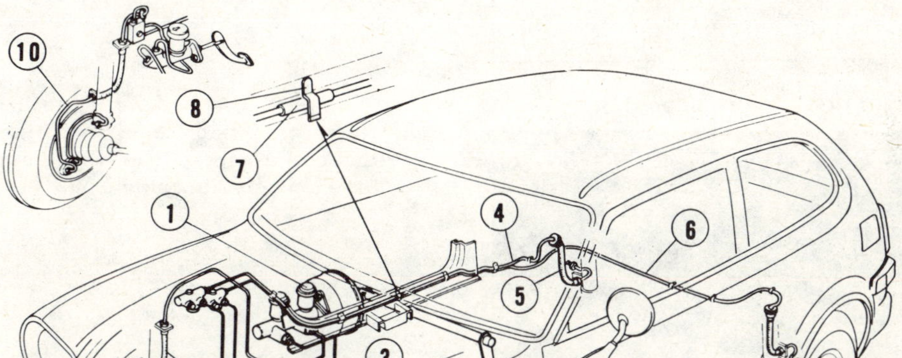

Fig. M-34

① Master cylinder
② Front brake hose
③ Brake pipe
④ Brake pipe
⑤ Rear brake hose
⑥ Brake pipe
⑦ Protector
⑧ Clip
⑨ For disc brake type
⑩ For drum brake type

Inspection before removal

- Check the brake hoses for twist or any other damage and for brake fluid leak.
- Check the brake pipes for collapse, bend, corrosion or any other damage.
- Carefully check the brake parts which are open to splash of mud or skipping stones.

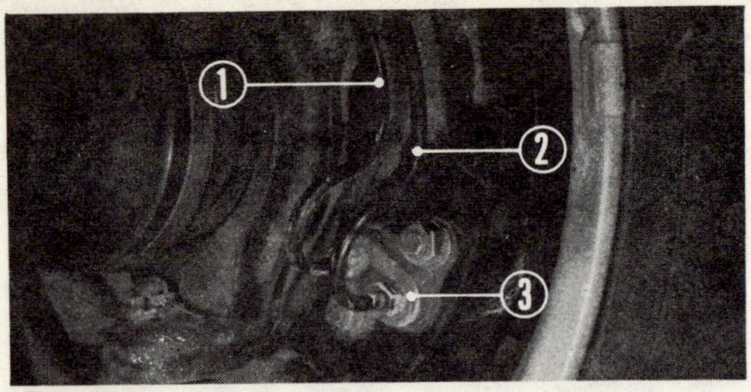

Fig. M-35
① Front wheel cylinder pipe ② Brake pipe G
③ Wheel cylinder

a. Disconnection

- When removing the flare nuts, be sure to use a flare nut wrench to prevent damage of the nuts.
- Disconnect the front wheel cylinder pipes, brake pipes G and front brake hoses.

1. Disconnect the brake pipe at the brake hose.

   NOTE:

   After disconnecting, plug the pipe to prevent the brake fluid from coming out.

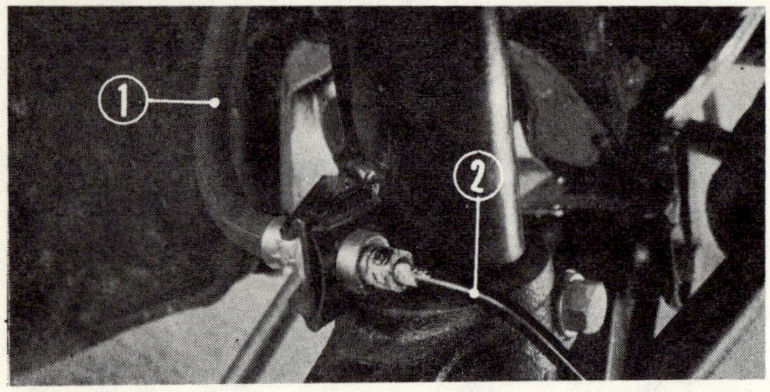

Fig. M-36
① Front brake hose ② Brake pipe

2. Disconnect the brake hoses at the brake pipes.
3. The pipes can easily be disconnected by loosening the right and left flare nuts first.

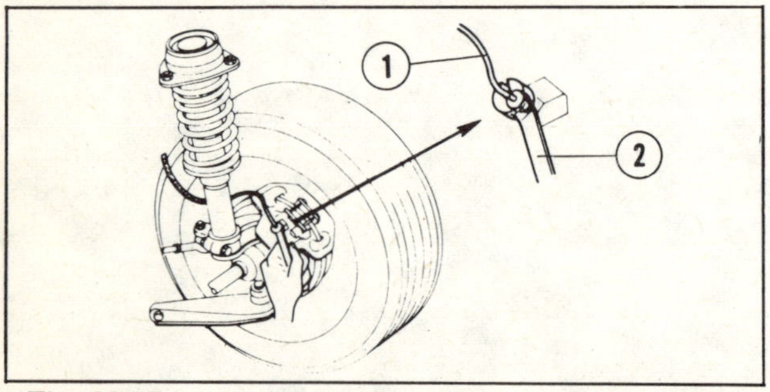

Fig. M-37
① Brake pipe ② Flare nut wrench

4. Remove the brake pipe proportioning valve. Disconnect the brake pipes. Remove the proportioning valve by loosening its attaching bolts.

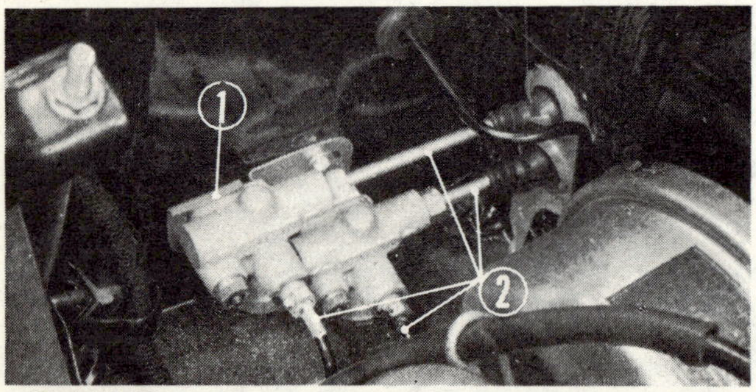

Fig. M-38
① Proportioning valve ② Brake pipe

b. Connection

To connect the brake hose and pipe, reverse the disconnecting procedures. Pay attention to the following points:

1. Connect the pipe securely with a flare nut wrench.
2. Connect the brake hoses so that they are not twisted or that they do not interfere with other parts. (Take care not to twist the product marking or white line on the hose.)

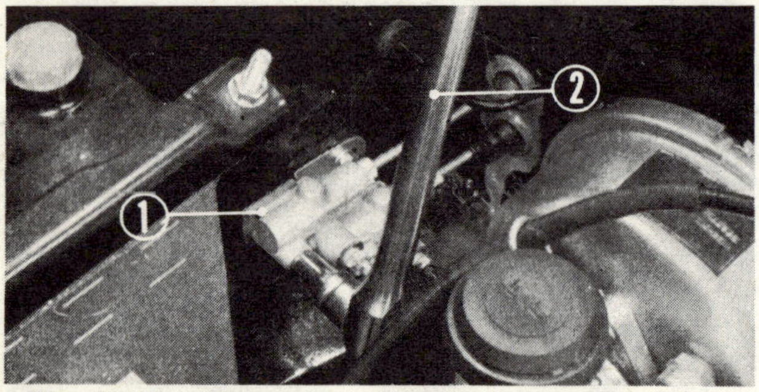

Fig. M-39 Tightening proportioning valve
① Proportioning valve ② Universal wrench

3. Turn the steering wheel in both direction and bounce the body to check if the brake hoses and pipes interfere with the wheel rims.
4. Bleed the brake lines.

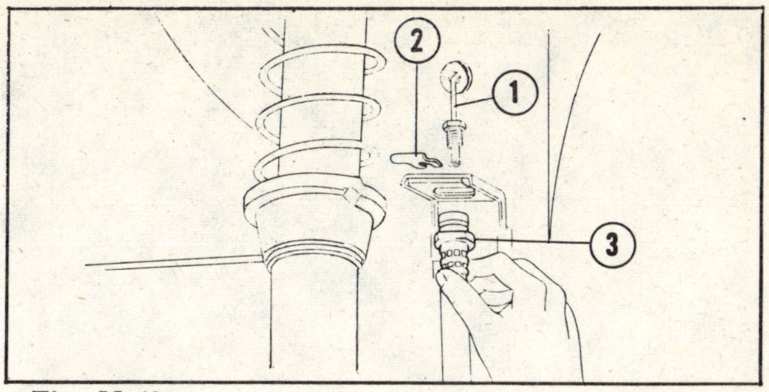

Fig. M–40
① Brake pipe A ② Brake hose clip
③ Front brake hose

Front Brakes
(On a car equipped with disc brakes)

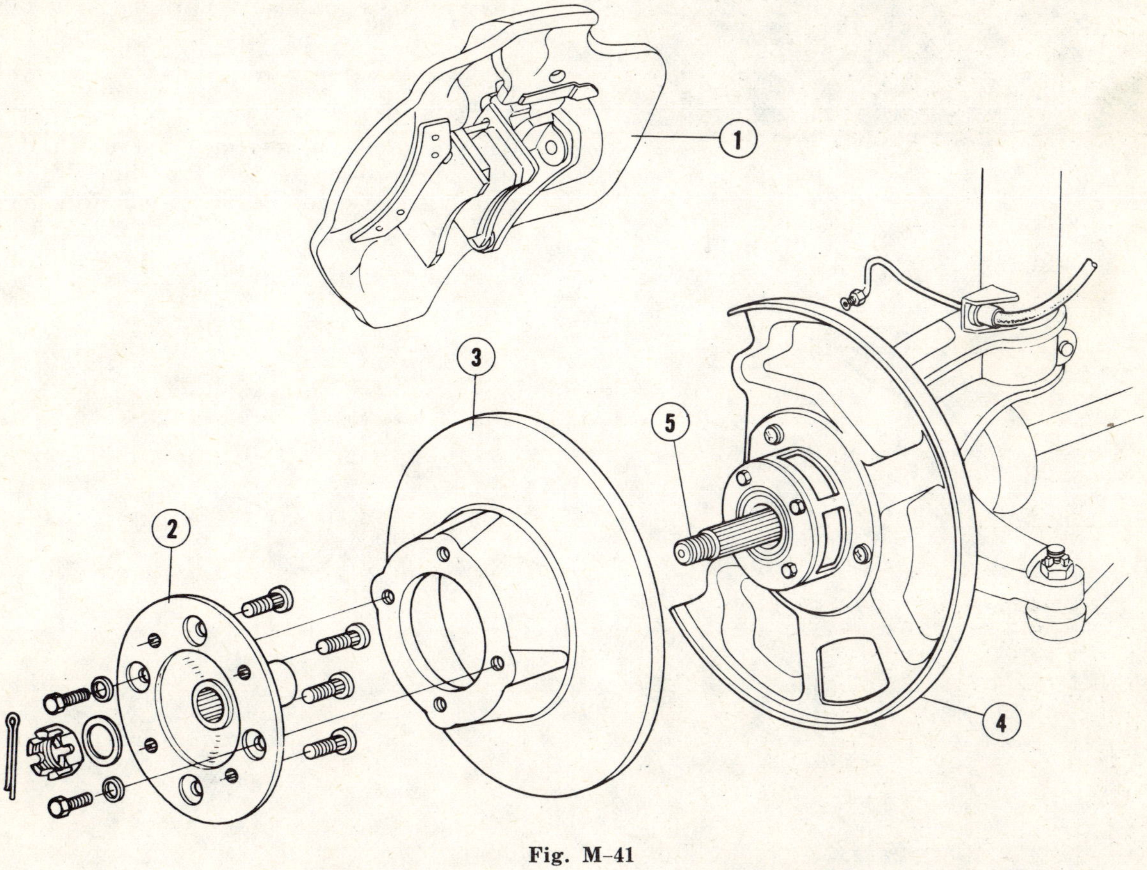

Fig. M–41

① Caliper assembly
② Front hub
③ Front brake disc
④ Back plate
⑤ Drive shaft

a. Removal and disassembly

1. Loosen the wheel clip nuts.
2. Remove the wheel center cap.
3. Pull out the cotter pin and loosen the spindle nut.

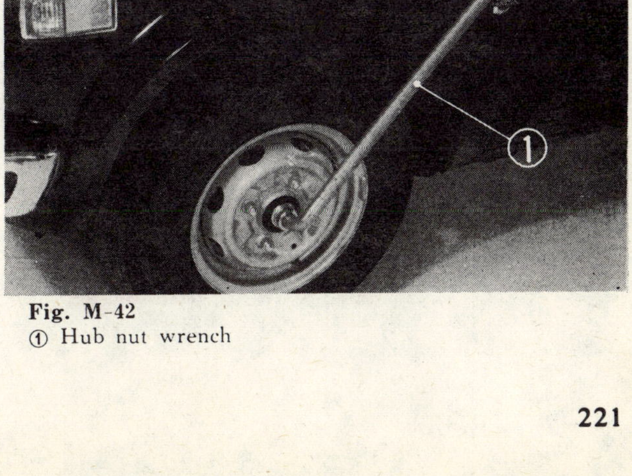

Fig. M–42
① Hub nut wrench

Fig. M-43
① Spindle nut

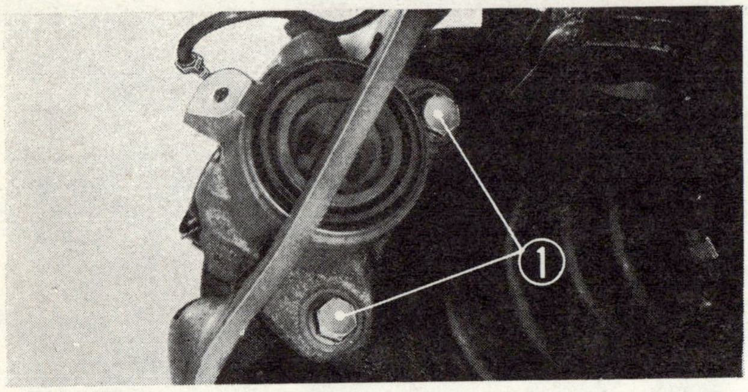

Fig. M-44
① Caliper attaching bolts

4. Jack up the car and support it with safety stands.

5. Remove the wheel clip nuts and remove the wheel.

6. Remove the spindle nut.

a-1. Removal of calipers

1. Remove the caliper assembly by loosening its attaching bolts.

2. Remove the pad retaining clip and pins. Then remove the pad set.

3. Move the yoke springs A and B, and remove the cylinder body.

4. Remove the piston and its seals from the cylinder body.

NOTE:

When it is difficult to remove the piston, direct a blast of compressed air to the piston through the brake pipe connection. Do not use pliers; otherwise the piston may be damaged.

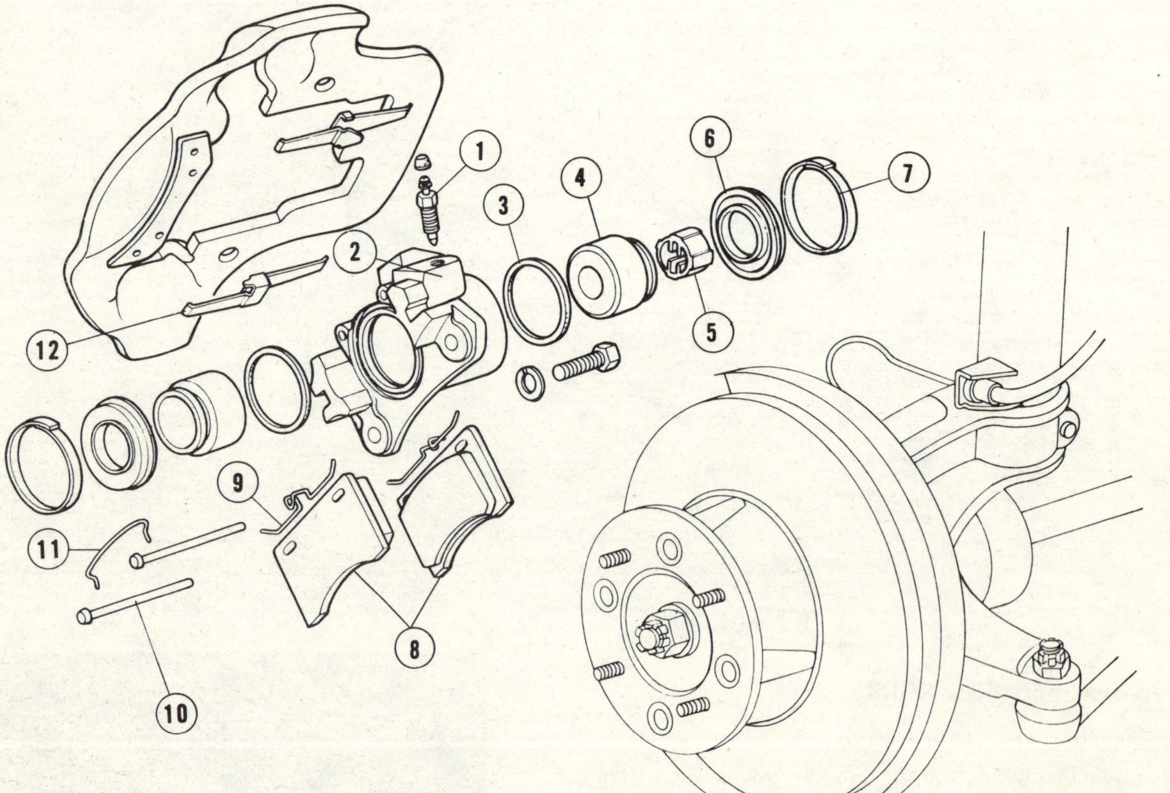

Fig. M-45

① Bleeder screw
② Cylinder
③ Piston seal
④ Piston
⑤ Bias ring
⑥ Boot
⑦ Retaining ring
⑧ Pad set
⑨ Pad spring
⑩ Pad retaining pin
⑪ Pad retaining clip
⑫ Yoke spring

a-2. Removal of brake discs

After removing the caliper assembly, remove the front hub and brake disc as a unit.

1. Install a hub puller attachment to the front hub.
2. Install a sliding hammer to the hub puller attachment and remove the front hub as shown.

   NOTES:

   1) Install the hub puller attachment to the front hub with the wheel clip nuts securely.
   2) When the front hub is disassembled, replace the wheel bearing.

3. Remove the brake disc from the front hub.

   NOTE:

   A small amount of oil or slight damage on the brake disc may affect the braking effect. Use soft, lint-free cloth to protect the disc from being damaged.

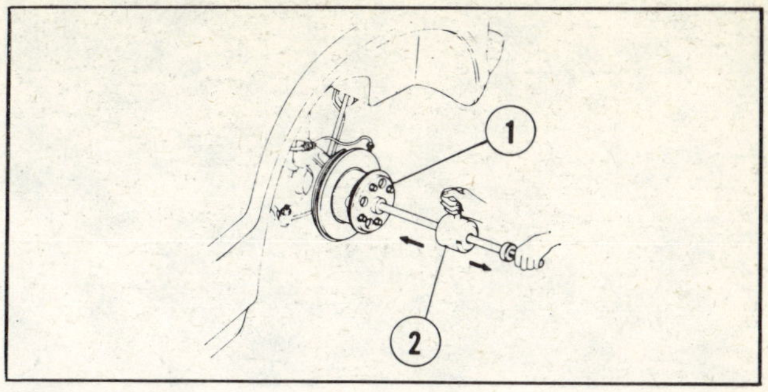

Fig. M-46
① Hub puller ② Sliding hammer

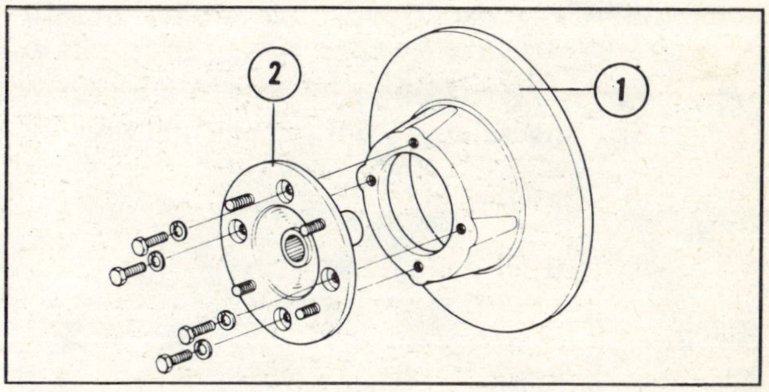

Fig. M-47
① Front brake disc ② Front hub

b. Inspection

See the Maintenance Standard.

Inspection item	Remedy
Wear or other damage of brake disc	Replace.
Wear of pads	Replace.
Damage of piston seal	Replace.
Breakage or other damage of boots	Replace.
Damage of retaining rings	Replace.
Damage of pad springs A and B	Replace.
Wear of pad retaining pins	Replace.
Piston-to-cylinder clearance	Replace.
Damage of yoke springs A and B	Replace.

c. Assembly and installation

To assemble the front brake, reverse the removal and disassembly procedures.

1. To install the wheel bearing to the knuckle, see page 230.
2. First install the brake disc to the front hub to the specified torque. Then install the front hub in the knuckle.
3. Install the caliper assembly to the knuckle to the specified torque.

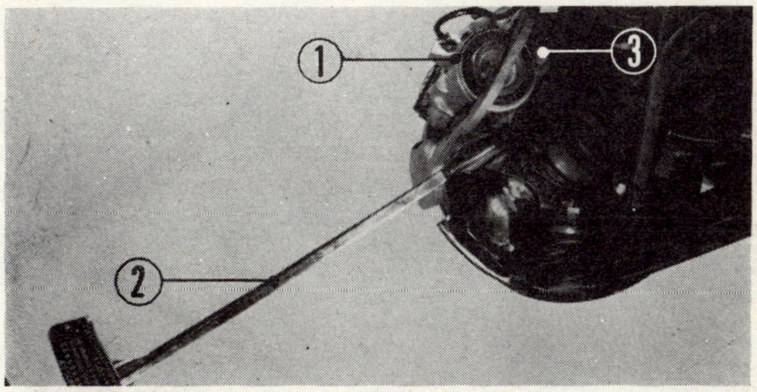

Fig. M-48
① Caliper assembly ② Torque wrench
③ Caliper attaching bolt

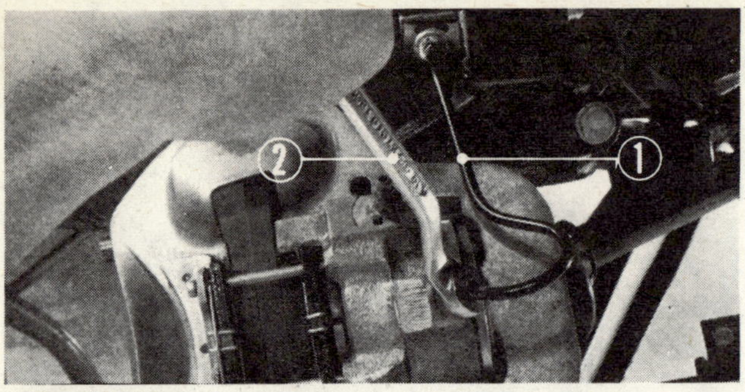

Fig. M-49
① Brake pipe ② Flare nut wrench

4. Check to be sure the pad set and disc are not damaged and free from oil or grease, and install them in place.
5. Connect the brake pipe to the brake cylinder body to the specified torque.
6. Bleed the brake lines.
7. Apply pressure to the pipe line and check the pipe joint for leak.
8. After installing, check for braking effect or abnormal braking by using a brake tester or by driving the car.

Front Brakes
(On cars equipped with drum brakes)

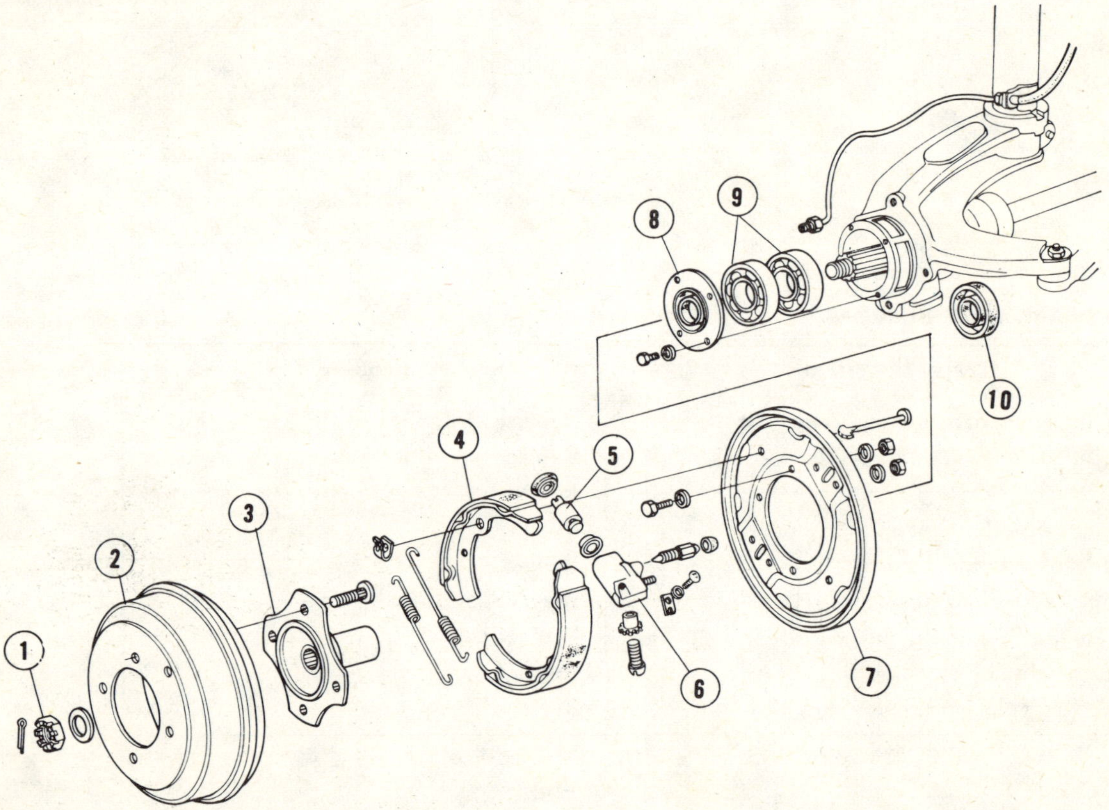

Fig. M-50 ① Spindle nut
② Front brake drum
③ Front hub
④ Brake shoe
⑤ Front piston
⑥ Front cylinder
⑦ Front back plate
⑧ Front bearing dust seal B
⑨ Front wheel bearings
⑩ Front bearing dust seal A

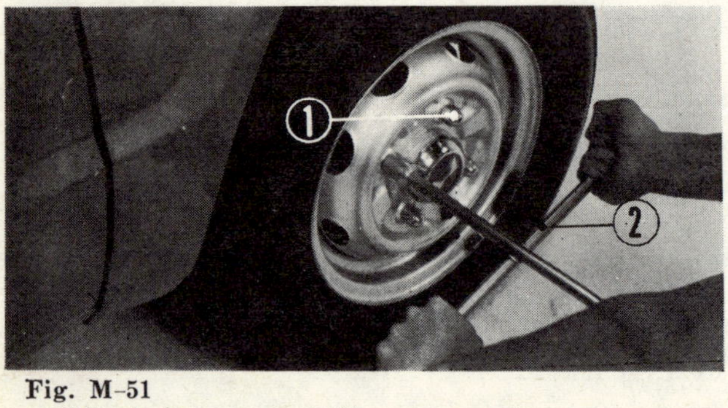

Fig. M-51
① Wheel clip nut ② Clip nut wrench

a. Removal and disassembly

1. Loosen the wheel clip nuts.
2. Jack up the front of the car and support it with safety stands.
3. Remove the clip nuts and then the wheel.
4. Remove the brake drum. Lightly tap the brake drum with a mallet. Then the drum can be removed with hands.

5. Remove the brake shoes. Remove the shoe clamp springs and pull out the tension pins. Then expand the brake springs and remove the brake shoes.
6. Disconnect the brake pipe at the wheel cylinder.

NOTE:

After disconnecting, plug the brake pipe to prevent the brake fluid from coming out.

Fig. M-52
① Brake shoe ② Shoe clamp spring
③ Tension pin ④ Brake spring

7. Remove the wheel cylinder by loosening its attaching nuts at the back of the backing plate.

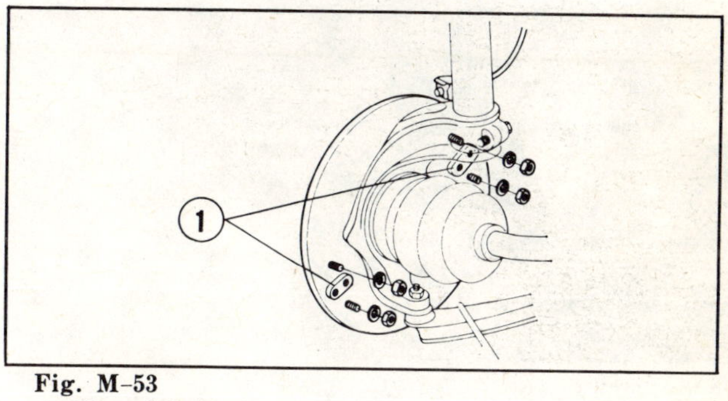

Fig. M-53
① Thread holes

8. Remove the dust cover. Remove the piston set from the wheel cylinder. Remove the wheel adjuster, and the adjusting screw can be removed.

NOTES:

1) Wash the parts disassembled in absolute alcohol (for fuel) quickly.
2) To remove the backing plate, remove the front hub.

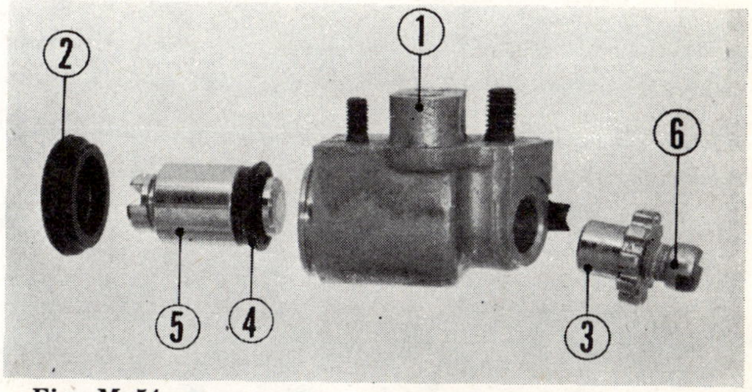

Fig. M-54
① Front wheel cylinder ④ Front piston cup
② Dust cover ⑤ Front piston
③ Wheel adjuster ⑥ Adjusting screw

b. Inspection

See the Maintenance Standard at the end of this manual.

Inspection item	Remedy
Deformation or other damage of piston cups and dust seal	Replace.
Outside diameter of piston	Replace.
Cylinder bore	Replace.
Damage of piston and cylinder	Replace.
Inside diameter and out of round of drum	Replace.
Lining thickness	Replace.

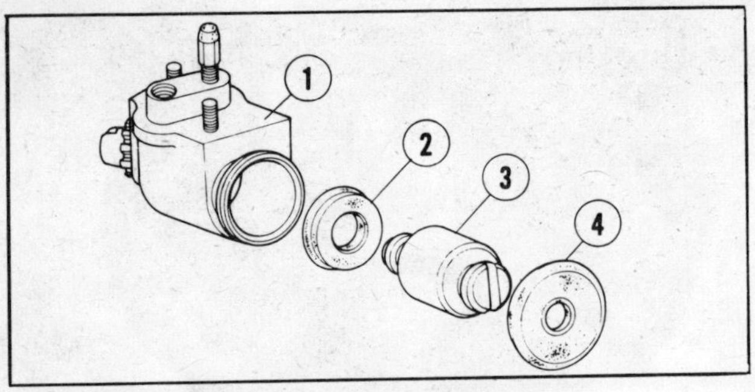

Fig. M-55
① Front cylinder ② Front piston cup
③ Front piston ④ Dust cover

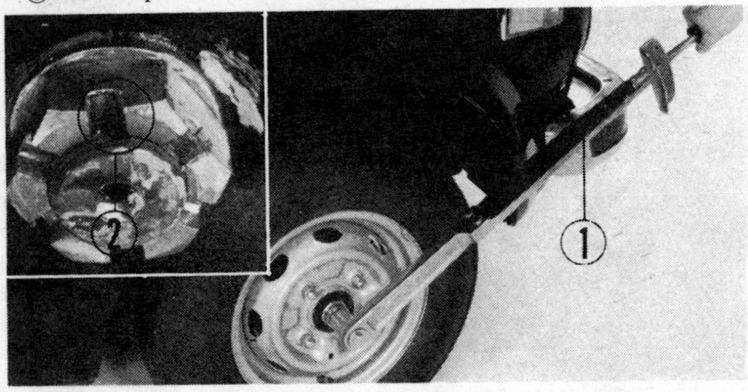

Fig. M-56
① Torque wrench ② Do not turn nut back

c. Assembly and installation

To assemble and install the front brake, reverse the removal and disassembly procedures. Pay attention to the following points :

1. Assemble the wheel cylinder. Apply a coat of fresh brake fluid to the piston, piston cups and cylinder bore. Put those parts together in the cylinder, taking care not to damage the piston cups.
2. Check to be sure the wheel adjuster stopper does not loosen and install the wheel cylinder assembly to the back plate.
   - Install the front hub, if necessary, to the knuckle refering to page to **page 230**.
3. Before installing the brake drum, check to be sure oil, grease or foreign materials are not accumulated on the drum sliding surface.
4. Tighten the front spindle nut with a torque wrench.

NOTES :

1) Turn the nut until the castelation and the pin hole are aligned. Be careful not to exceed the specified torque.
2) Do not align the holes by backing off the nut.
3) For the tightening torque, see the Torque Specifications.

Adjustment of brakes, see page 46.

Rear Brakes

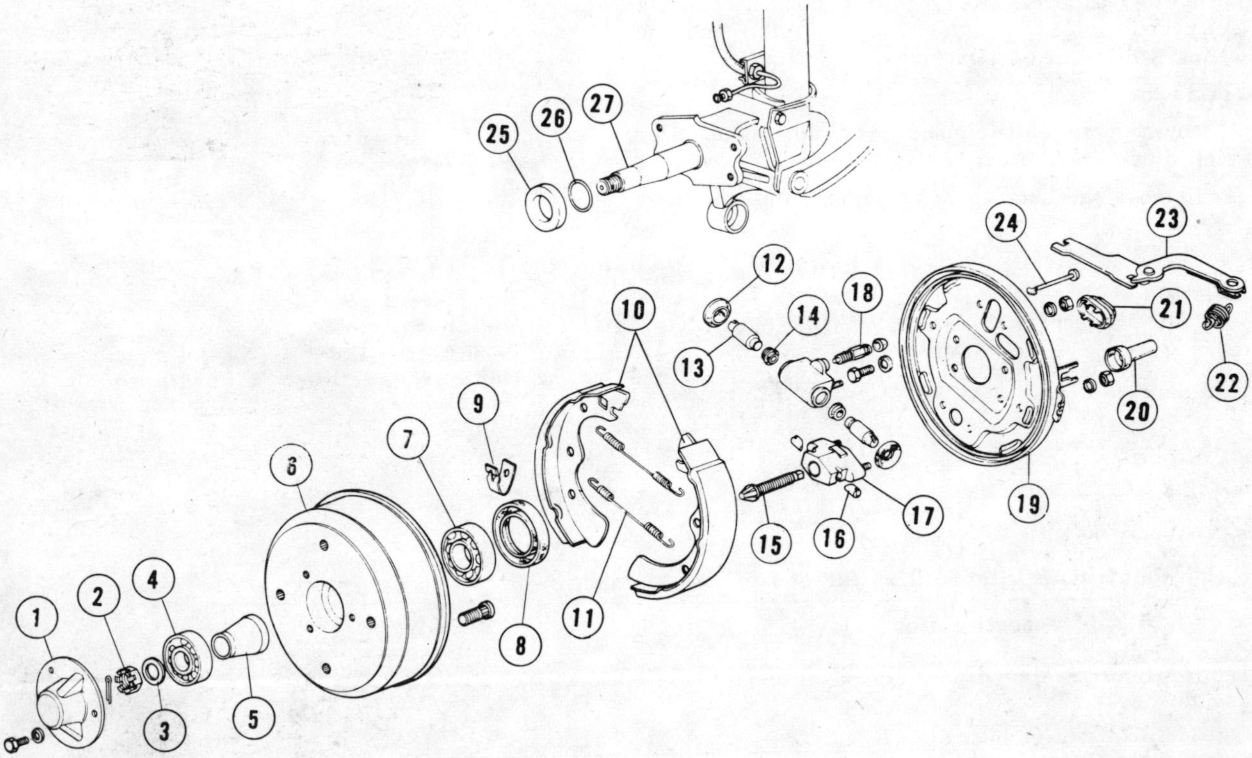

Fig. M-57

① Rear wheel bearing cap
② Rear axle nut
③ Rear hub washer
④ Rear wheel bearing A
⑤ Hub carrier spacer
⑥ Rear brake drum
⑦ Rear wheel bearing B
⑧ Rear wheel bearing oil seal
⑨ Shoe clamp spring
⑩ Shoe complete
⑪ Brake spring
⑫ Dust cover
⑬ Piston
⑭ Piston cup
⑮ Adjuster bolt
⑯ Brake adjuster piston
⑰ Brake adjuster
⑱ Bleeder screw
⑲ Rear back plate
⑳ Adjuster cap
㉑ Lever seal cap
㉒ Return spring
㉓ Lever
㉔ Tension pin
㉕ Collar
⑥ O-ring
㉗ Hub carrier

a. Removal and disassembly

1. Loosen the rear wheel clip nuts.
2. Fully pull up the parking brake lever. Jack up the rear of the car and support it with srfety stands.
3. Remove the wheel and bearing cap. Remove the cotter pin, rear axle nut and drum in this order.

NOTES:

1) When removing the drum, release the parking brake.
2) When it is difficult to remo:ve the drum, use a hub puller attachment and a sliding hammer.

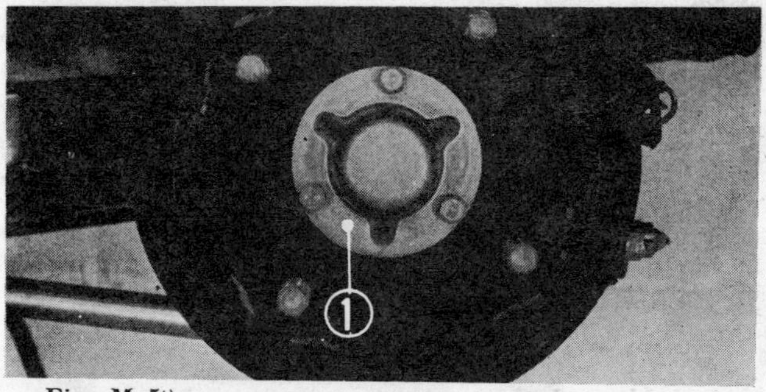

Fig. M-58
① Rear wheel bearing cap

4. To remove the brake shoes, remove the tension pins and expand the brake springs as in the case of the front brake.

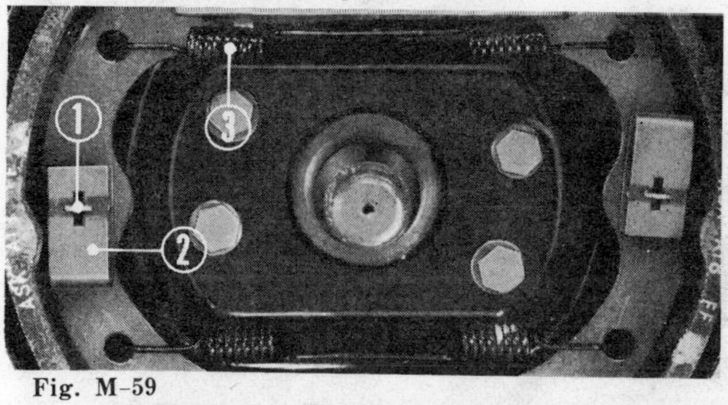

Fig. M-59
① Tension pin ② Shoe clamp spring
③ Brake spring

5. Disconnect the parking brake control cable and rear brake pipes at the back of the backing plate.
6. Loosen the backing plate attaching bolts and remove the backing plate together with the wheel cylinder from the rear hub carrier.

Fig. M-60
① Rear wheel cylinder ② Brake adjuster
③ Rear back plate

7. Remove the wheel cylinder by loosening its attaching nuts at the back of the backing plate.

NOTE:

The pistons can be removed by removing the dust covers in the grooves in the wheel cylinder.

Fig. M-61
① Rear wheel cylinder ② Piston ③ Dust cover
④ Piston cup ⑤ Attaching nut ⑥ Spring washer

8. Remove the brake adjuster by loosening its attaching nuts at the back of the backing plate. The adjuster piston set can be removed from the body by screwing in the adjuster bolt fully.

Fig. M-62
① Brake adjuster ② Adjuster bolt
③ Brake adjuster pin

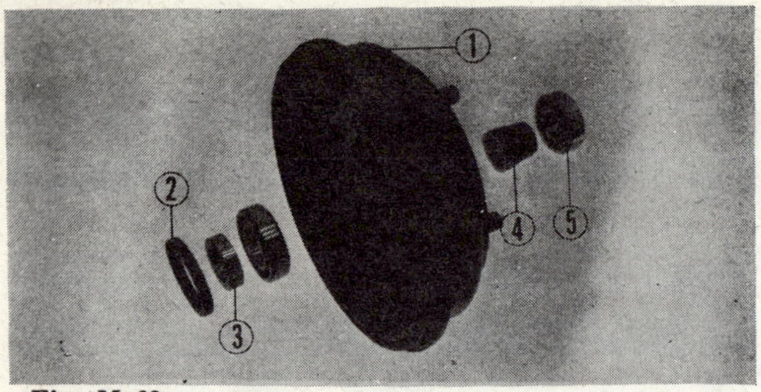

Fig. M-63
① Rear brake drum ② Rear wheel bearing oil seal
③ Hub carrier collar ④ Carrier spacer
⑤ Rear wheel bearing

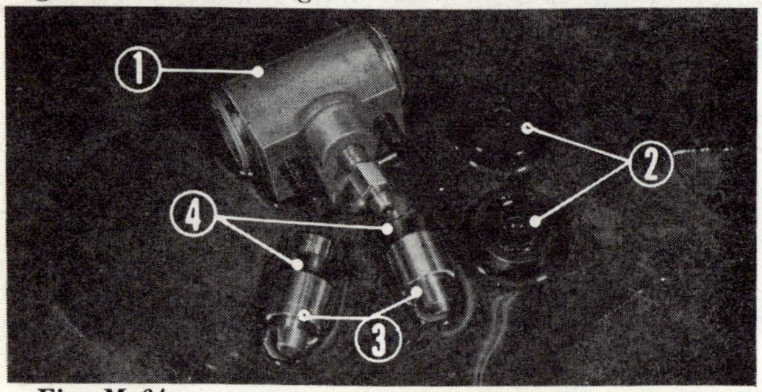

Fig. M-64
① Rear wheel cylinder ② Dust covers
③ Pistons ④ Piston cups

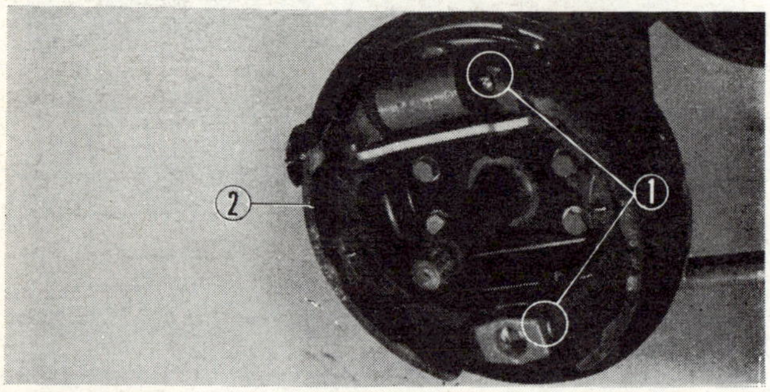

Fig. M-65
① Apply grease to these parts ② Brake lining

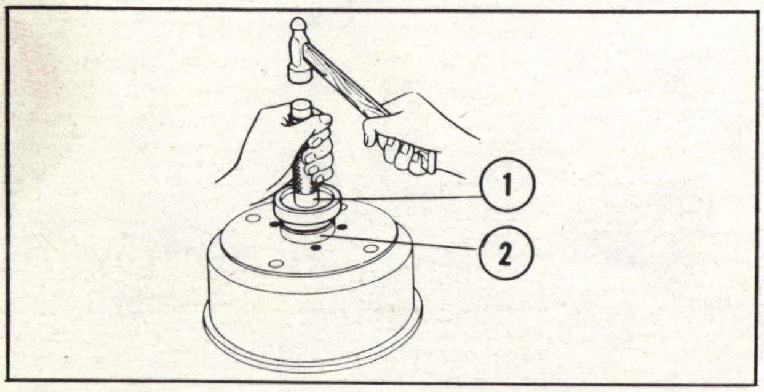

Fig. M-66
① Bearing installation jig ② Bearing

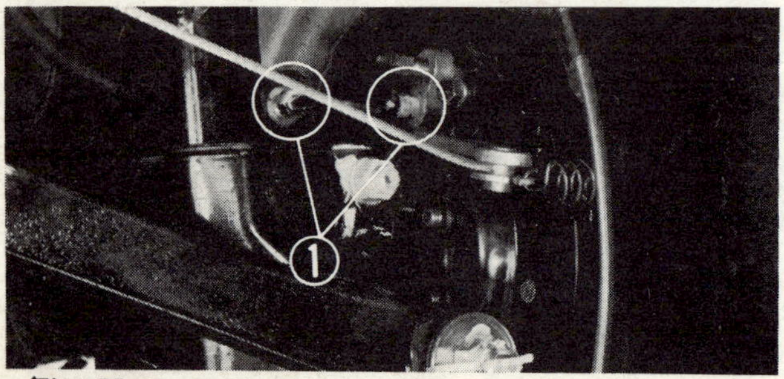

Fig. M-67
① Checking these parts for oil leak

9. Disassemble the drum. Remove the oil seal, bearings and distance collar in this order.

NOTE:

For the inspection items, see the paragraph for the front brakes.

b. Assembly and installation

To assemble and install the rear brake, reverse the removal and disassembly procedures. Pay attention to the following points:

1. Assemble the wheel cylinder. Apply a coat of fresh brake fluid to the pistons, piston cups and cylinder wall. Install those parts in the cylinder, taking care not to damage the piston cups.
2. Before installing, apply a thin coat of grease to the brake adjuster, grooves in the wheel cylinder pistons, and sliding surfaces of the backing plate and brake shoes.
3. Check to be sure the brake shoes are properly engaged in the grooves in the adjuster pistons.
4. Install the wheel bearings in place. Using the special tool, drive the outer race evenly. After installing, lubricate the bearings with grease. Drive the oil seal in the drum properly until it becomes flush with the drum surface.
5. Using a torque wrench, tighten the spindle nut.
6. Bleed the brake lines.
7. Adjust the rear brake.
8. Apply pressure to the pipe lines and check the joints of brake pipes for leak. Using a brake tester or driving the car, check the rear brake for proper operation.

Front Suspension

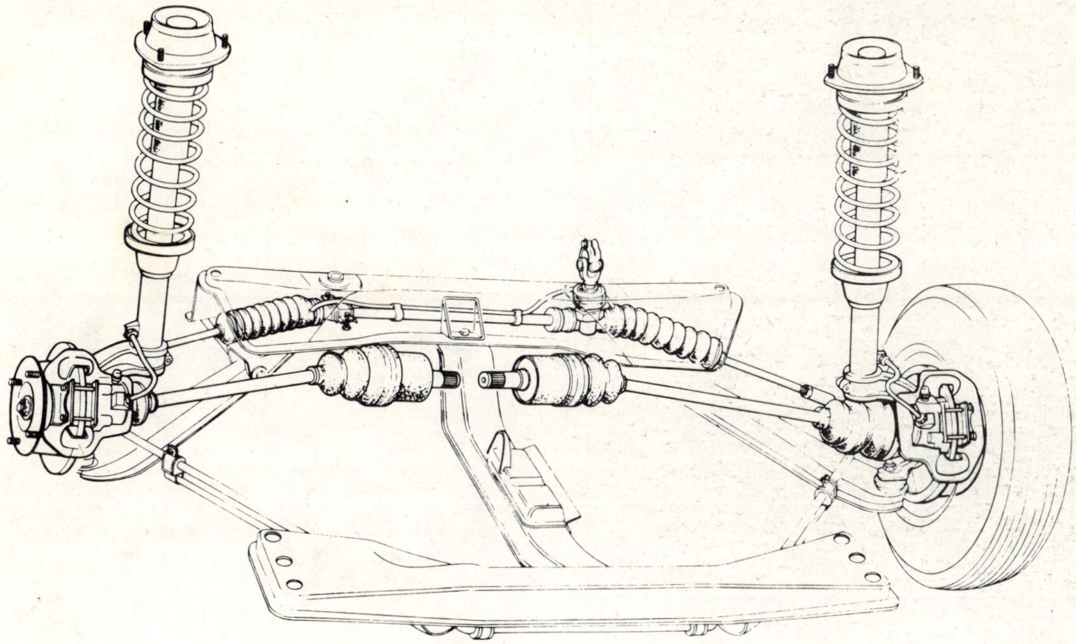

Fig. N–1 Front suspension

a. Knuckles

a–1. Removal

1. On a car equipped with drum brakes, remove the front wheel clip nuts, center wheel cap, spindle nut, wheel, brake drum, brake shoes and wheel cylinder.
2. Disconnect the brake pipe at the backing plate.
3. Using a ball joint puller, remove the tie rod end from the knuckle.
4. Remove the lower arm ball joint from the knuckle.
5. Remove the knuckle from shock absorber by loosening the absorber locking bolts.
6. Remove the knuckle together with the front hub from the drive shaft.

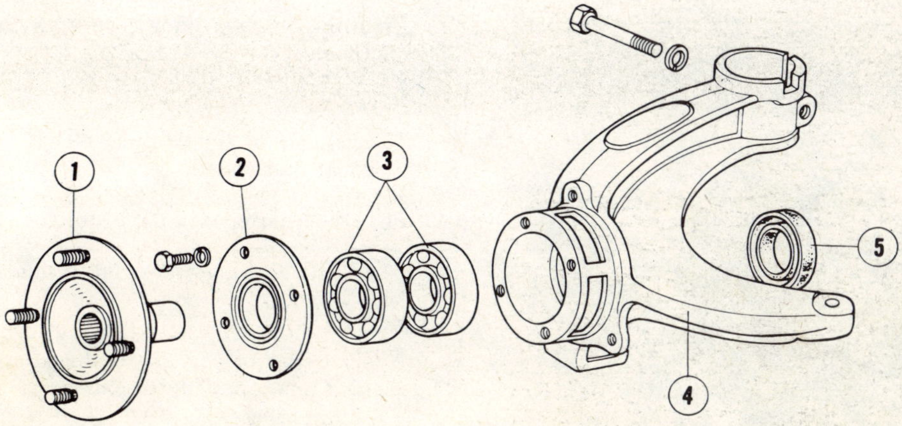

Fig. N–2 ① Front hub
② Front bearing dust seal B
③ Front wheel bearings
④ Knuckle
⑤ Front bearing dust seal A

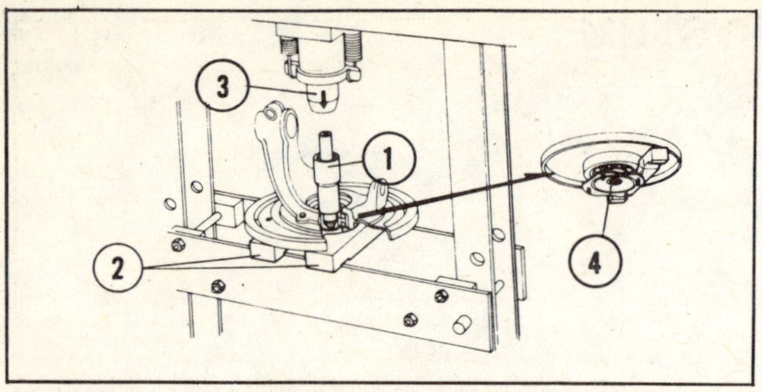

Fig. N-3
① Front hub disassembly jig pin A
② Front hub disassembly jig plate A
③ Hydraulic press ④ Front hub

a-2. Disassembly

● **Removal of front hub and wheel bearings**

1. Remove the wheel bearing dust seal A.
2. Position the plate A of the front hub disassembly jig, pin A and knuckle on a hydraulic press as shown. Carefully remove the front hub with the press.

   NOTE:

   Properly install the plate A in place.
3. Remove the four backing plate bolts.
4. Remove the bearing dust seal B.
5. Remove the wheel bearings.

   Install the plate A in place in such a way that it does not contact with the bearings.
6. Install the driver attachment to the driver handle and carefully remove the bearings as shown.

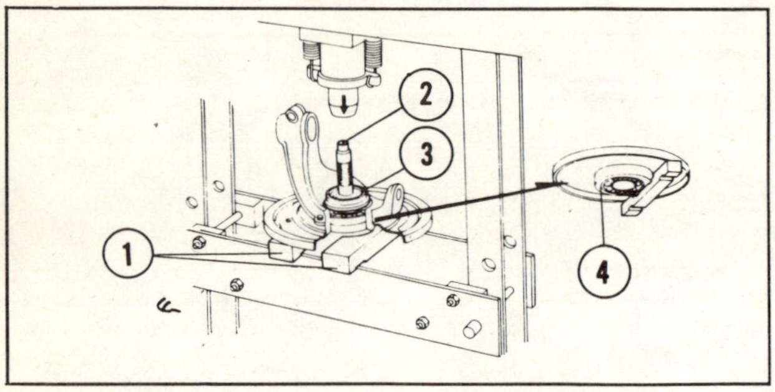

Fig. N-4
① Front hub disassembly jig plate A
② Driver handle ③ Driver attachment D
④ Front wheel bearing

a-3. Inspection

See Maintenance Standard at the end of this manual.

Inspection item.	Remedy
Cracks, scratches or other damage of knuckle	Replace.
Damage of bearings and bearing dust seal B	Replace.
Damage of bearing dust seal A	Replace.

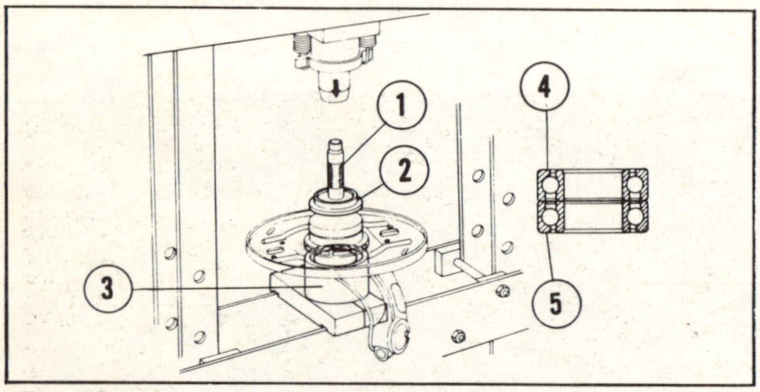

Fig. N-5
① Driver handle ② Driver attachment B
③ Front hub disassembly jig B ④ Product marking
⑤ Product marking

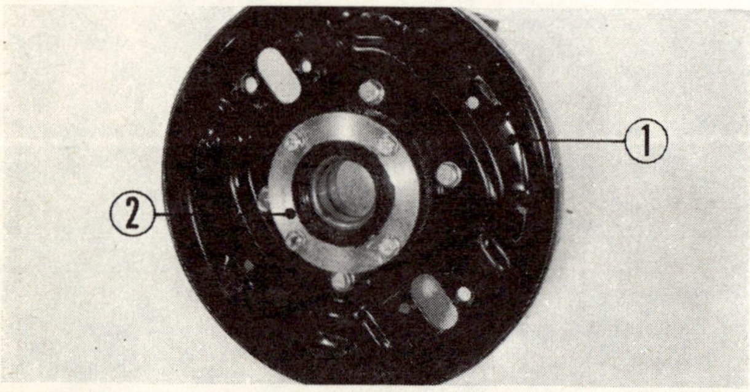

Fig. N-6
① Brake plate ② Dust seal B

a-4. Assembly and installation

To assemble and install the knuckle, reverse the removal and disassembly procedures. Pay attention to the following points:

1. Position the front hub disassembly jig plates A and B (facing the more deeply stepped side toward the knuckle) on a hydraulic press. Position the wheel bearings with their markings facing outward. Install the driver attachment B to the driver handle. Then, operate the press and install the bearings in place, as shown.

   Press the bearing outer race with the end having larger diameter.

   NOTE:

   The wheels of the Honda CIVIC use angular contact type ball bearings.
2. Fill each bearing with 80% full of recommended grease.
3. Install the backing plate and dust seal B to the knuckle.

   NOTE:

   Apply a coat of sealant to the attaching surfaces of the backing plate and dust seal.

4. Place the plate A and tool B on the hydraulic press and position the front hub as shown and press it down through the tool A.

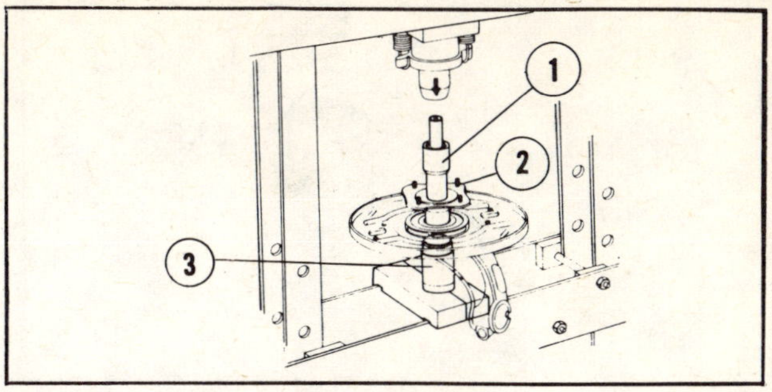

Fig. N–7
① Front hub disassembly jig pin A
② Front hub ③ Front hub disassembly jig plate B

b. Front shock absorbers

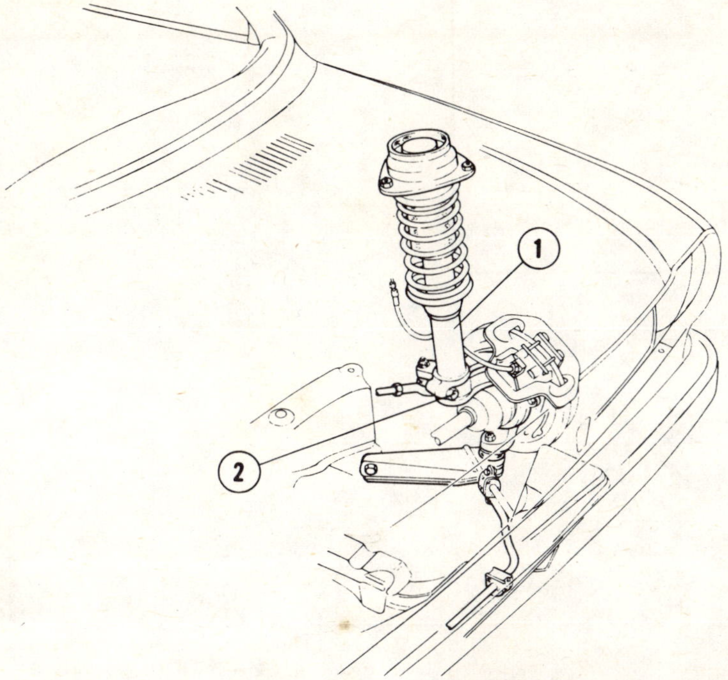

Fig. N–8 ① Front shock absorber ② Front shock absorber lock bolt

b–1. Removal

1. Remove the front wheel clip nuts, spindle nut and wheel.
2. Jack up the front of the car and support it with safety stands.
3. Disconnect the brake pipe at the shock absorber. Remove the hose clip and disconnect the hose from the pipe.
4. Plug the disconnected end of the brake hose.
5. Remove the knuckle from the shock absorber by loosening the absorber locking bolts.
6. Remove the front shock absorber by loosening the attaching nuts at the absorber rubber and on the body side.

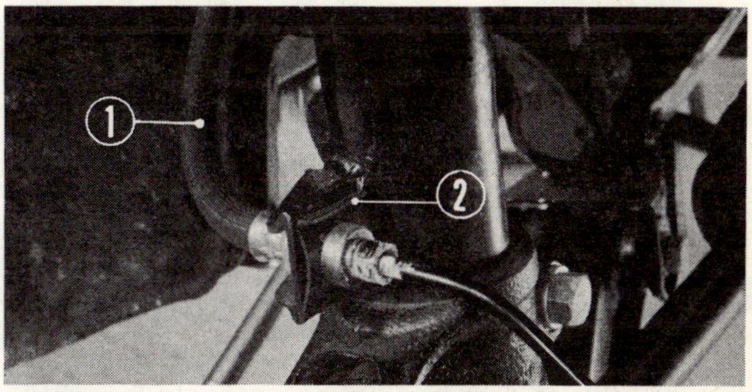

Fig. N–9
① Front brake hose ② Hose clip

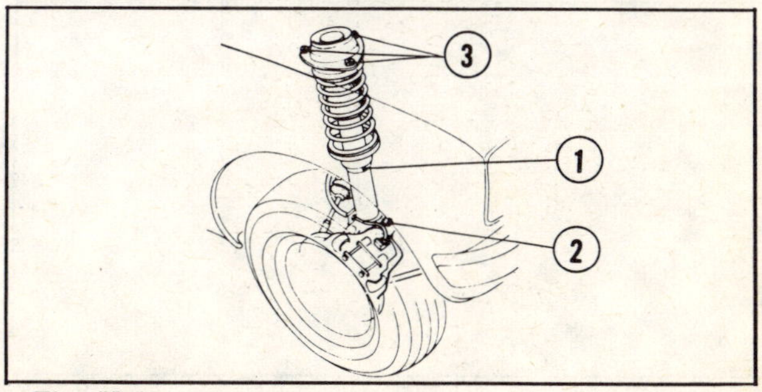

Fig. N–10
① Front shock absorber ② Lock bolt
③ Attaching nuts

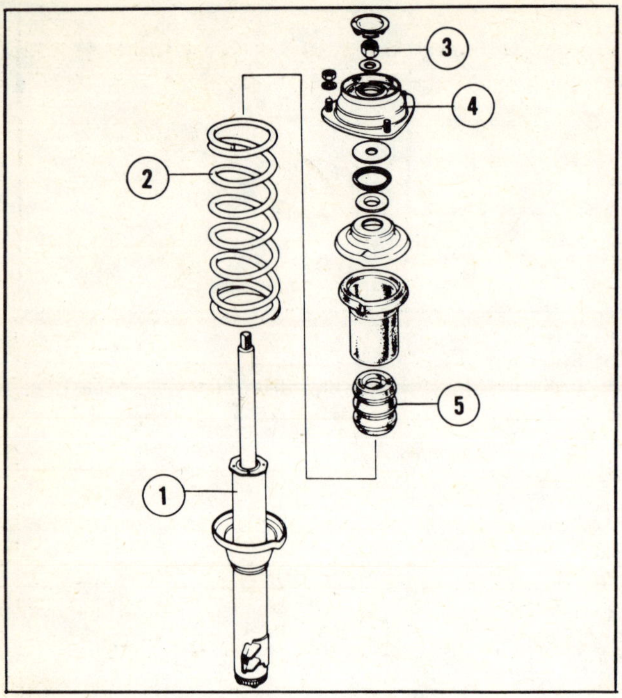

Fig. N–11
① Front shock absorber unit ② Spring
③ Self lock nut ④ Front shock absorber rubber
⑤ Bump stopper rubber

b–2. Disassembly

1. Assemble the front shock absorber compressor as shown. Slowly turn the handle and compress the shock absorber. Then remove the self-lock nut.

   NOTE:

   Use of any tool other than the front shock absorber compressor may cause the self-lock nut to jump out, resulting in an accident.

2. Turn back the handle and remove the front shock absorber unit. Then the shock absorber can be disassembled as shown in Fig. N–11.

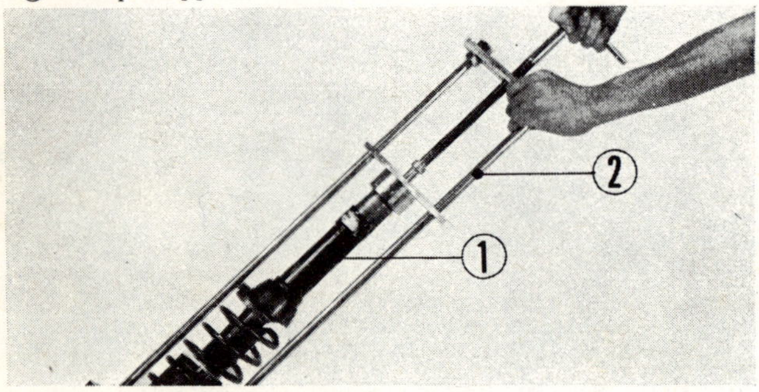

Fig. N–12
① Front shock absorber
② Front shock absorber compressor

b–3. Inspection

See Maintenance Standard at the end of this manual.

Inspection item	Remedy
Wear of needle roller bearing	Replace.
Damage of spring	Replace.
Oil leak from shock absorber	Replace.
Deterioration of front shock absorber rubber	Replace.
Deterioration or other damage of front bump stopper rubber	Replace.

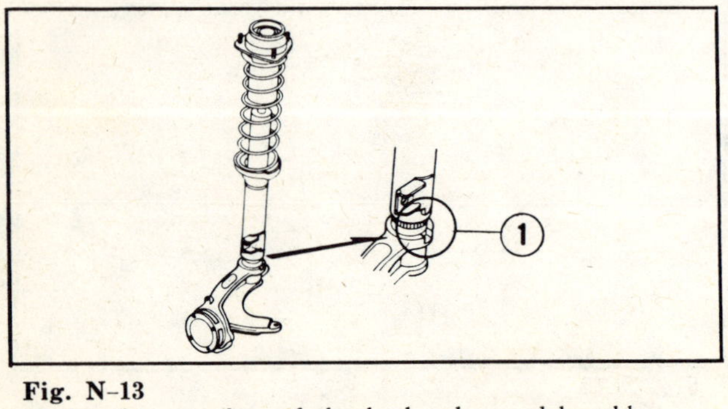

Fig. N–13
① Attaching surface of shock absorber and knuckle

b–4. Assembly and installation

To assemble and install the front shock absorber, reverse the removal and disassembly procedures. Pay attention to the following points:

1. Tighten the shock absorber nuts to the specified torque.
2. Install the shock absorber to the knuckle. Fit the locating tab of the shock absorber in the recess in the knuckle and tighten the locking bolt to the specified torque.
   - If the locating tab of the shock absorber does not fit properly or if the shock absorber is not inserted in the knuckle properly, the locking bolt will not be tightened securely.
3. Connect the brake hose to the pipe, taking care not to twist the hose.

b-5. Checking

After assembling, check the following points:

1. Turn the steering wheel all the way to right and left, making sure the brake hose does not contact with the wheel rim.
2. Bleed the brake lines. Apply pressure to the brake lines and check the brake hose and pipe joints for leak.
3. Inspect the wheel alignment. For the specifications, see the specification at the end of this manual.

c. Lower arms and stabilizer

c-1. Removal

1. Remove the front wheel.
2. Remove the lower arm ball joint.

NOTE:

Take care not to damage the lower arm and knuckle.

3. Disconnect the lower arm at the body.
4. Disconnect the radius rod at the front.
5. Remove the ball joint seal by removing the snap ring.
6. Disconnect the stabilizer by loosening the four band bolts. Rest the center of the stabilizer on a jack. Remove the two center bands and then the right and left bands.

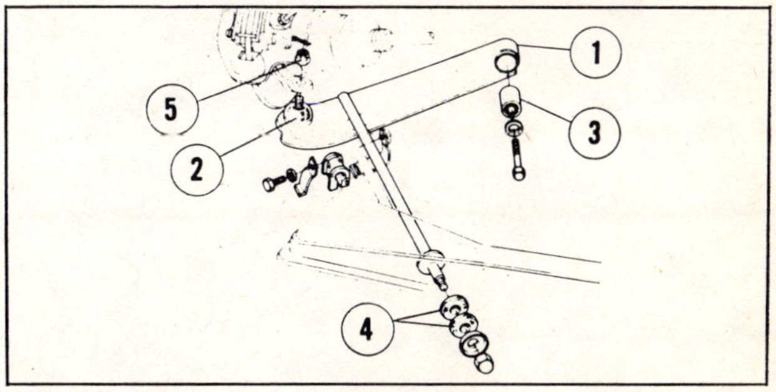

Fig. N-14
① Lower arm ② Ball joint seal
③ Lower arm bushing ④ Radius rod bushings
⑤ Castle nut

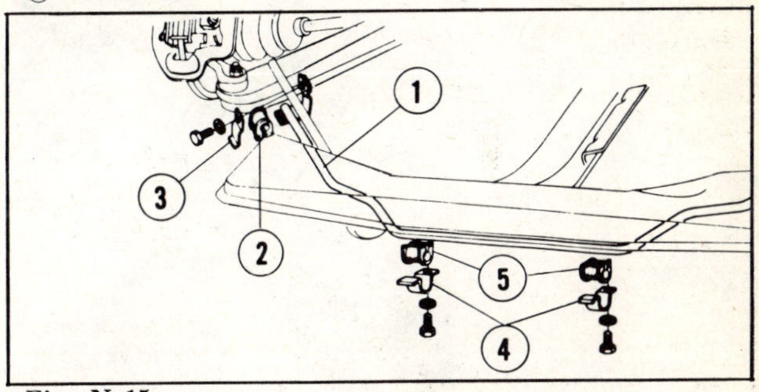

Fig. N-15
① Stabilizer spring ② Stabilizer bushing A
③ Stabilizer band B ④ Stabilizer bands
⑤ Stabilizer bushings

c-2. Inspection

Inspection item	Remedy
Deformation or other damage of lower arm	Replace.
Cracks or breakage of ball joints	Replace.
Deterioration of lower arm and radius rod bushings	Replace.
Deformation or other damage of stabilizer	Replace.
Deterioration of stabilizer bushings	Replace.

c-3. Installation

To install, reverse the removal procedures. Pay attention to the following points:

1. Fill the lower arm ball joint seal with the recommended grease when installing.
2. Tighten all bolts and nuts to the specified torque.
   See the Torque Specifications.
3. Connect both ends of the stabilizer. Rest the center of the stabilizer on a jack and secure the stabilizer with the two center bands.
4. Inspect the wheel alignment.
   For the specifications, see the Maintenance Standard.

N-2. Rear Suspension

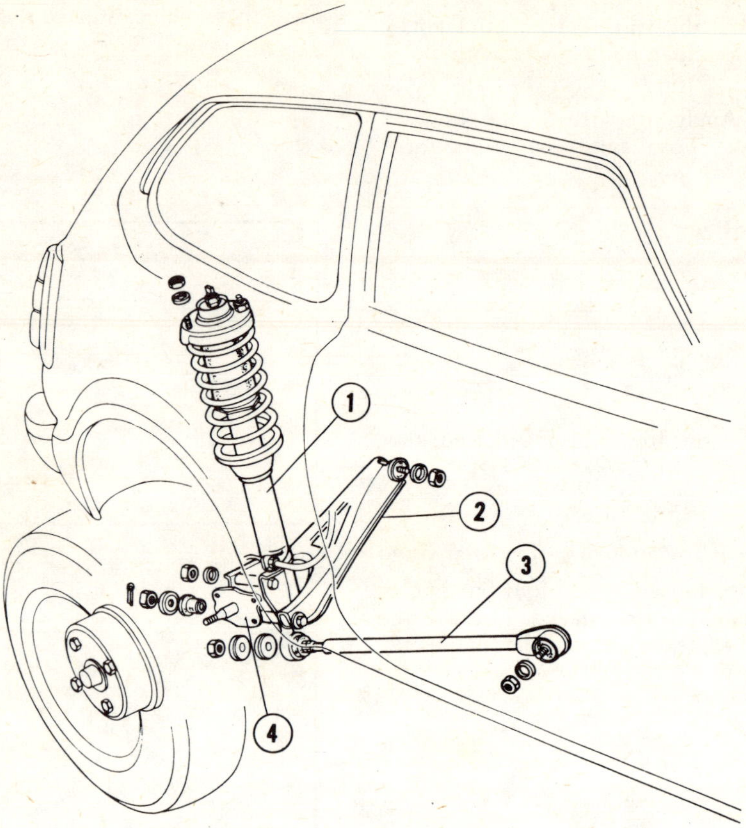

Fig. N-16

① Rear shock absorber
② Rear lower arm
③ Rear radius rod
④ Rear hub carrier

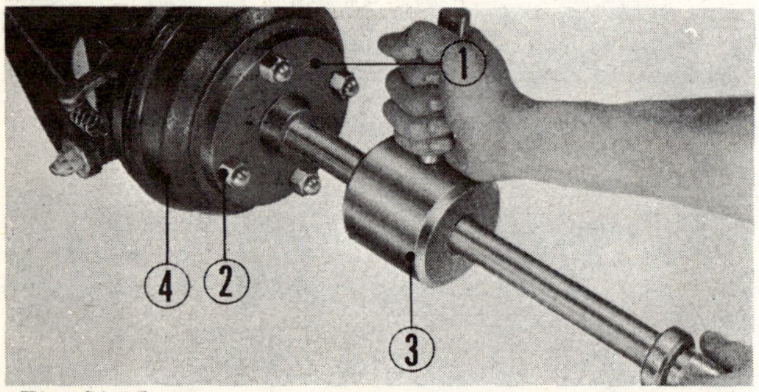

Fig. N-17
① Hub puller attachment ② Wheel clip nut
③ Sliding hammer ④ Brake drum

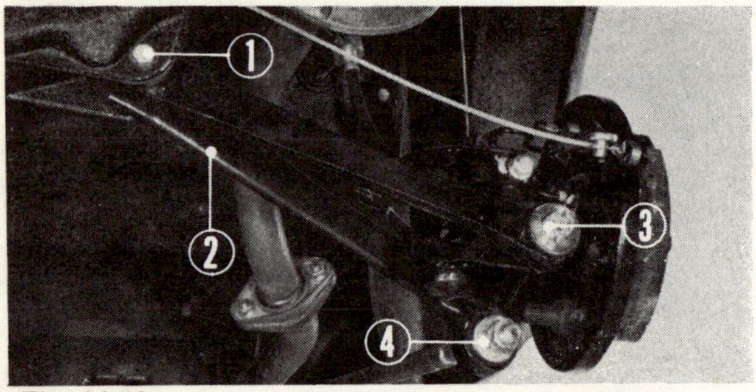

Fig. N-18
① Rear lower arm bolt ② Rear lower arm
③ Hub carrier castle nut
④ Rear radius rod self lock nut

a. Removal

1. Jack up the rear of the car and support it with safety stands.
2. Remove the rear wheel and brake drum.

   NOTE:

   When it is difficult to remove the brake drum, use the hub puller attachment and a sliding hammer.
3. Disconnect the brake pipe at the rear wheel cylinder and the parking brake control cable at the brake lever.

   Plug the disconnected end of the brake pipe to prevent the brake fluid from coming out.
4. Remove the rear backing plate.
5. Remove the hub carrier cotter pin and bolt.
6. Remove the rear shock absorber and hub carrier bolts.

   Remove the rear hub carrier from the radius rod by loosening the radius rod self-lock nuts.

7. Remove the rear shock absorber assembly from the body by loosening the attaching nuts.
8. Disconnect the rear radius rod at the body by loosening the attaching nuts.
9. Disconnect the rear lower arm at the body by loosening the attaching nuts.

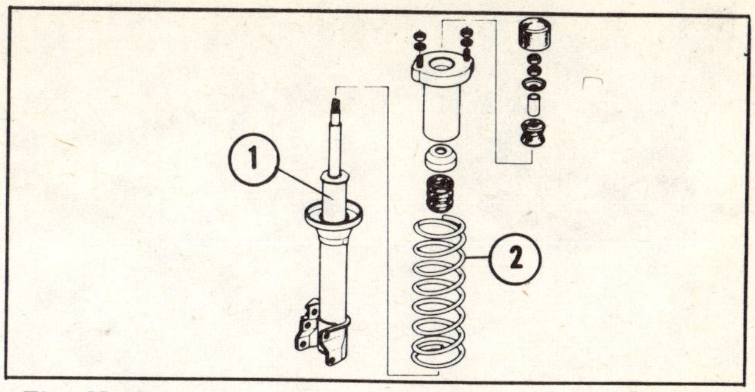

Fig. N–19
① Rear shock absorber unit ② Spring

b. Inspection

See Maintenance Standard at the end of this manual.

Inspection item	Remedy
Oil leak from rear shock absorber	Replace.
Fatigue of rear shock absorber springs	Replace.
Deterioration or other damage of rear bump stopper rubber	Replace.
Deterioration or other damage of rear lower arm bushing and bushings A	Replace.
Deformation, cracks or other damage of lower arm	Replace.
Bend, cracks or other damage of rear radius rod	Replace.
Deterioration or other damage of radius rod bushings A and B	Replace.

c. Installation

To install, reverse the removal procedures. Pay attention to the following points:

1. To connect the rear lower arm and radius rod, first tighten the attaching nuts temporarily with the car cleared off the ground. Then retighten the nuts to the specified torque with the car on a level ground. See the Torque Specifications at the end of this manual.
2. After connecting the brake pipes, bleed the brake lines.

• After bleeding, check for oil leak.

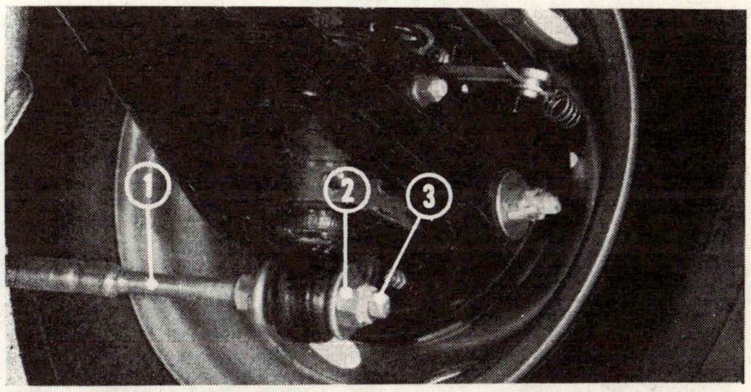

Fig. N–20
① Rear radius rod ② Self lock nut ③ 14 mm nut

c–1. Adjustment of toe-in

Measure the wheel alignment with a side-slip tester.

Unit: mm (in.)

	Specification
Side-slip	0

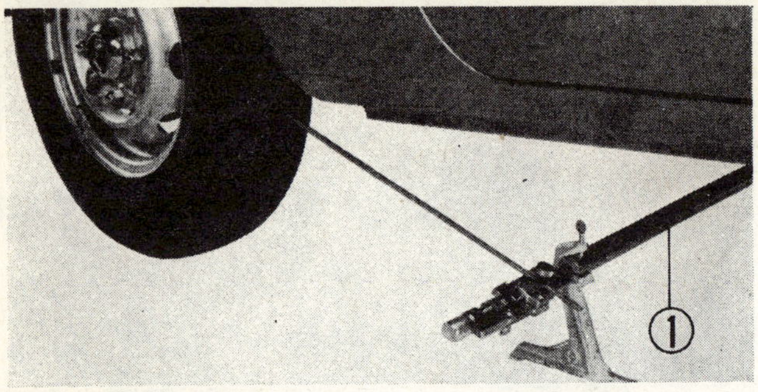

Fig. N-21
① Toe-in gauge

If out of specification, check and adjust the toe-in without a driver and passengers or load in or on the car.

1. Place the car on level ground.
2. Make chalk marks on the center of the tires at the height of the axle center.

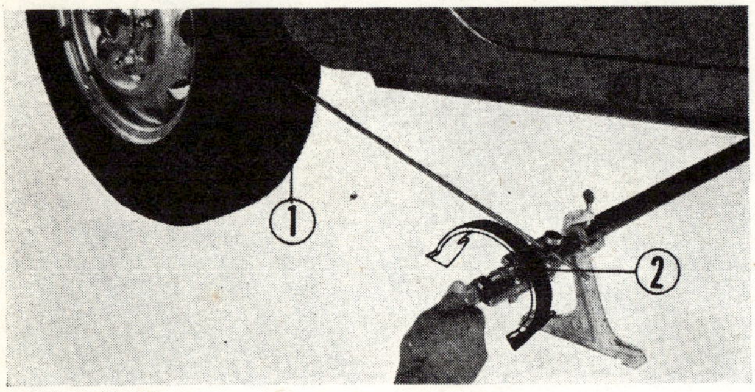

Fig. N-22
① Tire center ② Measure

3. Align the toe-in gauges with the marks on the tires as shown.
4. Read the scales of the gauges.
5. Slowly reverse the car and rotate the wheels 180°.
6. Measure the toe-in on the rear part of the tires at the same point.

Unit : mm (in.)

	Specification
Toe-in	0

7. If out of specification, adjust it by changing the length of the rear radius rods.

NOTE :

Maladjustment of toe-in may result in unstable driving or uneven wear of tires.

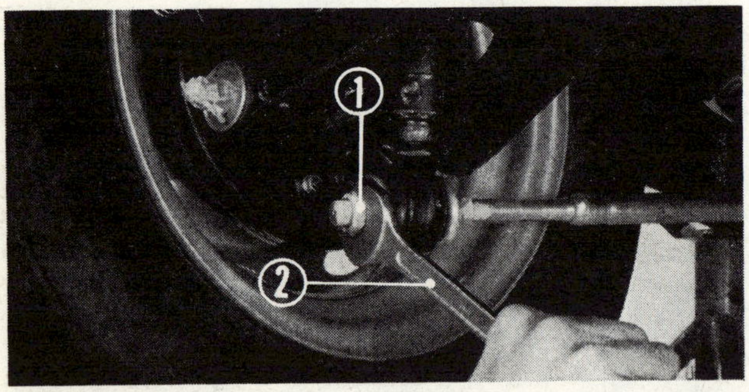

Fig. N-23
① Self lock nut ② Wrench

Adjustment

1. Raise the rear wheels.
2. Turn in or out the rear radius rod self-lock nuts until the correct adjustment can be made.
3. After adjusting, retighten the self-lock nuts.

NOTE :

If the car is still unstable even after the toe-in adjustment has been made properly, the rear lower arms and body are suspected to be misaligned.

EXHAUST MANIFOLD, PIPE AND SILENCER

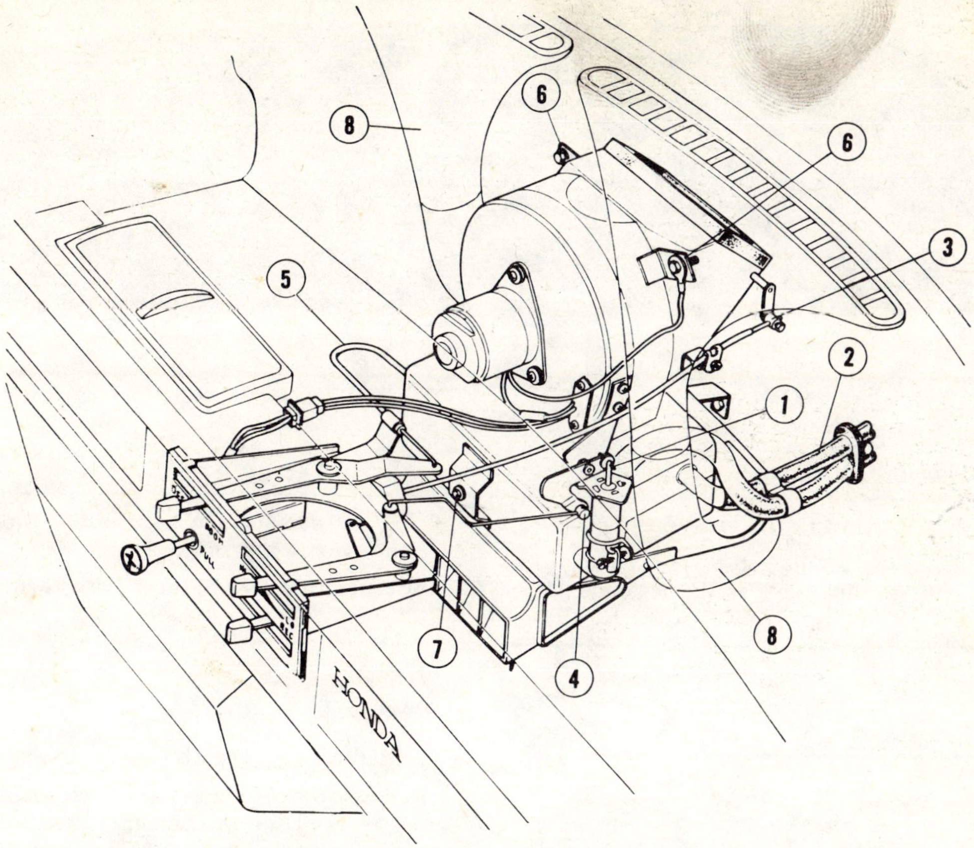

Fig. O-1

① Hot air cover	⑤ Silencer mount
② Exhaust manifold	⑥ Exhaust silencer
③ Exhaust pipe	⑦ Exhaust manifold
④ Exhaust pipe clamp B	⑧ Exhaust pipe

a. Removal

1. Remove the front grille.
2. Remove the exhaust pipe clamps A and B.
3. Disconnect the exhaust pipe at the manifold.
4. Remove the hot air cover and remove the exhaust manifold attaching nuts.
5. Loosen the exhaust silencer attaching bolts, and remove the exhaust pipe and silencer as a unit.
6. Disconnect the exhaust pipe at the silencer.

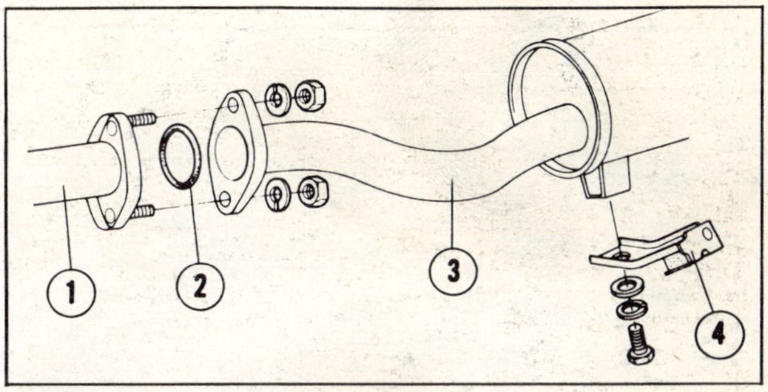

Fig. O-2

① Exhaust pipe	③ Exhaust silencer
② Gasket	④ Silencer mount

b. Installation

To install, reverse the removal procedures. Pay attention to the following points:

1. Use new gaskets.
2. Tighten the exhaust manifold and pipe bolts and nuts evenly.
3. After installing, check each joint for leak of exhaust gas. Also check to see if the exhaust manifold and pipe do not interfere with other parts.

P. INSTRUMENT PANEL

Fig. P–1

Precautions for handling

1. Take care not to damage the instrument panel.
2. When removing or installing the panel, be sure to disconnect the ground cable at the terminal of the battery.
3. Do not damage the wire leads. Connect the cables and wire leads correctly.
4. Secure the wire harness with their clips properly.

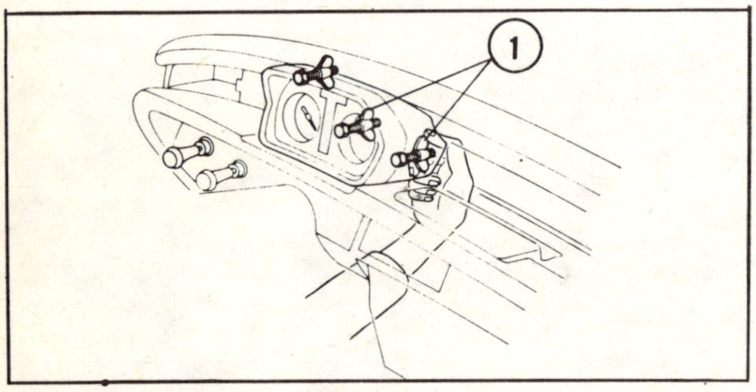

Fig. P–2 Removing meter case
① Meter case wing nut

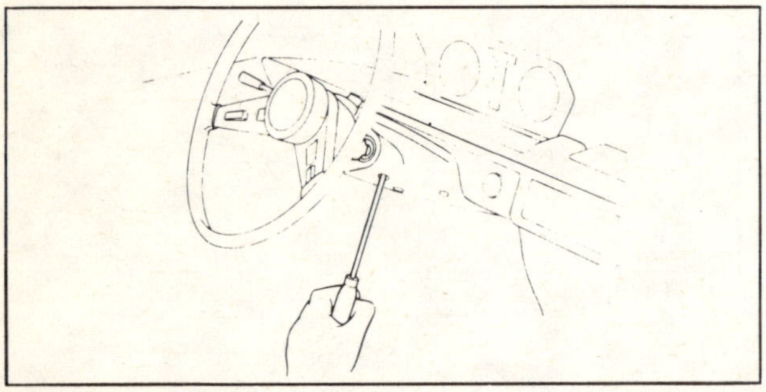

Fig. P–3 Removing column cover

a. Removal

a–1. Meter case assembly

1. Remove the two meter case mounting wing nuts from the rear of the instrument panel with hands.
2. Disconnect the speedometer and tachometer drive cables at the engine.
3. Pull the meter case forward. Disconnect the speedometer and tachometer drive cables at the meter case by loosening their attaching nuts. Disconnect the meter wires at the couplers and connectors to separate them from the harnesses.
4. Remove the meter case.

a–2. Switch panel

1. Loosen the five steering wheel column cover screws and remove the upper and lower covers.

2. Remove the four steering column bolts and rest the steering wheel assembly on the floor.

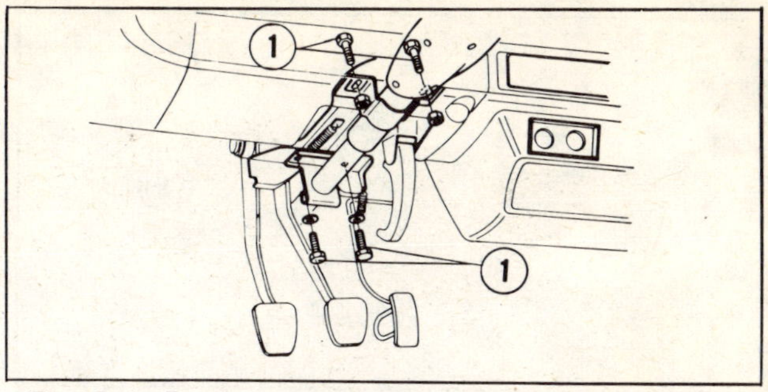

Fig. P-4 Removing steering column
① Steering column attaching bolts

3. Remove the four switch panel screws from the rear of the instrument panel and remove the switch panel.

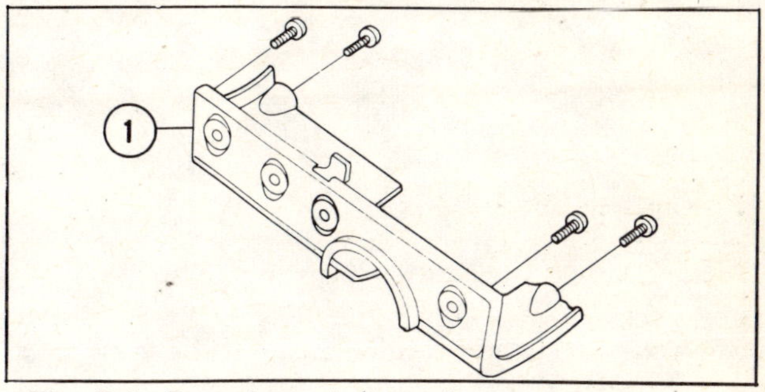

Fig. P-5 Removing switch panel
① Switch panel

a-3. Instrument panel assembly

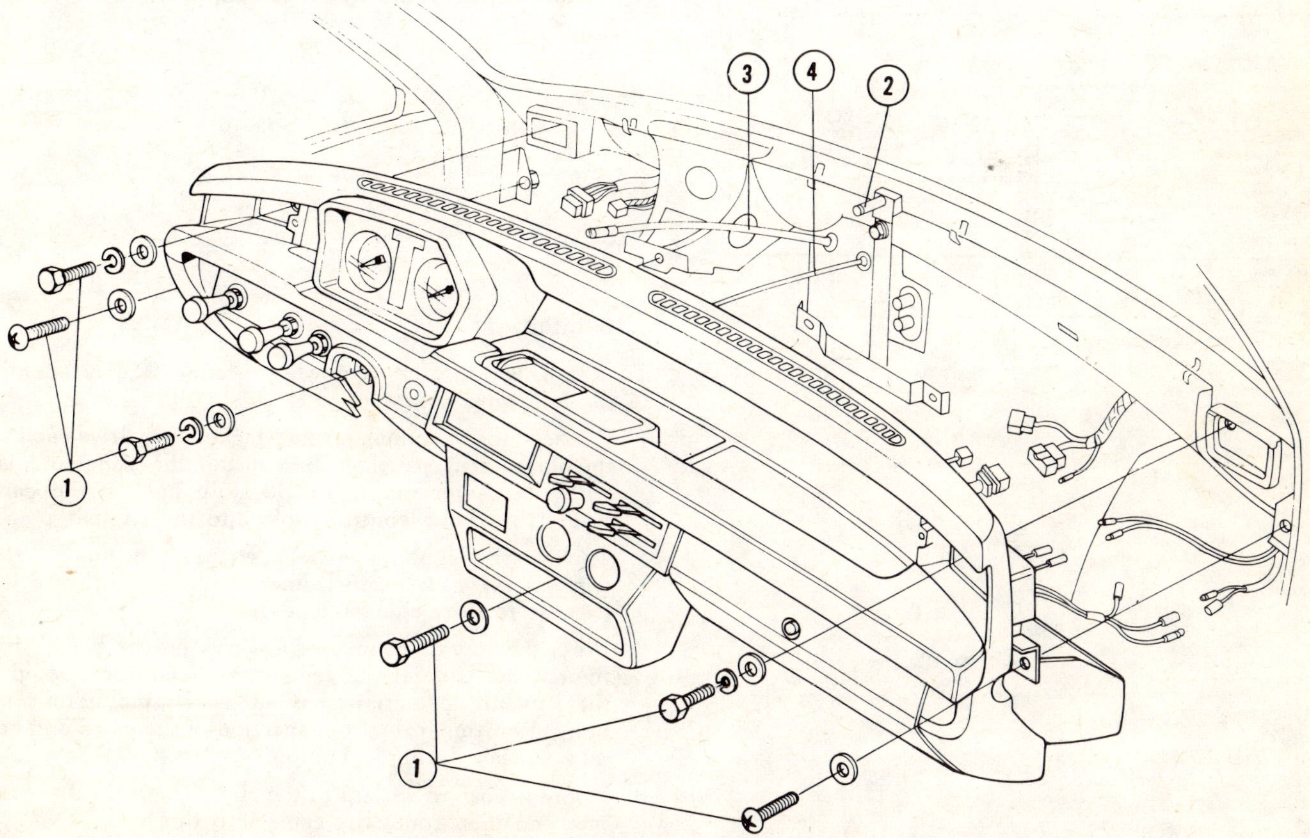

Fig. P-6 Removing instrument panel

① Attaching bolts
② Center pin
③ Meter cable
④ Choke control cable

1. Remove the steering column covers and column bolts. Rest the steering assembly on the floor.

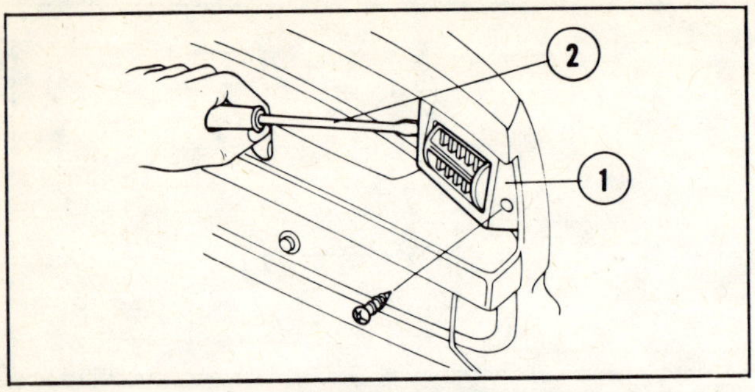

Fig. P–7 Removing fresh air panel
① Fresh air panel ② Screwdriver

2. Remove one of the fresh air panel screws. Pry off the air panels with a standard-type screwdriver as shown and remove the air panels. Then the two more panel bolts will come out from behind them.

3. Disconnect the instrument panel wire harnesses from their cabin harnesses by removing the connectors and couplers.
4. Disconnect the speedometer and tachometer drive cables at the engine.
5. Disconnect the choke control cable at the carburetor.
6. Disconnect the heater control cables (valve control cable, hot air control cable, cool air control cable, etc.) at the heater and valve.

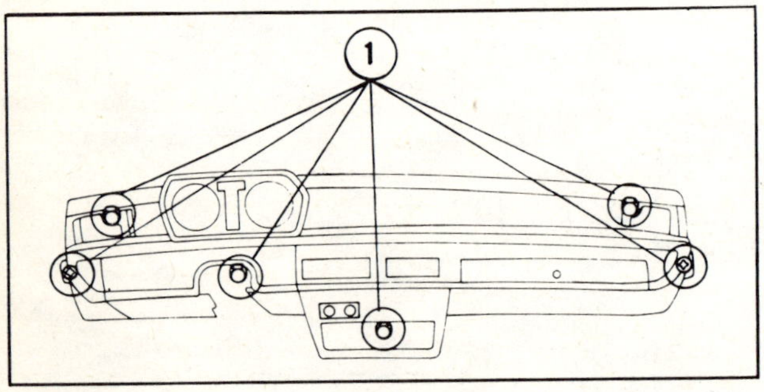

Fig. P–8
① Instrument panel attaching bolts

7. Remove the heater motor cord coupler.
8. Remove the instrument panel by loosening the five bolts.

b. Installation

To install, reverse the removal procedures. Pay attention to the following points:

1. Connect the speedometer and tachometer drive cables to the instrument panel. Then install the panel in place with the center pin in the locating hole in the panel. Route the choke control cable into the engine room.
2. Temporarily tighten the bolts securing the upper, right and left sides of the instrument panel. Check the wire harness for any sign of binding.
3. Press the instrument panel against the car body. First tighten the panel bolts temporarily and then retighten them evenly in a criss-cross pattern, beginning on center bolt. Remember, loose connection of the panel and body may squeak during operation.
4. Apply a coat of sealant to the choke control cable grommet and then route the cable into the body.
5. After installing, check the instrument panel parts for correct installation and the switches and levers for proper operation.
6. Install the steering column securely.

HEATER ASSEMBLY

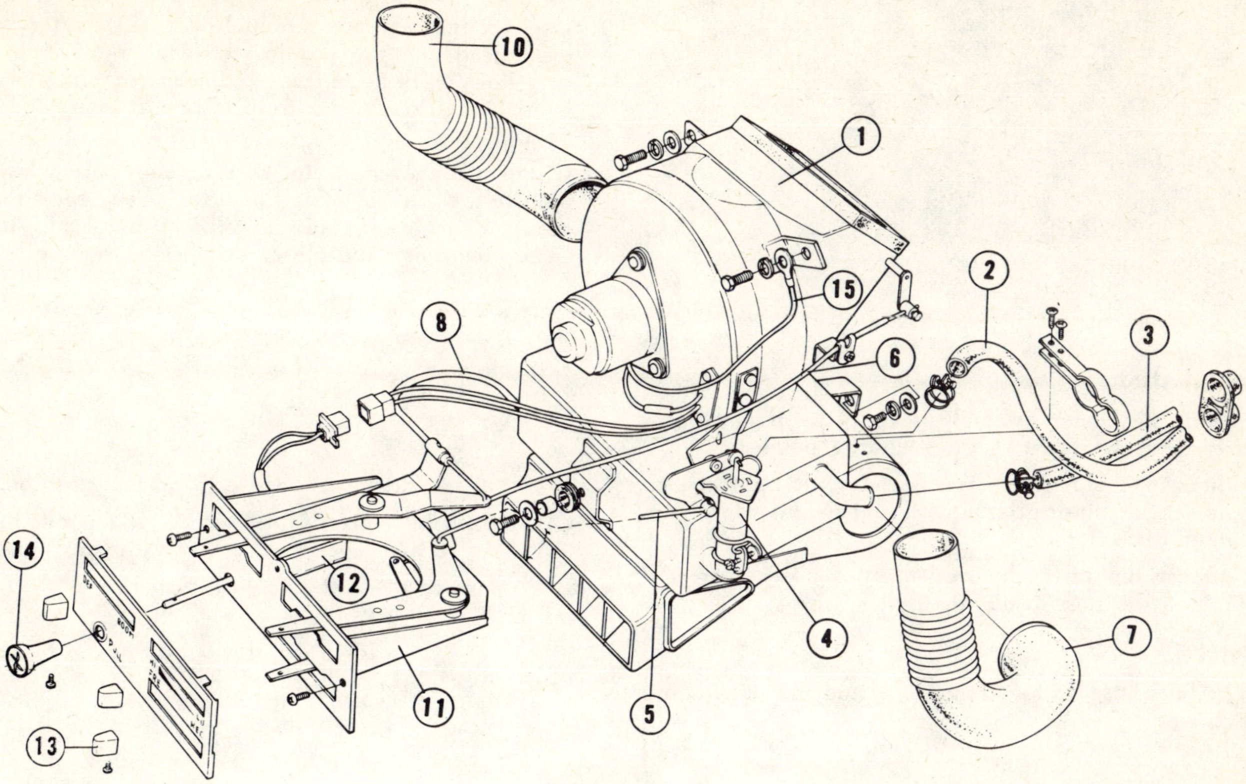

Fig. Q-1

① Heater body	⑥ Cool air control cable	⑪ Heater control assembly
② Inlet pipe	⑦ Right defroster hose	⑫ Heater switch
③ Outlet pipe	⑧ Hot air control cable	⑬ Lever knob
④ Water valve	⑨	⑭ Heater switch knob
⑤ Water control cable	⑩ Left defroster hose	⑮ Ground cord

a. Removal

1. Drain the lower radiator tank by loosening the drain cock.
2. Disconnect the right and left defroster hoses.
3. Disconnect the inlet and outlet water hoses at the heater assembly by loosening their bands.

   NOTE:

   When disconnecting the hoses, do not spill the cooling water over the floor mat. Cover the mat with a waste cloth.

4. Disconnect the hot air, cool air and valve control cables at the heater assembly.
5. Remove the heater motor cord coupler.
6. Remove the heater assembly by loosening the four attaching bolts.

   NOTE:

   First remove the upper attaching bolts.

7. Remove the three control lever knobs and heater control switch knob by loosening their attaching screws.
8. Peel off the heater control garnish plate and remove the heater control assembly by the three attaching screws through the opening.

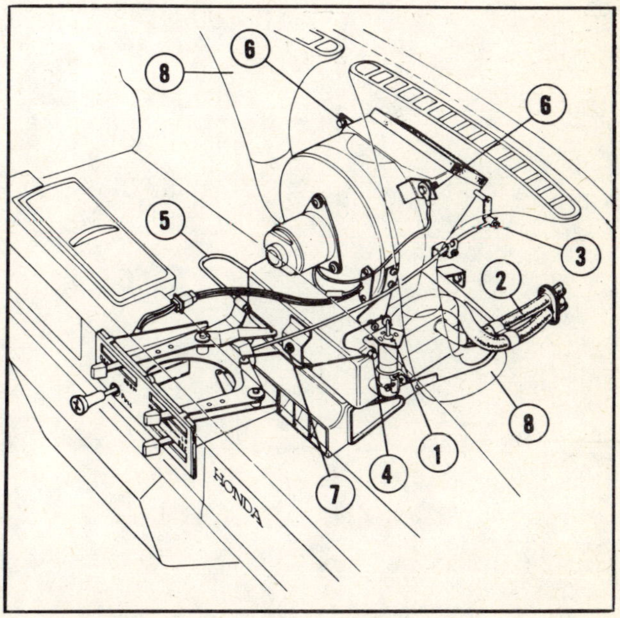

Fig. Q-2 Removing heater assembly

① Inlet hose band	⑤ Hot air control cable
② Outlet hose band	⑥ Upper attaching bolts
③ Cool air control cable	⑦ Lower attaching bolts
④ Valve Control cable	⑧ Defroster hoses

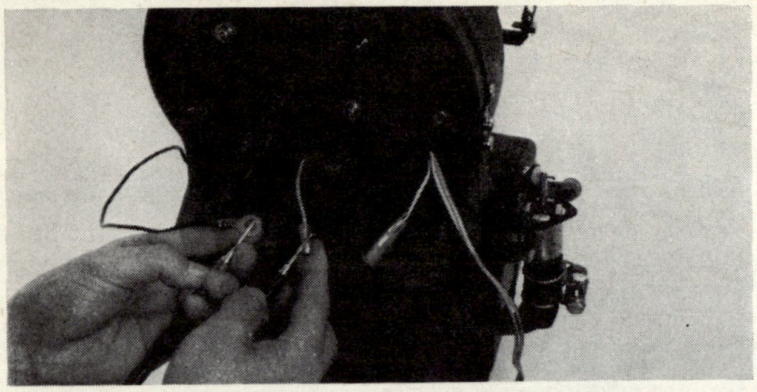

Fig. Q-3

b. Inspection

b-1 Inspect the motor

1. Connect the negative terminal of a 12V battery to the black lead and the positive terminal to the yellow/red or yellow/black terminals. If the motor turns smoothly at high and low speeds respectively, it is in good condition.
2. Disconnect the connector of the yellow/black lead running to the motor. Using a tester, check for continuity between the black lead and yellow/black lead. If there is no continuity, the motor coil has an open circuit.

c. Installation

To install the heater assembly, reverse the removal procedures. Pay attention to the following points:

1. When installing the heater assembly, do not forget to connect the motor ground cord to the right side of the center bracket.
2. Connect the inlet and outlet water hoses correctly. Tighten the hose bands tightly.

   NOTE:

   The inlet hose is of a straight type and the outlet hose a L-type.

3. Install the defroster nozzles in correct position.
4. Connect the control cables securely. Operate the control valve and lever to check for proper operation.
5. Be sure to bleed the cooling system.

NOTES

DOORS

Door Linings

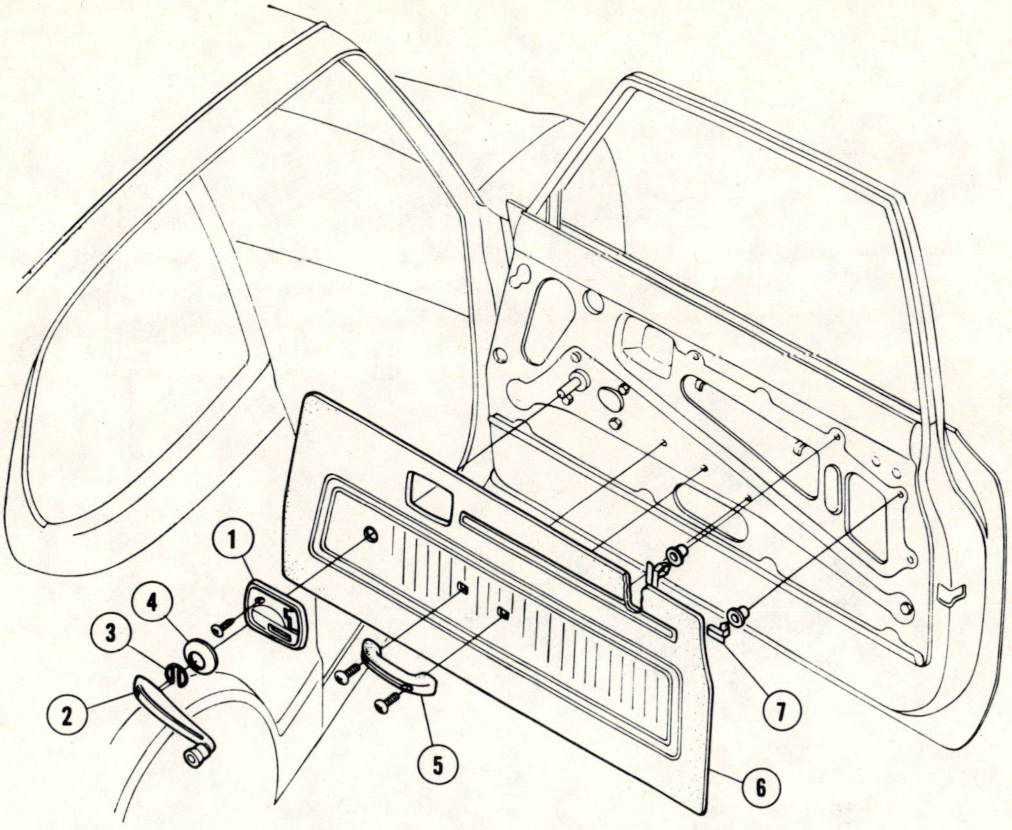

Fig. R-1

① Inside handle case
② Regulator handle
③ Regulator handle pin
④ Escutcheon
⑤ Arm rest
⑥ Door lining
⑦ Mount clip

a. Removal

1. Remove the inside handle case.

   Pull the inside handle forward and remove its case.

2. Remove the regulator handle.

   Push out the handle pin with a screwdriver. Then the handle can be removed together with the escutcheon.

3. Remove the arm rest.

4. Remove the door lining. The lower side of the lining is inserted in the panel rail. The upper, front and rear sides are secured with the mount clips. First pull out the clips and then remove the lining.

   NOTE:

   The door lining is made up of resin. Be careful not to damage the lining when removing it.

b. Installation

To install the door lining, reverse the removal procedures. Pay attention to the following points:

1. To install the lining, first insert its lower side into the panel rail. Then, engage the lining clips in their holes and secure the lining properly.
2. Install the regulator handle so that its knob comes to the front side when the door glass is closed completely.

Door Glasses

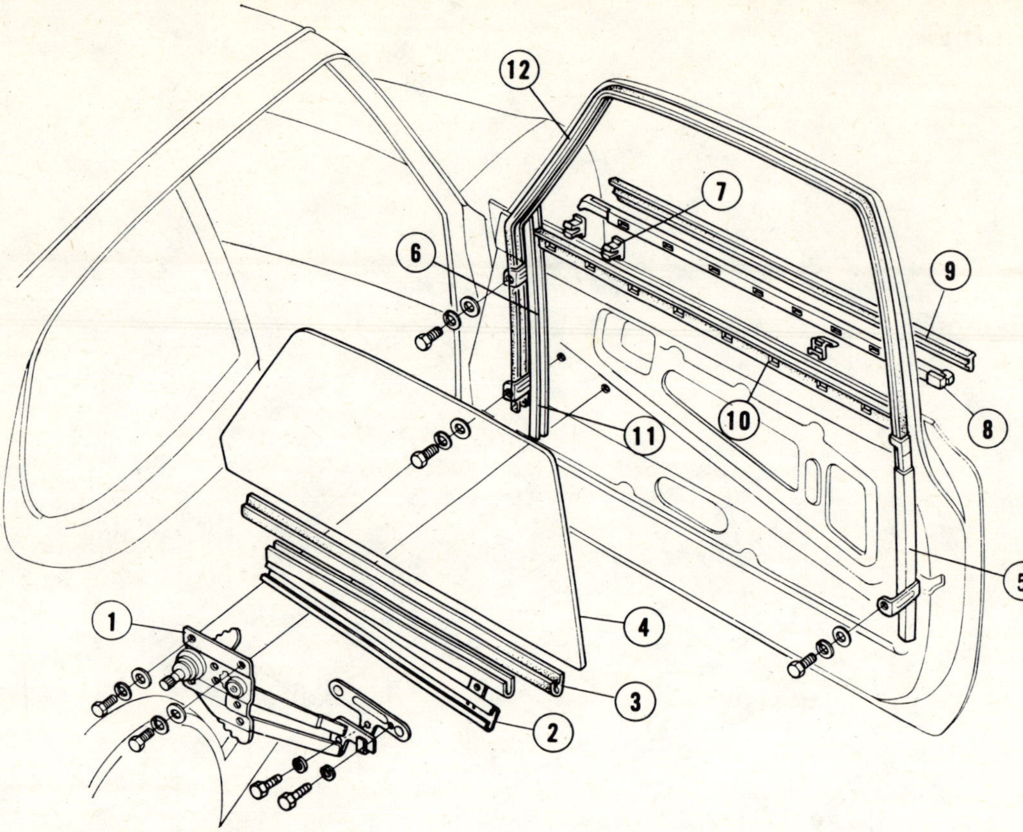

Fig. R-2

① Door glass regulator	⑦ Door mole clip
② Door glass holder	⑧ Door mole
③ Door glass front cushion	⑨ Door outer weather strip
④ Front door glass	⑩ Front door inner weather strip
⑤ Door rear sash	⑪ Door front run channel
⑥ Door front sash	⑫ Door run channel

a. Removal

1. Remove the door lining. (See page C–153.) Remove the door panel seals.
2. Remove the inner and outer weather strips with a standard-type screwdriver.

NOTES:

1) When removing the inner weather strip, take care not care not to drop the clips inside the door.
2) Do not apply a force to the weather strips more than necessary when removing; otherwise they may be damaged.
3) Operate the regulator handle until the door glass holder can be removed. Remove the glass by loosening its attaching bolts.

   Loosen four regulator bolts and the regulator can be removed through the door panel seal attaching hole.

b. Installation

To install the door glass, reverse the removal procedures. Pay attention to the following points:

1. Apply a coat of grease to the sliding surface of the regulator.
2. Install the door glass to its holder by tightening its attaching bolts temporarily.

   Operate the regulator handle and raise the glass, making sure it is in correct position. Then retighten the glass attaching bolts.
3. Correctly position the inner weather strip clips; otherwise they will be damaged.
4. Install the door mole and outer weather strip as a unit to the door.
5. After installing, check to be sure the glass moves smoothly without any sign of binding.

Door Locks

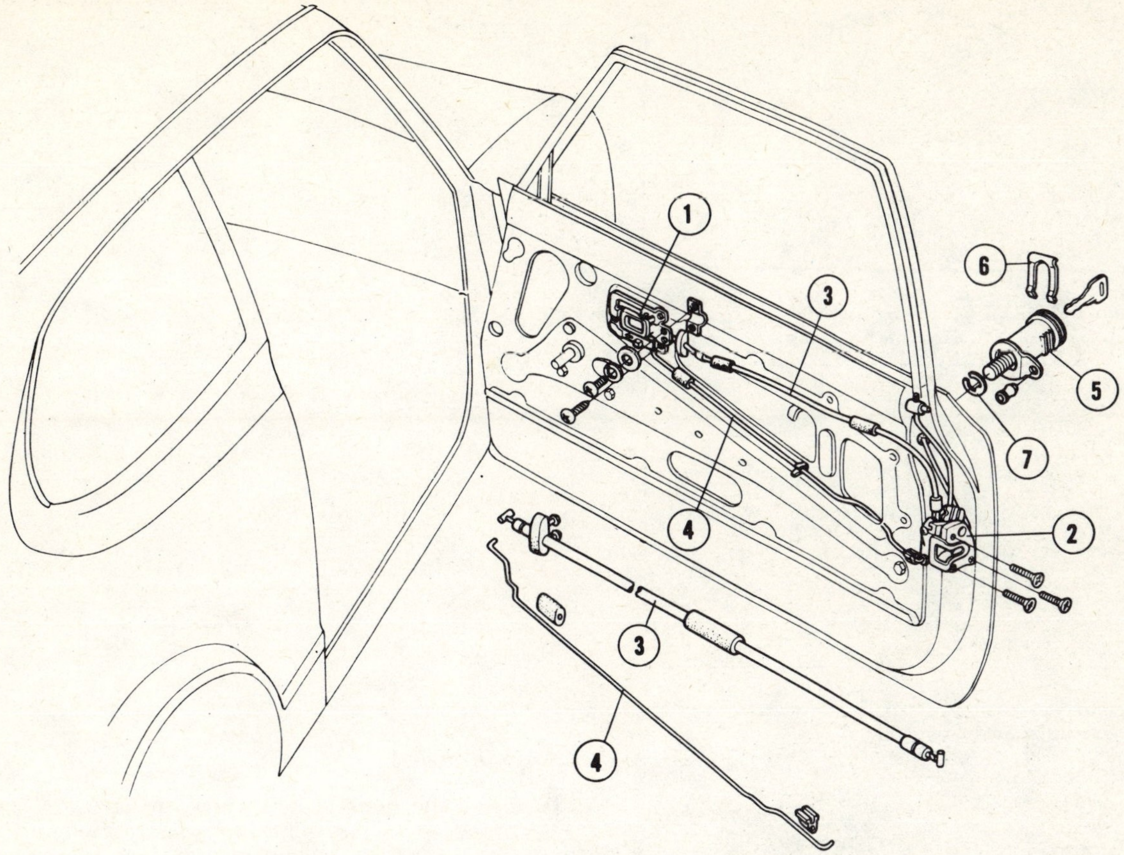

Fig. R-3

① Inside handle assembly
② Door lock
③ Inside handle cable
④ Lock knob rod
⑤ Door cylinder
⑥ Door lock key set spring
⑦ Cylinder rod clip

a. Removal

- Remove the door lining and glass. (See page C-153.)

1. Remov the inside handle case. Remove the inside handle tapping screws.
2. Remove rear sash bolts. Using a standard-type screw-driver, push down the sash lightly. Then the sash can be removed. Keep the sash raised as shown.
3. Using a standard-type screwdriver, remove the door lock from the outside handle.

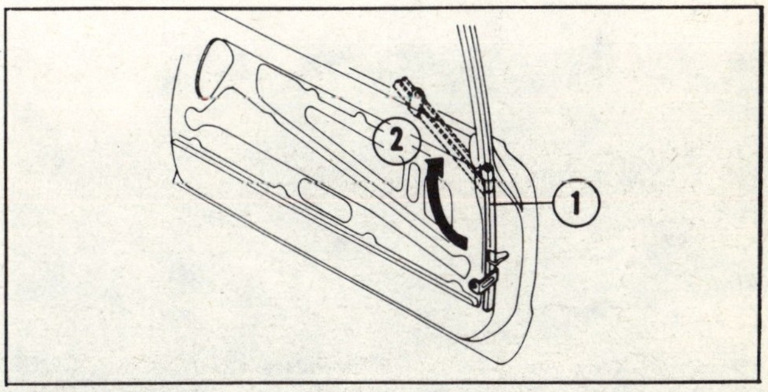

Fig. R-4
① Rear sash ② Raising sash

4. Remove the door lock from the door cylinder by removing the cylinder rod clip with a screwdriver.

NOTE:

Take care not to drop the clip inside the door.

5. Remove the lock knob rod clip on the door lock side. Then, disconnect the rod at the door look by pulling it forward.

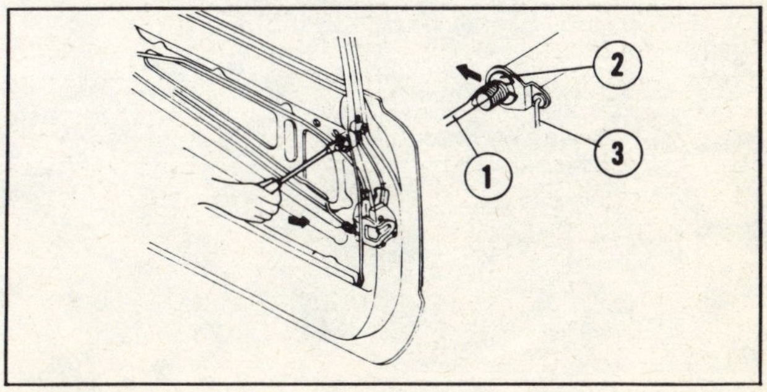

Fig. R-5
① Standard type screwdriver ③ Door lock rod
② Cylinder rod clip

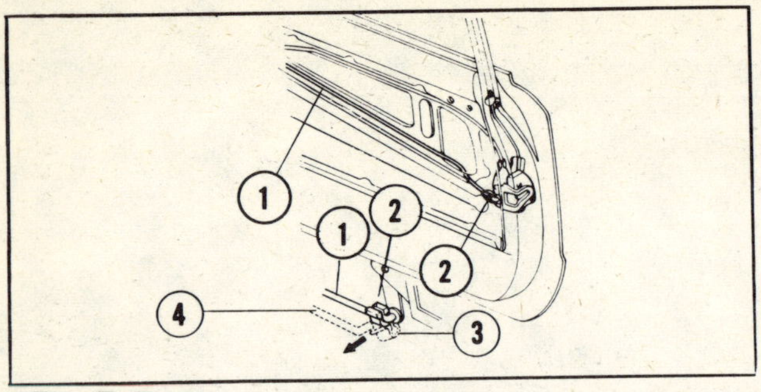

Fig. R-6
① Lock knob rod ③ Removing clip
② Resin clip ④ Disconnecting lock knob rod

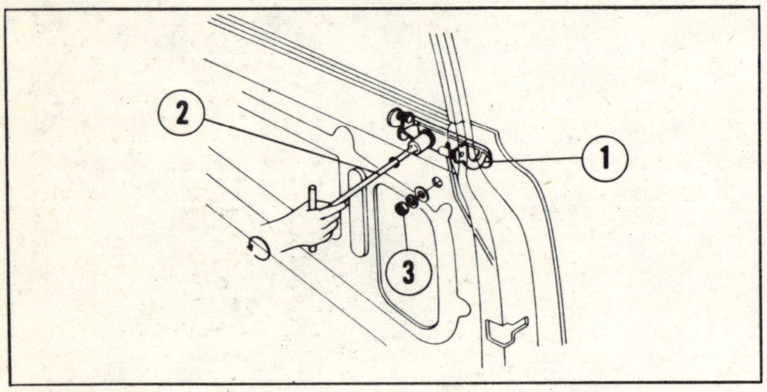

Fig. R-7
① Outside handle assembly ② T-type socket wrench
③ Outside handle attaching nut

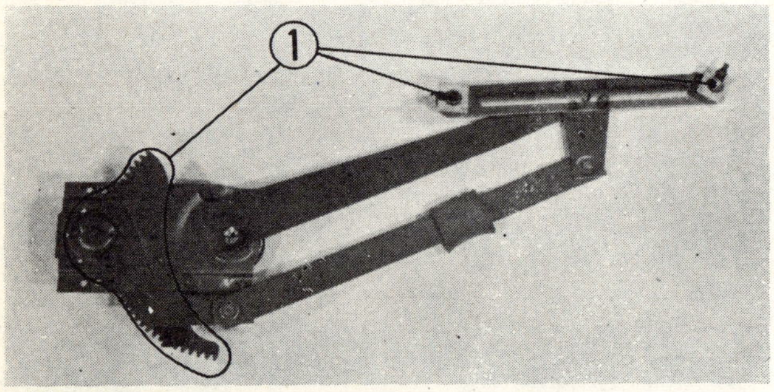

Fig. R-8
① Apply grease to these parts

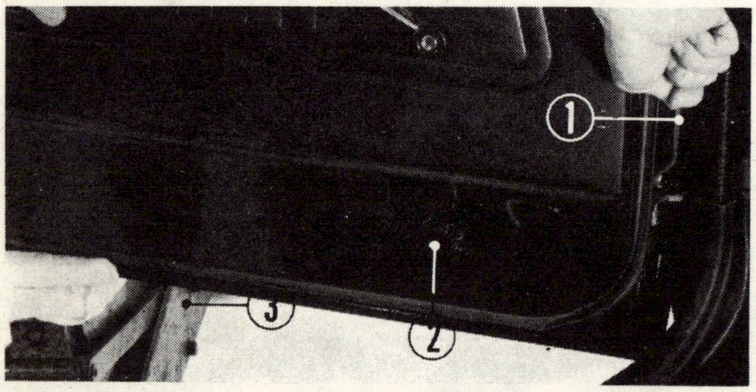

Fig. R-9
① Box wrench ③ Garage jack
② Door

6. Straighten the inside handle cable clip and remove the door lock screws. Then the door lock and inside handle assembly can be removed.

7. Remove the outside handle assembly by loosening the attaching nuts with a T-socket wrench.

b. Installation

To install the door lock, reverse the removal procedures. Pay attention to the following points:

1. When disassembling, apply a coat of grease to the rod, arm and linkage.
2. Check the door lock, outside handle, door cylinder and lock knob rod for proper operation and adjust them if necessary.

R-4. Door Assemblies

a. Removal

1. Place a jack under the door assembly. Loosen the door hinge bolts.
2. Remove the hinge bolts, beginning on the lower bolts, while supporting the door.

   NOTE:

   Put a rubber mat or waste cloth on the jack to prevent the door from being damaged.

b. Installation

- To install the door assembly, reverse the removal procedures. Pay attention to the following points:

1. After installing, check the door lock and striker for proper engagement. Lock the door and check for rattle. If the door fails to be locked properly, it may drop by itself. Adjust it by loosening the door hinge bolts or by inserting shims into between the hinge and body.

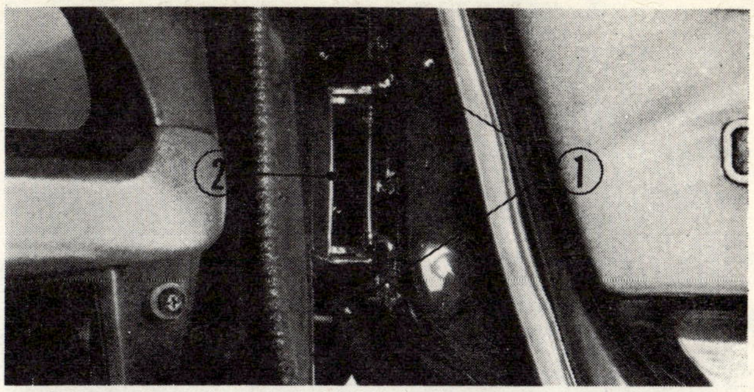

Fig. R–10
① Door hinge ② Door hinge attaching bolts

2. Adjust the door lock striker. Loosen the lock striker bolts and move the striker in either direction to check for proper engagement. Adjust the striker so that the door is opened or closed smoothly.

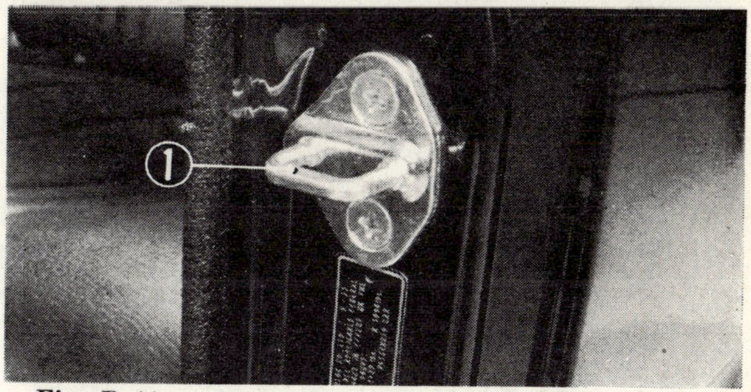

Fig. R–11
① Door lock striker

NOTES

S. FUEL TANK

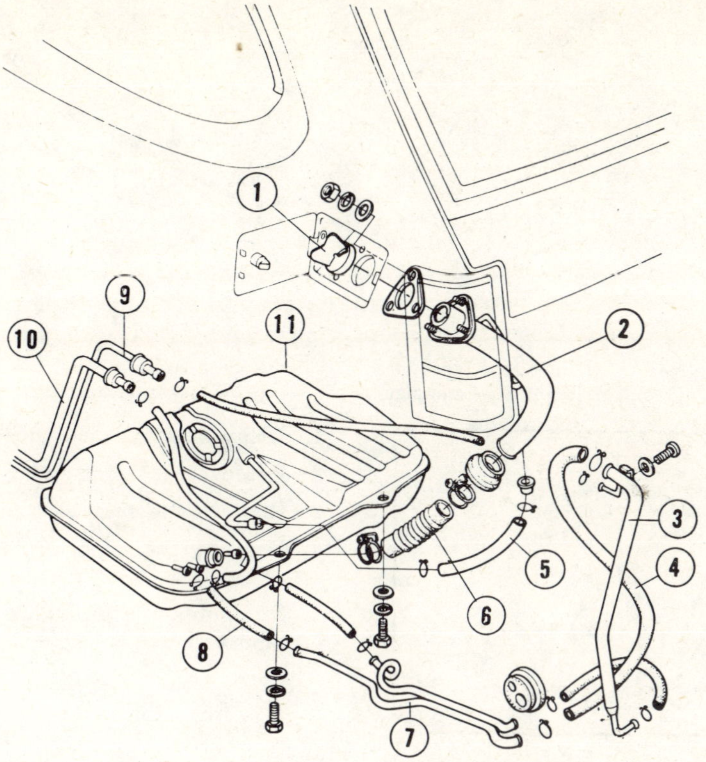

Fig. S–1

① Fuel filler cap
② Fuel filler pipe
③ Separator pipe
④ Fuel tube (5.7×750)
⑤ Vinyl tube
⑥ Filler neck connecting tube
⑦ Bend pipe B
⑧ Fuel tube (7.5×130)
⑨ Bend pipe
⑩ Fuel feed pipe
⑪ Fuel tank

a. Removal

a–1. Fuel tank

1. Drain the tank by loosening the tank drain bolt.

NOTE:

Catch fuel in a clean, safe container.

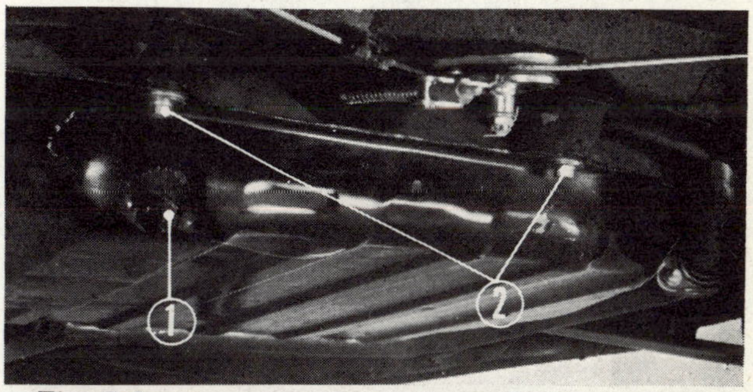

Fig. S–2
① Drain bolt ② Attaching bolts

2. Disconnect the fuel tubes, filler neck connecting tube and vinyl tubes.

NOTE:

Disconnect the fuel tubes by removing clips, taking care not to damage the tubes.

3. Disconnect the fuel meter unit cord at its connection.
4. Remove the fuel tank by loosening its attaching bolts.
5. Remove the fuel meter unit from the tank, taking care not to bend the rod.

Fig. S–3
① Fuel tubes ③ Filler neck connecting tube
② Vinyl tube

a–2. Fuel feed pipe and bends

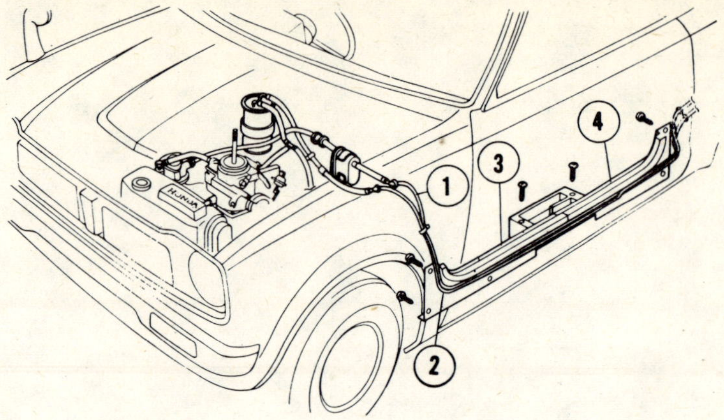

Fig. S–4

① Fuel feed pipe
② Bend pipe complete
③ Side front garnish
④ Side rear garnish

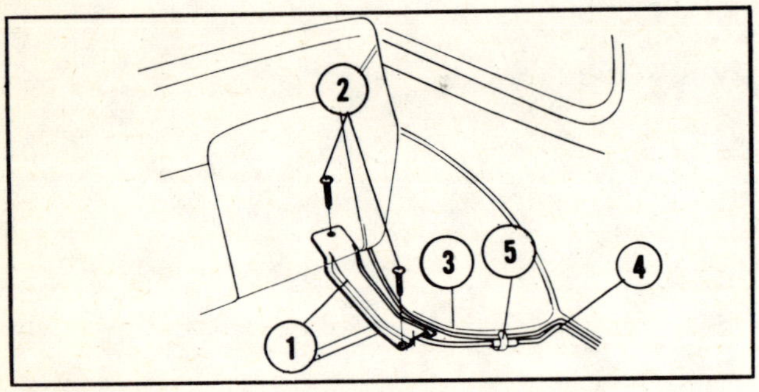

Fig. S–5

① Pipe protector
② Tapping screws
③ Bend complete
④ Fuel feed pipe
⑤ Pipe clip

1. Disconnect the fuel tubes in the engine room.
2. Remove the front and rear side garnishes.
3. Remove the rear seat cushion.
4. Remove the pipe protector by loosening its tapping screws.

   Straighten the pipe clips, and disconnect the fuel feed pipe and bends.

b. Inspection

See the Maintenance Standard.

Inspection item	Remedy
Collapse, deformation, corrosion or other damage of fuel feed pipe, bends and separator pipe	Replace.
Cracks in fuel tubes and filler connecting tube	Replace.
Damage of tube clips	Replace.
Cracks or deformation of fuel tank and filler pipe	Repair or replace.
Clogging or other damage of fuel strainer, one-way valve and check valve	Repair or replace.
Breakage or other damage of vinyl tubes and grommets	Replace.

c. Installation

To install the fuel tank, reverse the removal procedures. Pay attention to the following points:

1. When the fuel meter unit has been removed, apply a coat of sealant to the attaching surface of the tank and to the screws when installing.
2. After connecting the fuel feed pipe and bends, apply a sealant to their grommets to prevent entry of rain into the body.
3. Connect the right pipes and tubes in the correct manner. Use new clips.
4. Install the fuel strainer correctly.
5. After installing, check the pipes and tubes for leak.

MISCELLANEOUS

Front Grille

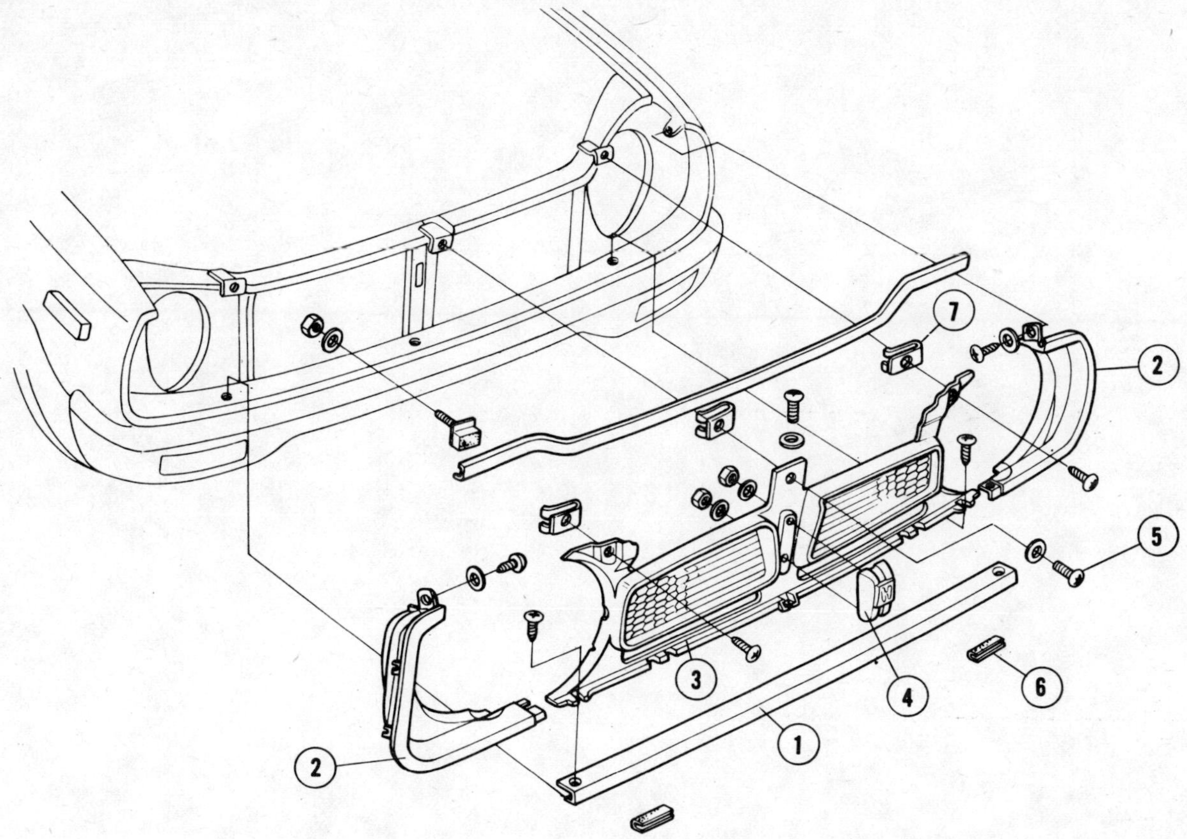

Fig. T-1

① Front grille lower mole
② Head light garnishes
③ Front grille
④ Front grille emblem
⑤ Front grille attaching screw
⑥ Rubber patch
⑦ Clip nut

a. Removal

1. Remove the front grille lower mole by loosening the tapping screws at the lower sections of the right and left head light garnishes.
2. Remove the front grille by loosening the four attaching screws.
   - Loosen the tapping screws at the upper sections of the right and left head light garnishes, and remove the head light garnishes.
   - Remove the front grille emblem and remove the clip nuts and rubber patches. Then the front grille can be removed as shown in Fig. T-1.

b. Installation

To install the front grille, reverse the removal procedures. Pay attention to the following points:

1. Tighten the tapping screws and screws, taking care not to damage the front grille and its moles.
2. To install the front grille lower mole, engage both ends of the grille in the grooves in the rubber patches properly.

Windshield and Rear Window Glass

a. Front windshield

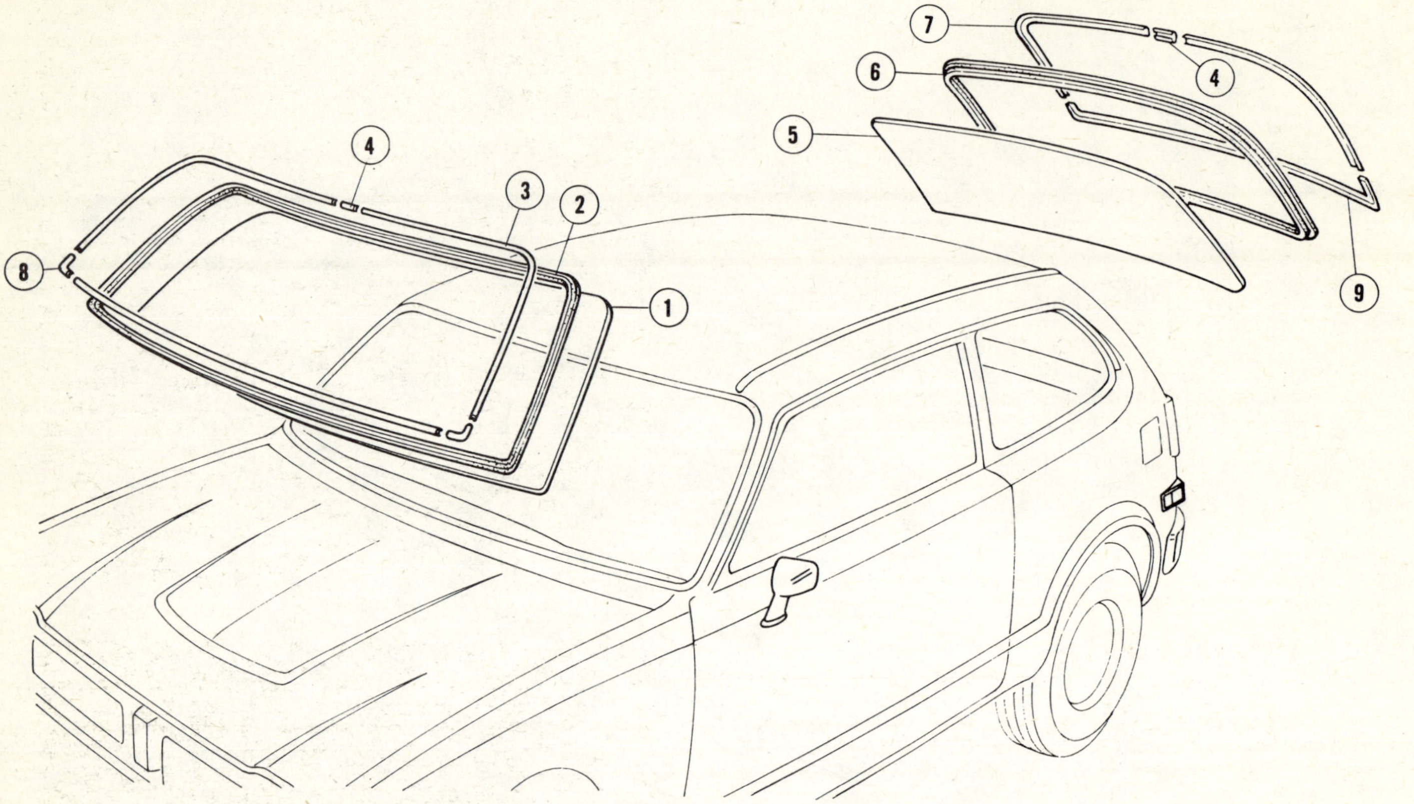

Fig. T-2 ① Front windshield
② Front windshield weather strip
③ Front mole
④ Window mole joint
⑤ Rear windshield
⑥ Rear windshield weather strip
⑦ Rear mole
⑧ Front corner joint
⑨ Rear corner joint

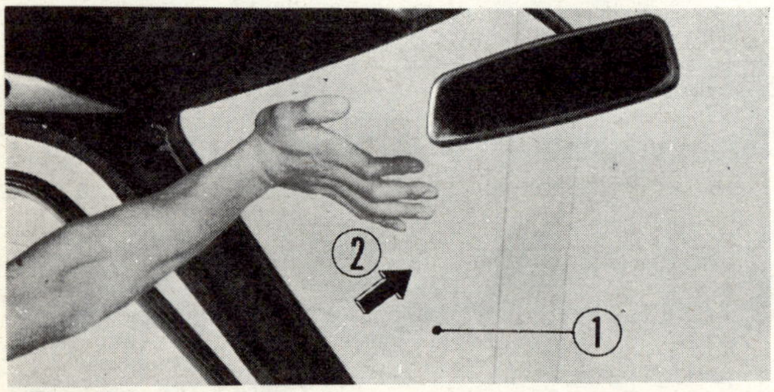

Fig. T-3
① Front windshield ② Pushing front windshield

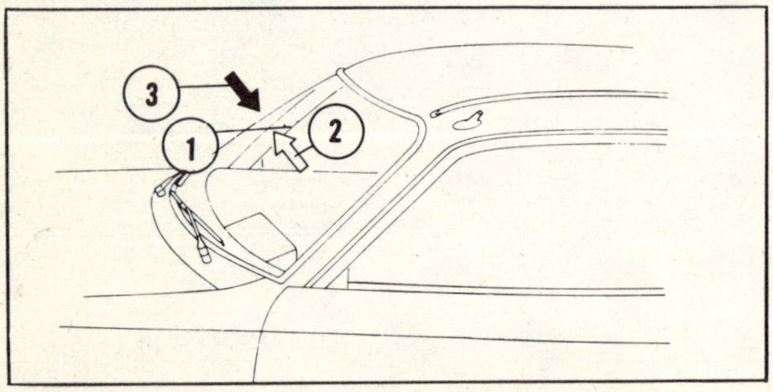

Fig. T-4
① Front windshield ③ Strong
② Weak

a-1. Removal

1. Push the upper corner of the front windshield from inside the car as shown.
2. Hold the windshield with hands outside the car and pull it outward.

   Then the windshield can be removed.

NOTES:

1) It is advisable that this removal be performed by two persons; one person inside the car and ther other outside the car.
2) First remove the wiper arms to facilitate this job.
3) Do not apply a strong force to the front windshield and rear window glass from inside the car to prevent breakage. Carefully handle them.

a-2. Installation

1. Install the weather strip to the windshield correctly.

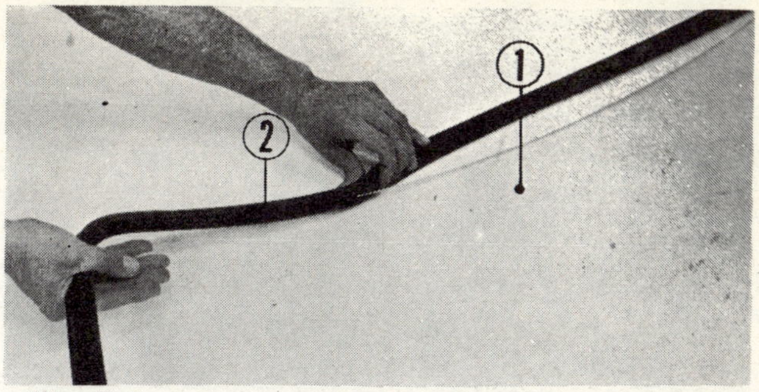

Fig. T-5
① Front windshield ② Front windshield weather strip

2. Apply a coat of soap solution to the weather strip and put a rope in the groove in the strip as shown.
3. Place the windshield in correct position, and put both ends of the rope inside the body.
4. While pushing the windshield from outside, slowly pull the rope and install the windshield to the body.

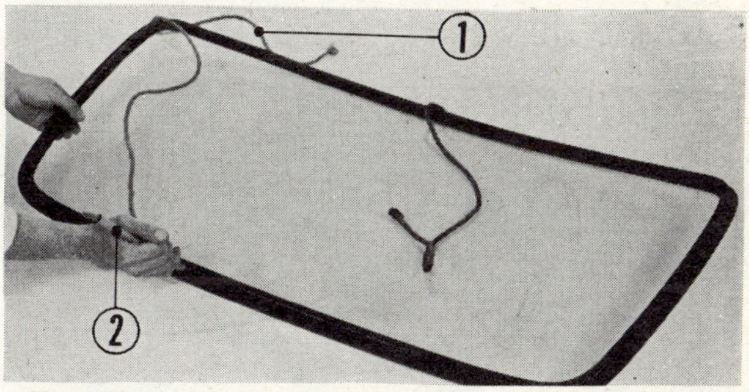

Fig. T-6
① Rope ② Copper pipe

- Do not pull the rope as shown in ②; otherwise the strip will be cut into saw teeth. Be sure to pull the rope in parallel to the weather strip as shown in ①.
- After installing, check the weather strip for proper fit. If necessary, correct with a mallet.

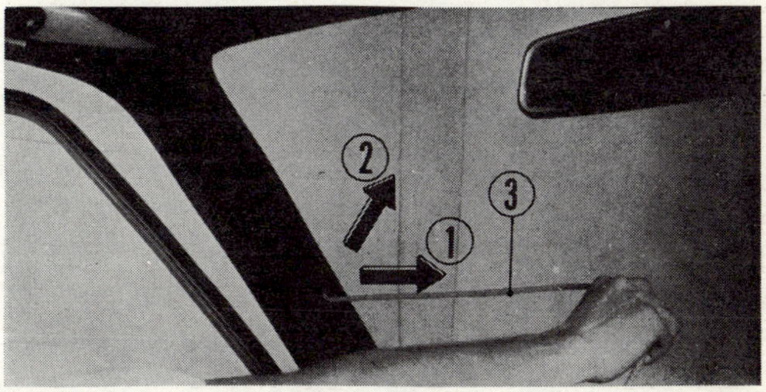

Fig. T-7
① Pull rope in parallel to weather strip ③ Rope
② Rubber is cut in form of saw teeth

5. Apply a coat of soap solution to the mole attaching surface and install the moles in place.

   The mole installer (special tool) must be used at correct angle; otherwise the tool may slip off or the weather strip may be out.

   Install the mole joint in place.

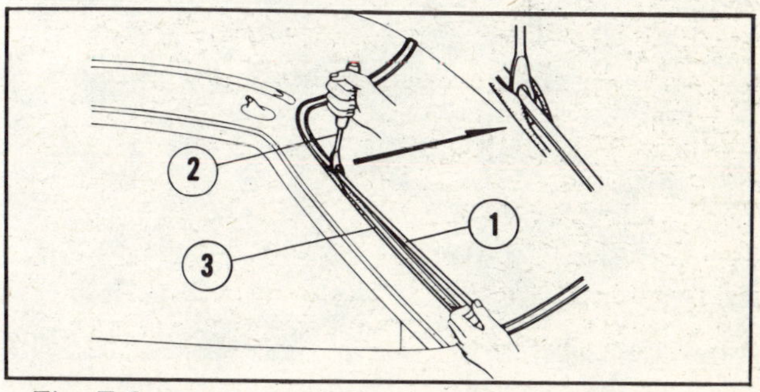

Fig. T-8
① Mole ③ Windshield weather strip
② Special tool

6. The rear window glass can be removed or installed in the same manner as for the front windshield.

   After installing, check the windshields for water leak.

Roof Lining

- When removing or installing, take care not to stain or break the roof lining.

a. Removal

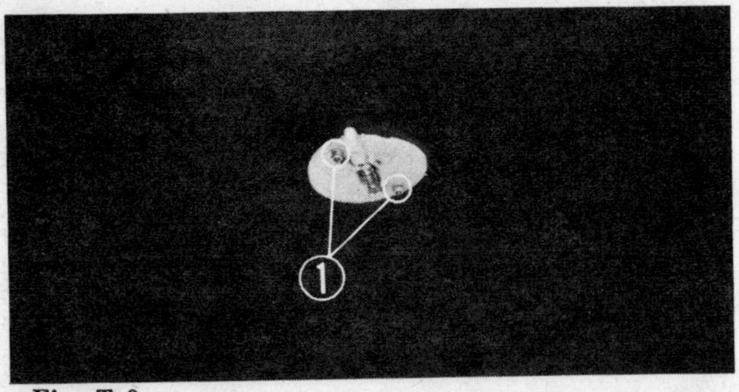

Fig. T-9
① Room light attaching nut

1. Remove the room light assembly. Remove the lens and loosen the attaching nuts.

   Disconnect the cords at their connections. Do not pull the cords strongly; otherwise the roof lining may be broken. Disconnect the cords while holding them with hands.

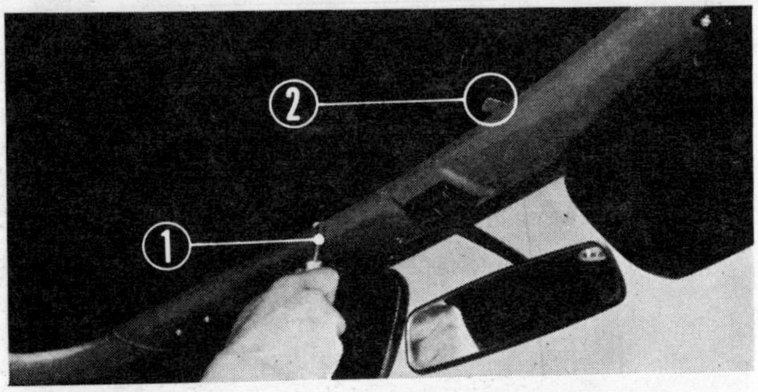

Fig. T-10
① Special tool ② Roof lining clip

2. Attach a lining installer (special tool) to the lining clip as shown in Fig. T-10 and lightly tap the installer with right hand. Then the clip can be removed as shown.

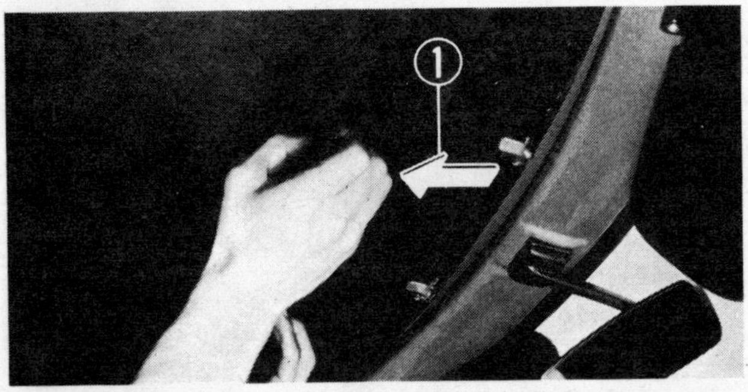

Fig. T-11
① Pull out

3. After removing the four clips, pull the roof lining downward.

   The lining can easily be pulled with the clips removed.

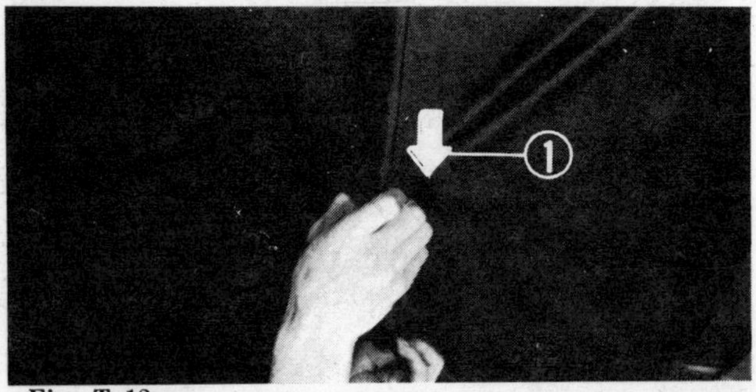

Fig. T-12
① Pull downward

4. After removing the front side of the lining, hold its center part and pull it downward.

   Repeat this step on the rear side.

5. After removing the front and rear clips, remove the roof lining.

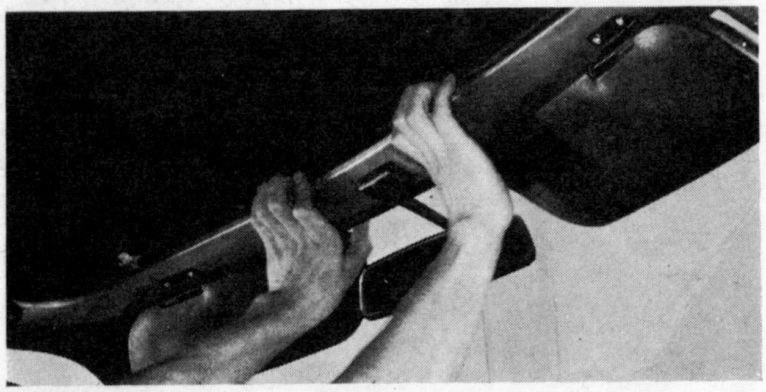

Fig. T-13 Stretch roof lining evenly

b. Installation

To install the roof lining, reverse the removal procedures. Pay attention to the following points:

1. Put the roof lining correctly in the attaching position.

   NOTES:

   1) Take care not to stain or break the lining.

   2) Do not damage the inside surfaces of the body.

2. Face the lining clip as shown in Fig. T-14, attach a special tool to the clip rivet and push the clip downward. Then the clip can be installed properly.
   - Correctly attach the tool to the rivet; otherwise the tool may slip off to result in damage to the lining.
3. After installing the lining, only in the fore and aft directions, remove wrinkles if any, by repositioning the locations of steel wires with the special tool.

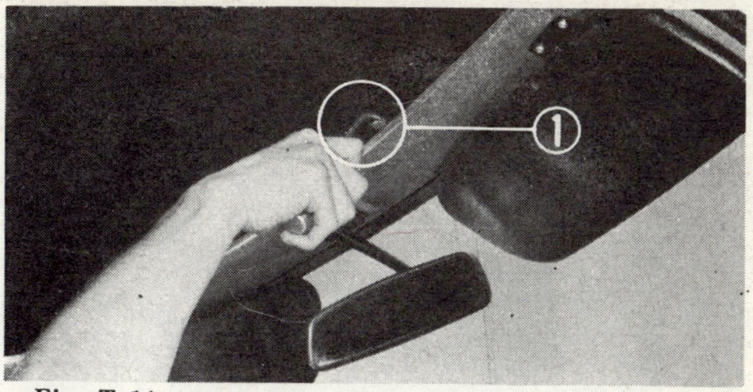

Fig. T-14
① Clip rivet

4. Make sure of the positions of the room light bolts and permit the bolts to come through the lining.
5. Make a hole in the lining to permit the room light cords to come out.

NOTE:

Do not cut the lining more than necessary.

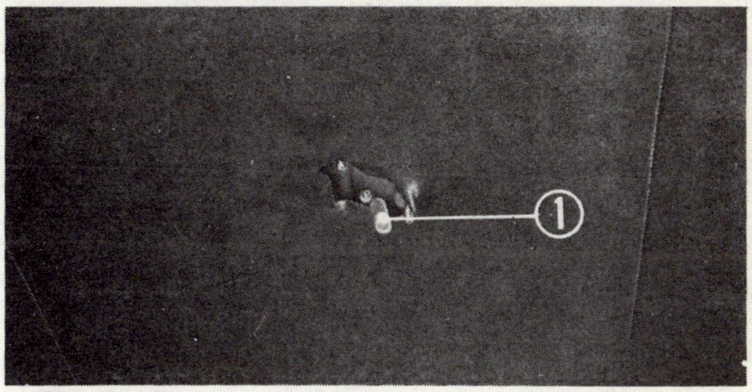

Fig. T-15
① Room light cord

Engine Hood

a. Removal

- It is advisable to remove the engine hood by two persons to prevent damaging the hood.

1. Disconnect the vinyl tube from the washer nozzle, and straighten the tube clips.

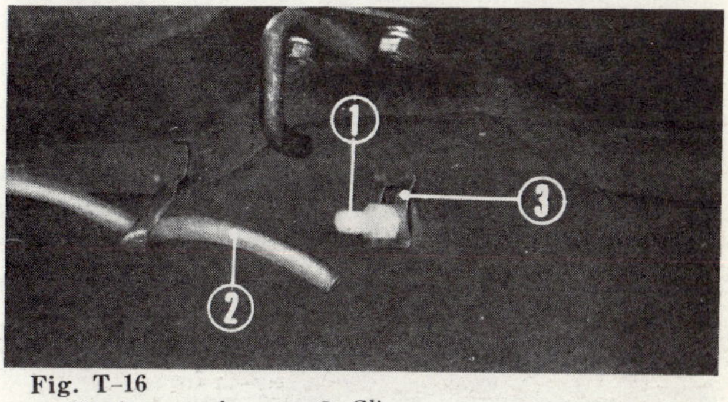

Fig. T-16
① Washer nozzle ③ Clip
② Vinyl tube

2. Remove the engine hood assembly by loosening the hinge bolts.

NOTE:

Store the removed engine hood, exercising care not to damage it.

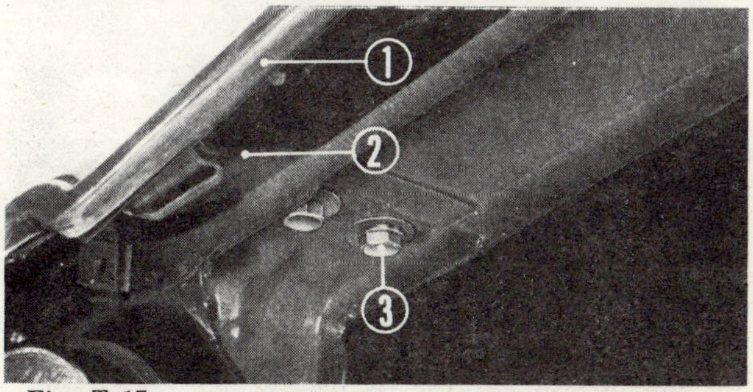

Fig. T-17
① Engine hood assembly ③ Hinge bolt
② Engine hood hinge

b. Installation

To install the engine hood, reverse the removal procedures. Pay attention to the following points:

1. Correctly install the engine hood by adjusting with the hinges.

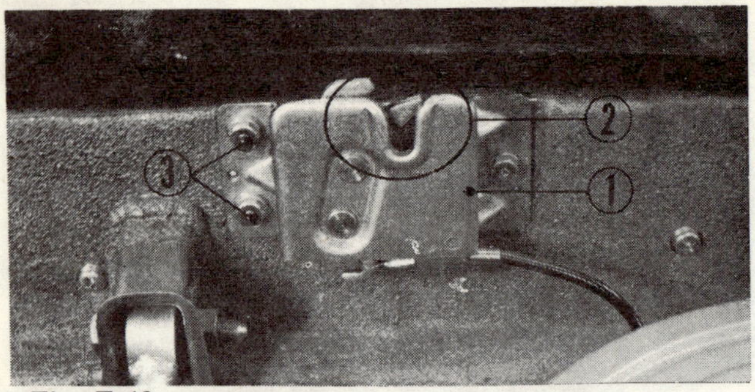

Fig. T-18
① Engine hood lock ③ Attaching bolts
② Apply grease to this part

2. Apply a coat of grease to the engine hood lock.
   - Check to see if the hood can be opened or closed smoothly. If not, adjust it by loosening the striker bolts. Move the hood to right and left on the striker side and up and down on the lock side.

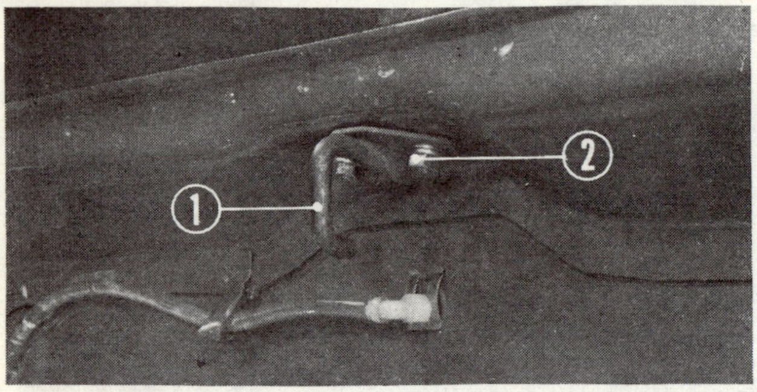

Fig. T-19
① Striker ② Attaching bolt

T-5. Trunk Lid

a. Removal

1. Disconnect the licence number plate light cord.
2. Remove the trunk hinge screws at the body side.
3. Separate the trunk lid from the open stay.
4. Remove the trunk lid seal if damaged.

NOTE:

Where it is difficult to remove the seal, pour a small amount of gasoline into between the seal and trunk lid.

5. Separate the trunk hinges from the lid.
6. Remove the lock cylinder set spring.
7. Remove the licence number plate light assemblies by loosening the 6 mm bolts.
8. Remove the licence number plate by loosening the 5 mm nuts.

NOTE:

Carefully remove the speed nuts from the emblem; otherwise the pins may be broken.

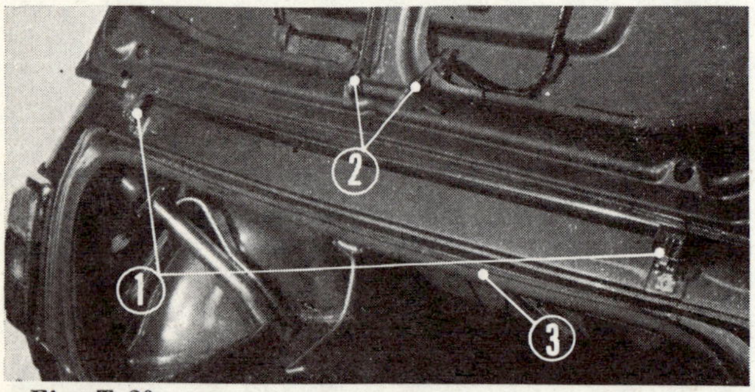

Fig. T-20
① Trunk hinge ③ Trunk lid seal
② Side license number plate light cords

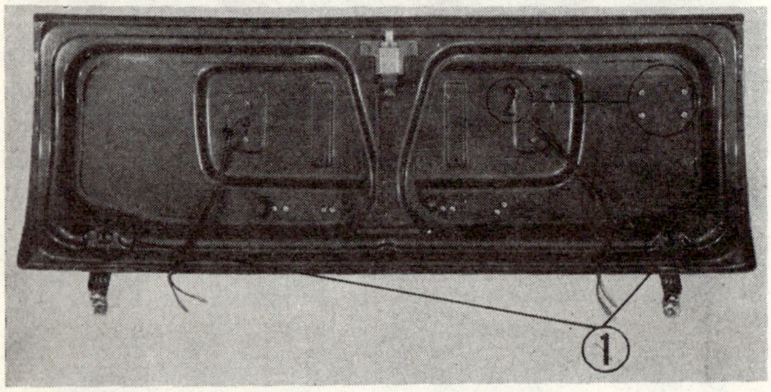

Fig. T-21
① Trunk hinge screws (Trunk lid side)
② Emblem pins

b. Installation

To install the trunk lid, reverse the removal procedures. Pay attention to the following points:

1. Install the trunk lid seal in place. Apply a uniform coat of sealant to the seal attaching surface of the lid and press-fit the seal in place.
2. After installing the lid, check for correct installation. If necessary adjust it.

NOTE:

When the trunk lid is removed, the engagement of the lid and striker may be disturbed. Adjust it by loosening the lid attaching bolts.

Tail Gate

The tail gate means the rear door on the three door type.

Precautions for service work:

Service work on the tail gate must be performed by two persons to prevent the body and gate from being damaged.

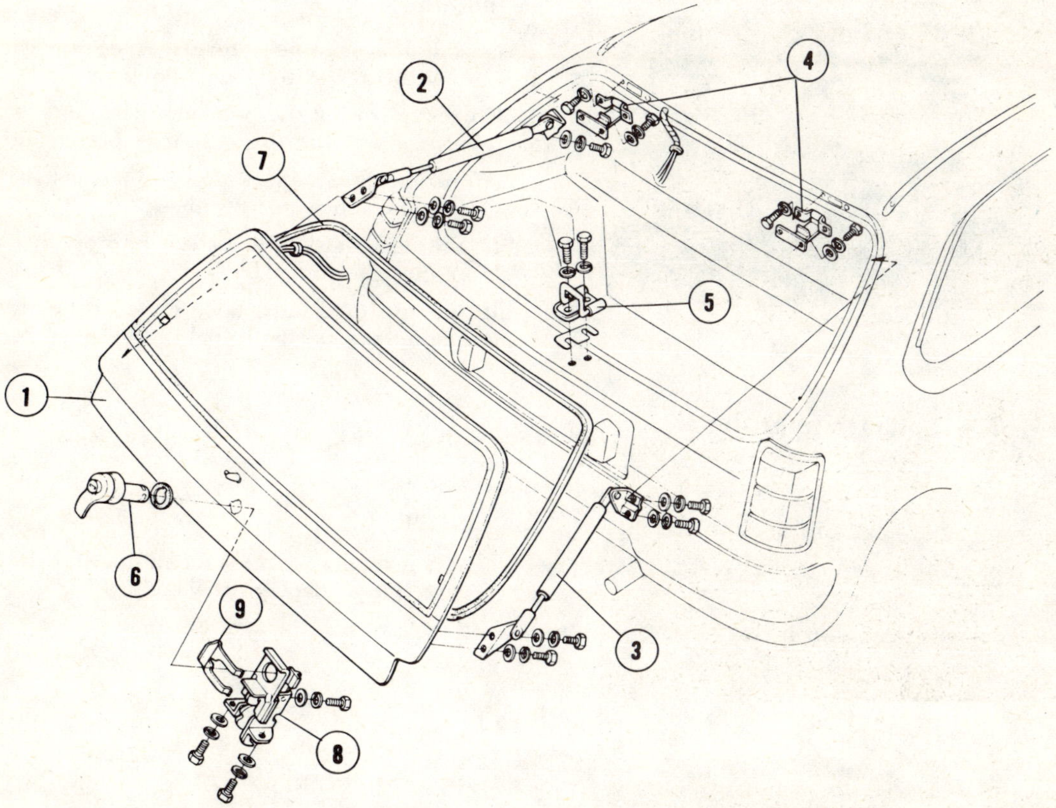

Fig. T-22

① Tail gate
② Tail gate open stay (left)
③ Tail gate open stay (right)
④ Tail gate hinges
⑤ Tail gate striker
⑥ Tail gate lock cylinder
⑦ Tail gate weather strip
⑧ Tail gate lock
⑨ Tail gate lock cylinder spring

a. Removal

1. Open the tail gate and remove the lining by loosening the tapping screws.
2. Disconnect the right and left tail gate open stays at the tail gate.
3. Remove the bolts securing the right and left tail gate hinges to the body.

NOTE:

Take care not to damage painted surfaces.

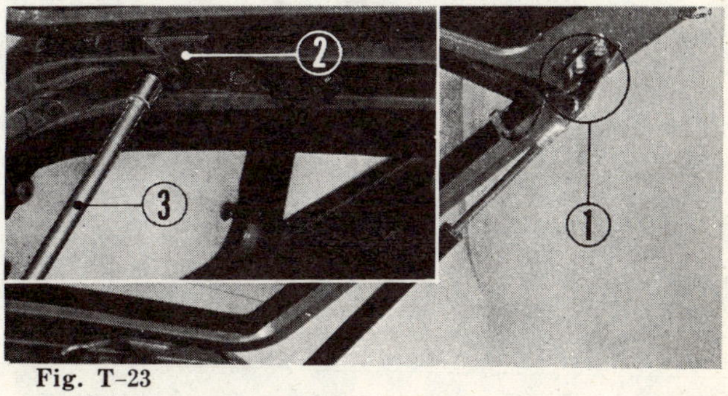

Fig. T-23

① Tail gate open stay bolts
② Tail gate hinge
③ T-type socket wrench

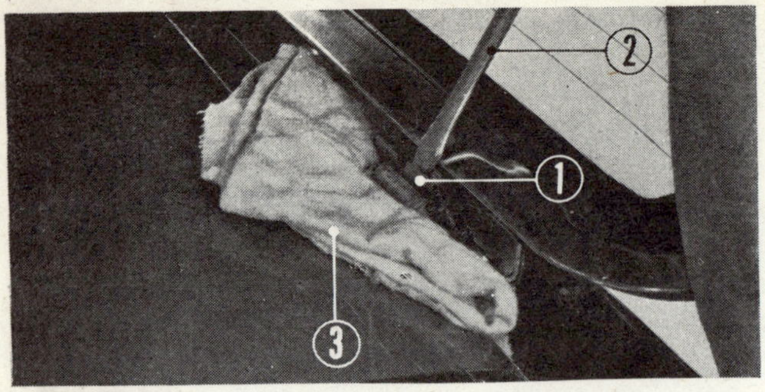

Fig. T-24
① Clip ③ Waste cloth
② Standard type screwdriver

Fig. T-25
① Tail gate lock ② Striker

4. Straighten the tail gate wire harness clip and remove the wire harness from the tail gate carefully.

NOTE:

When straightening the clip, do not damage the tail gate.

b. Installation

To install the tail gate, reverse the removal procedures. Pay attention to the following points:

1. Correctly route the wire harness into the tail gate; otherwise the wire harness may be damaged.
2. After installing, check for proper installation, smooth movement and water tightness.
   - If the tail gate is installed improperly, adjust it by loosening the hinge bolts.
   - If the tail gate fails to be closed properly, adjust it by loosening the tail gate lock and striker bolts.

T-7. Roof Moles and Drip Moles

a. Removal

1. Remove the roof mole joint, and remove the moles.
2. Remove the roof mole ends and remove the weather strip.
3. Remove the front and rear drip moles.

NOTE:

When removing the roof moles and drip moles, take care not to damage the body.

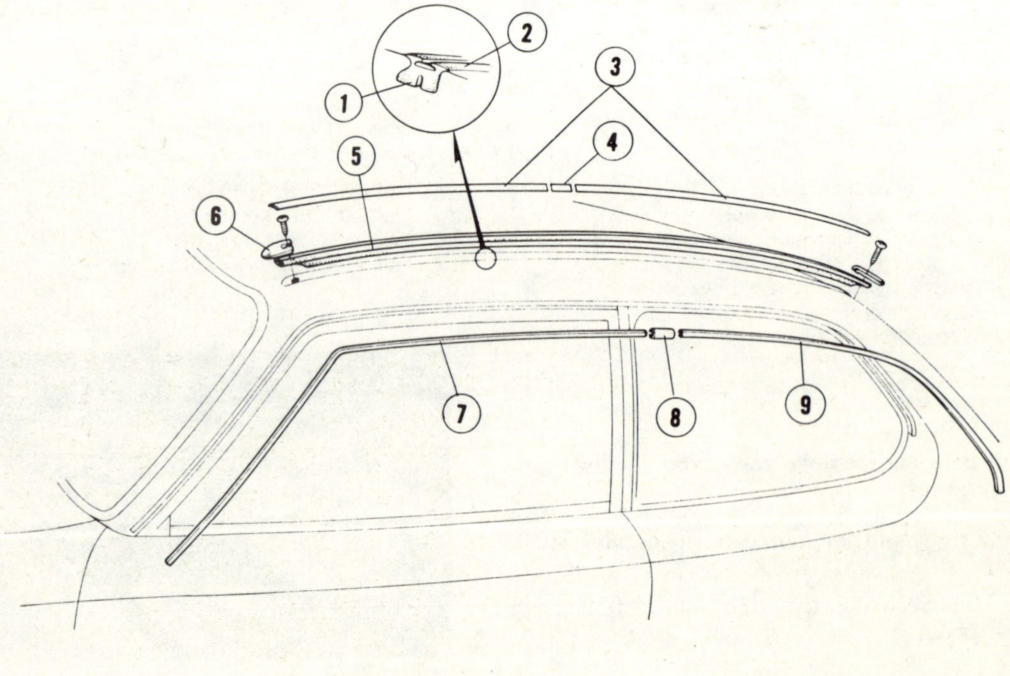

Fig. T-26

① Taper ④ Roof mole joint ⑦ Front drip mole
② Press in ⑤ Roof weather strip ⑧ Drip mole joint
③ Roof moles ⑥ Roof mole end ⑨ Rear drip mole

b. Installation

1. As shown in Fig. T–26, the part ① is slightly tapered. Tuck the roof weather into the part ① by pressing the part ② with a standard-type screwdriver, taking care not to damage the body.
2. Fit both weather strip ends to the mole ends, and secure the mole ends with tapping screws.
3. Install the roof moles in place. Use the front windshield mole installer (special tool).

NOTE:

To facilitate installation, apply a coat of soap solution to the mole attaching surfaces.

4. Correctly install the drip moles with a mallet.

Frame Repair Chart

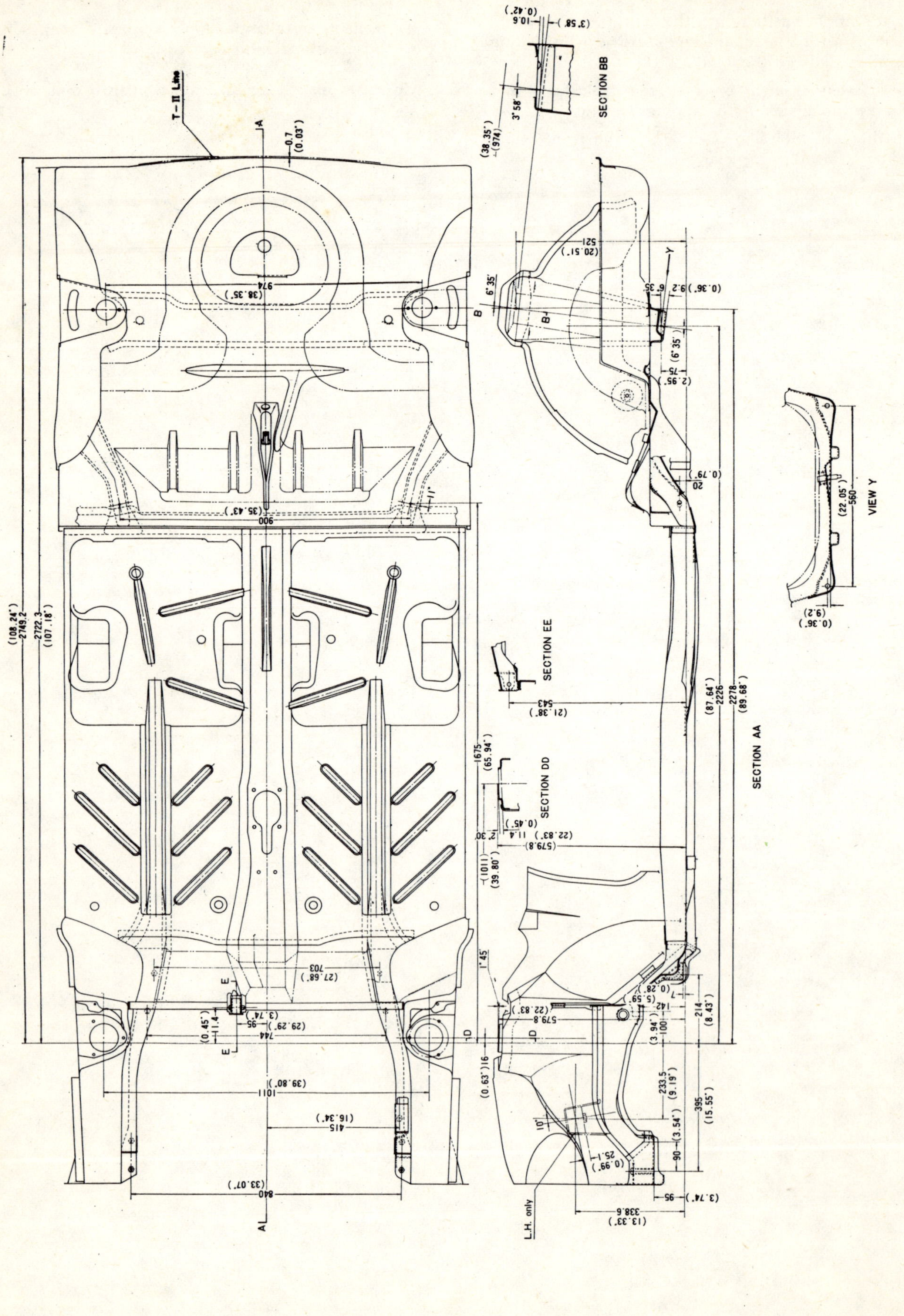

NOTES

U.S.A. Type

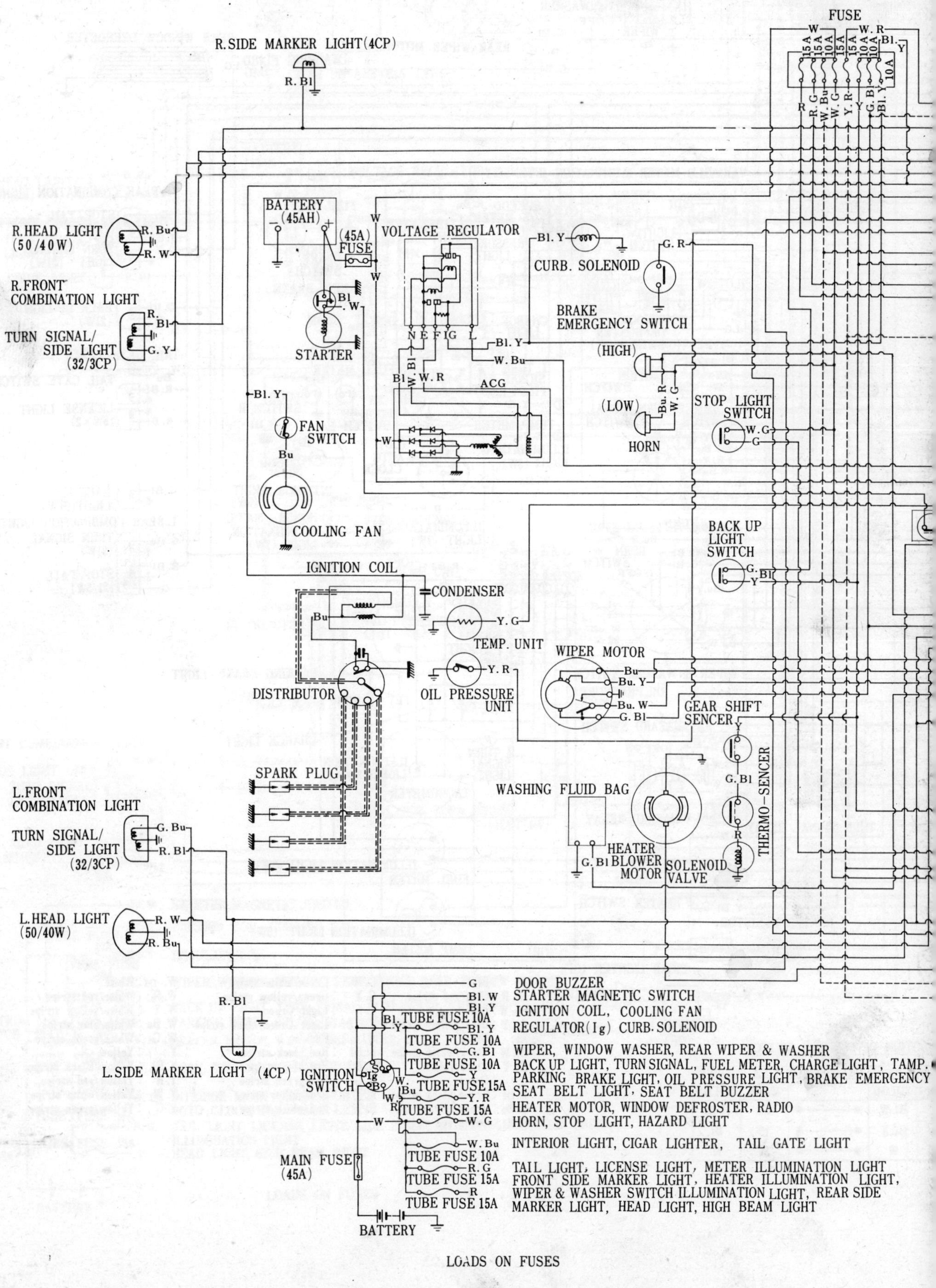
R.SIDE MARKER LIGHT (4CP)
R.HEAD LIGHT (50/40W)
R.FRONT COMBINATION LIGHT
TURN SIGNAL/ SIDE LIGHT (32/3CP)
BATTERY (45AH)
(45A) FUSE
VOLTAGE REGULATOR
STARTER
FAN SWITCH
COOLING FAN
ACG
CURB. SOLENOID
BRAKE EMERGENCY SWITCH
(HIGH)
(LOW)
HORN
STOP LIGHT SWITCH
BACK UP LIGHT SWITCH
FUSE
IGNITION COIL
CONDENSER
TEMP. UNIT
OIL PRESSURE UNIT
DISTRIBUTOR
WIPER MOTOR
GEAR SHIFT SENCER
THERMO-SENCER
SPARK PLUG
WASHING FLUID BAG
HEATER BLOWER MOTOR
SOLENOID VALVE
L.FRONT COMBINATION LIGHT
TURN SIGNAL/ SIDE LIGHT (32/3CP)
L.HEAD LIGHT (50/40W)
L.SIDE MARKER LIGHT (4CP)
IGNITION SWITCH
MAIN FUSE (45A)
BATTERY
TUBE FUSE 10A
TUBE FUSE 10A
TUBE FUSE 10A
TUBE FUSE 15A
TUBE FUSE 15A
TUBE FUSE 10A
TUBE FUSE 15A
TUBE FUSE 15A
DOOR BUZZER
STARTER MAGNETIC SWITCH
IGNITION COIL, COOLING FAN
REGULATOR (Ig) CURB. SOLENOID
WIPER, WINDOW WASHER, REAR WIPER & WASHER
BACK UP LIGHT, TURN SIGNAL, FUEL METER, CHARGE LIGHT, TAMP.
PARKING BRAKE LIGHT, OIL PRESSURE LIGHT, BRAKE EMERGENCY
SEAT BELT LIGHT, SEAT BELT BUZZER
HEATER MOTOR, WINDOW DEFROSTER, RADIO
HORN, STOP LIGHT, HAZARD LIGHT
INTERIOR LIGHT, CIGAR LIGHTER, TAIL GATE LIGHT
TAIL LIGHT, LICENSE LIGHT, METER ILLUMINATION LIGHT
FRONT SIDE MARKER LIGHT, HEATER ILLUMINATION LIGHT,
WIPER & WASHER SWITCH ILLUMINATION LIGHT, REAR SIDE
MARKER LIGHT, HEAD LIGHT, HIGH BEAM LIGHT
LOADS ON FUSES

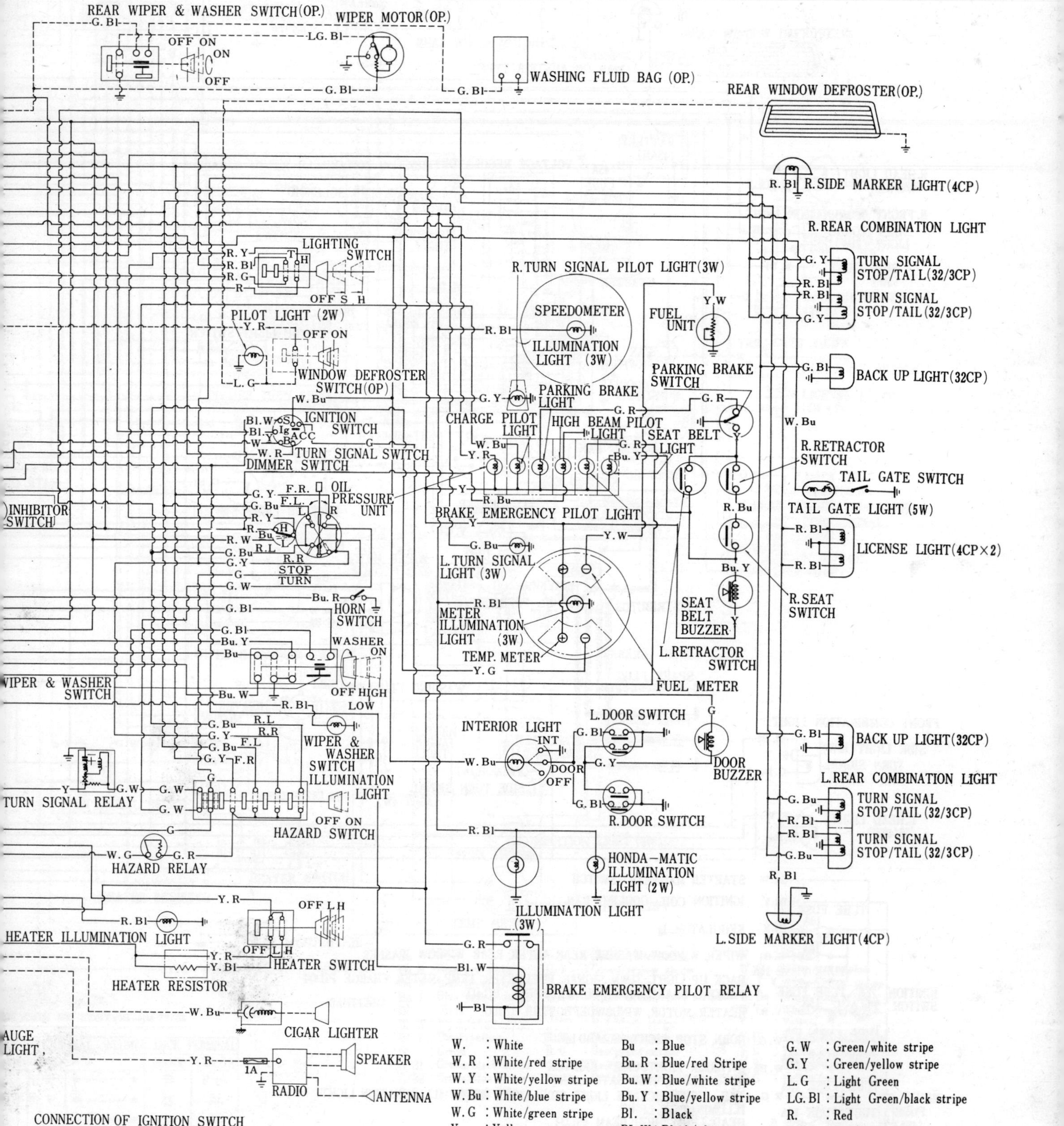

W.	: White	Bu	: Blue	G. W	: Green/white stripe
W. R	: White/red stripe	Bu. R	: Blue/red Stripe	G. Y	: Green/yellow stripe
W. Y	: White/yellow stripe	Bu. W	: Blue/white stripe	L. G	: Light Green
W. Bu	: White/blue stripe	Bu. Y	: Blue/yellow stripe	LG. Bl	: Light Green/black stripe
W. G	: White/green stripe	Bl.	: Black	R.	: Red
Y	: Yellow	Bl. W	: Black/white stripe	R. Bl	: Red/black stripe
Y. Bl	: Yellow/Black stripe	Bl. Y	: Black/yellow spripe	R. Bu	: Red/blue stripe
Y. R	: Yellow/red stripe	G.	: Green	R. G	: Red/green stripe
Y. W	: Yellow/white stripe	G. Bl	: Green/black spripe	R. Y	: Red/yellow stripe
Y. G	: Yellow/green stripe	G. Bu	: Green/black stripe	R. W	: Red/white stripe
		G. R	: Green/red stripe		

CONNECTION OF IGNITION SWITCH

SWITCH TERMINAL			NEUMAN LOCK TERMINAL	
MARK	COLOR		MARK	COLOR
S	BI.W		50	R.BI
Ig	BI.Y		15	BI
B	W		30	R
ACC	W.R		×	Y

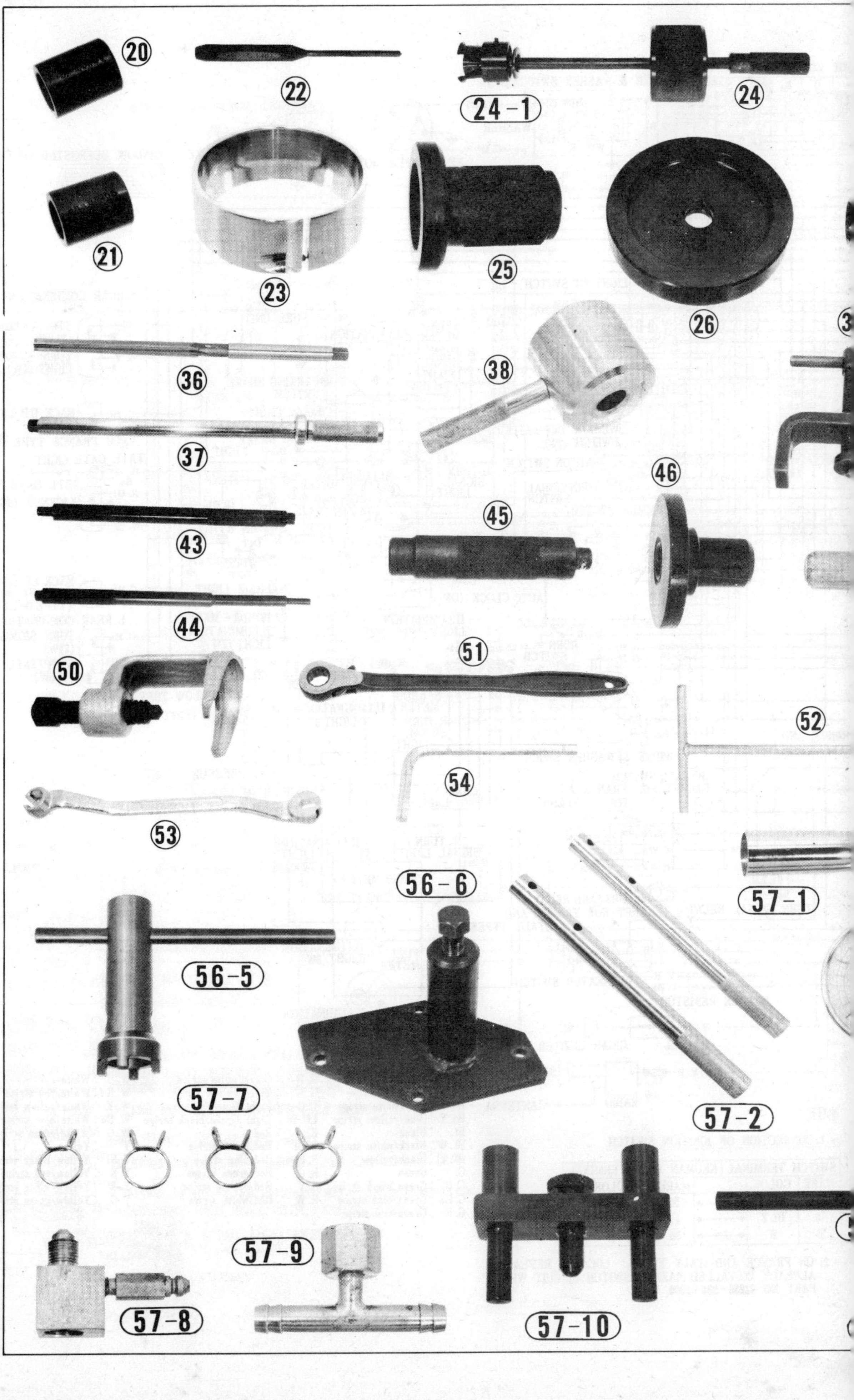
20
22
24-1
24
21
23
25
26
36
38
37
46
45
43
44
51
50
52
54
53
56-6
57-1
56-5
57-7
57-2
57-9
57-8
57-10

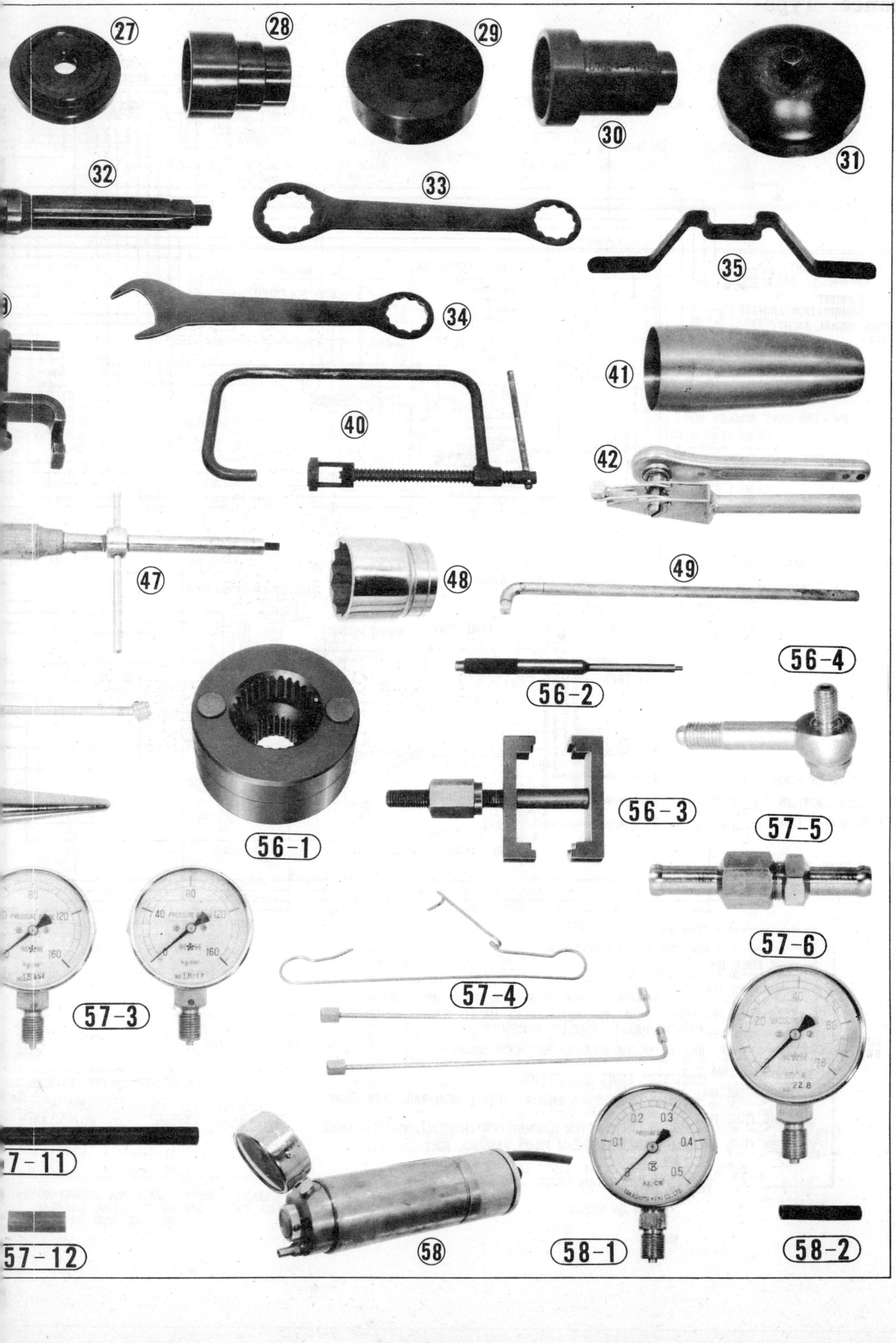
27
28
29
30
31
32
33
34
35
40
41
42
47
48
49
56-1
56-2
56-3
56-4
57-3
57-4
57-5
57-6
7-11
57-12
58
58-1
58-2

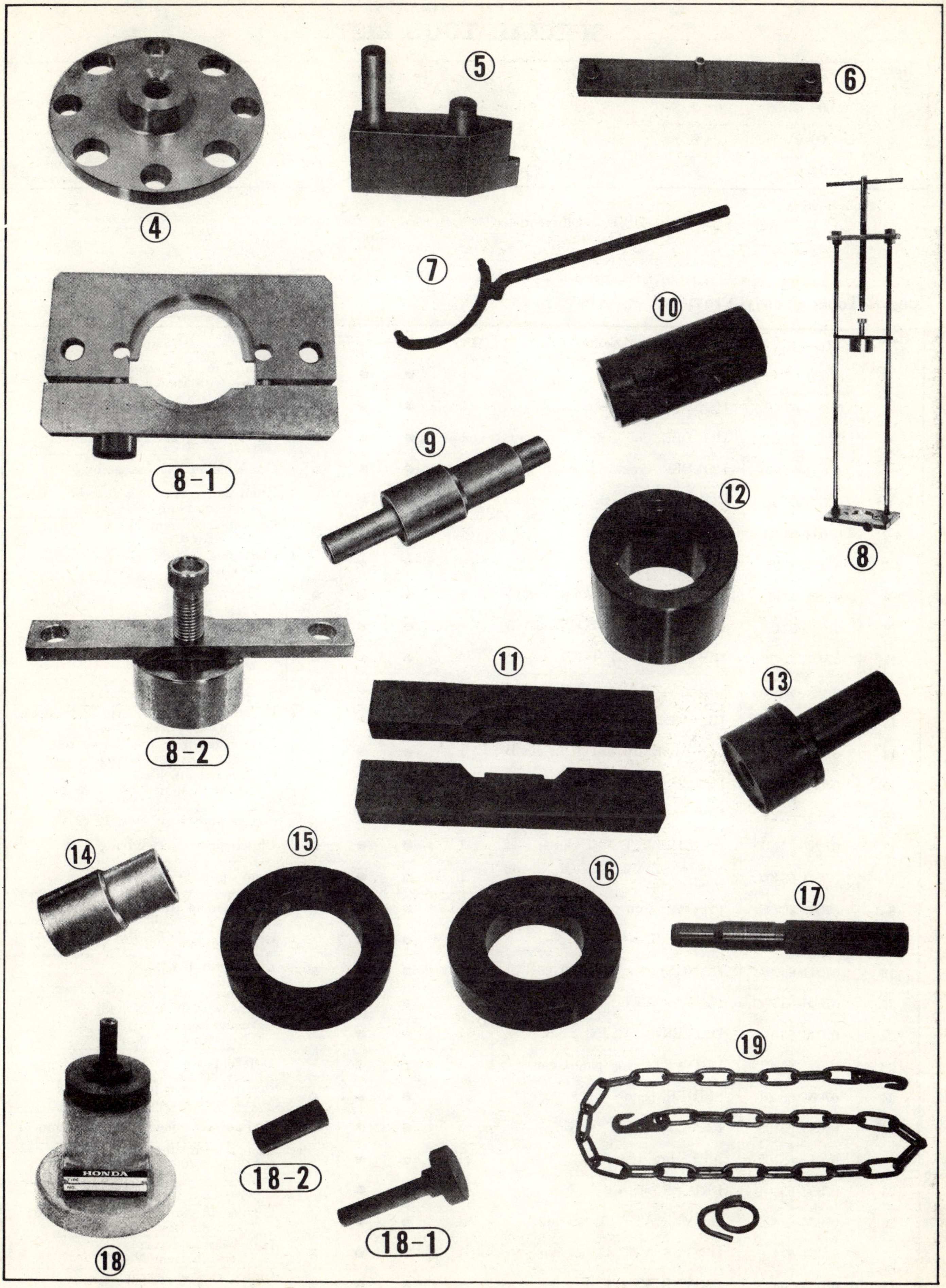
④
⑤
⑥
⑦
⑧
8-1
⑨
⑩
⑫
8-2
⑪
⑬
⑭
⑮
⑯
⑰
⑲
HONDA
18-2
18-1
⑱

SPECIAL TOOL LIST

Item No.	Tool No.	Tool Name	Q'ty.	Remarks	Applicable Model
1	07000-63400	A Set	1	All special tools set	Manual-shift transmission car Hondamatic transmission car
2	07000-63405	B Set	1	All special tools excluding all sets of optional special tools kit	Manual-shift transmission car Hondamatic transmission car
3	07000-63410	C Set	1	Indespensable special tools picked up of B set	Manual-shift transmission car Hondamatic transmission car

Abbreviation

ATT.: Attachment DIFF.: Differential

Special Tools (Newly Provided)

Item No.	Tool No.	Tool Name	Q'ty.	Set Contained			Remarks
				A	B	C	
4	07009-63401	ATT., hub puller	1	●	●	●	Front hub Dis/assembling (Use with Item Nos. 37 & 38)
5	07023-63401	HOLDER, ring gear	1	●	●	●	Ring gear Dis/assembling
6	07023-63405	HOLDER, def. gear	1	●	●	●	Differential gear Dis/assembling
7	07023-63410	HOLDER, crank pulley	1	●	●	●	Crank pulley Dis/assembing
8	07034-63410	COMPPESSOR, shock absorber	1	●	●		Front and rear shock absorber Dis/assembling
8-1	07034-63401	ATT. A, compressor	(1)	●	●		Contained in Item No. 8 (07034-63410)
8-2	07034-63405	ATT. B, compressor	(1)	●	●		Contained in Item No. 8 (07034-63410)
9	07036-63401	DIS/ASSEMBLY TOOL, pin A	1	●	●	●	Front hub, knuckle and hub bearing Dis/assembling
10	07036-63405	DIS/ASSEMBLY TOOL, pin B	1	●	●	●	
11	07036-63410	DIS/ASSEMBLY TOOL, base A	2	●	●	●	
12	07036-63415	DIS/ASSEMBLY TOOL, base B	1	●	●	●	
13	07036-63420	DIS/ASSEMBLY TOOL, pin A	1	●	●		Frontand rear lower-arm and engine mount Dis/assembling
14	07036-63425	DIS/ASSEMBLY TOOL, pin B	1	●	●		Rear radius rod and torque rod mount Dis/assembling
15	07036-63430	DIS/ASSEMBLY TOOL, base A	1	●	●		Radius rod att. base (Use with Item Nos. 12 & 14
16	07036-63435	DIS/ASSEMBLY TOOL, base B	1	●	●		Torque rod att. base (Use with Item Nos. 12 & 14)
17	07037-63401	ASSEMBLY TOOL, clutch	1	●	●	●	Clutch plate Assembling
18	07043-63400	DIS/ASSEMBLY TOOL, piston pin	1	●	●		Piston pin Dis/assembling
18-1	07043-63405	PILOT, piston pin	(1)	●	●		Contained in Item No. 18
18-2	07043-63410	COLLARS, piston pin pilot	(1)	●	●		Contained in Item No. 18
19	07044-63401	HANGER, engine block	1	●	●	●	Engine block hanger
20	07046-63401	DRIVER ATT., In. guide	1	●	●		In. valve guide (Use with Item No. 43)
21	07046-63405	DRIVER ATT., Ex. guide	1	●	●		Ex. valve guide (Use with Item No. 43)
22	07047-63401	DRIVER, spring pin, 4 mm	1	●	●		Differential gear Dis/assembling
23	07051-63401	SLIDER, piston	1	●	●	●	Inserting piston in cyl bore (Adjustable type)
24	07053-63400	REMOVER SET, bearing	1	●	●		Countershaft needle bearing remover
24-1	07053-63405	ATT., brg. remover	(1)	●	●		Contained in item No. 24 (07053-63400)
25	07054-63401	DRIVER, oil seal	1	●	●		Crankshaft and camshaft seal driver (35508-43279 seal)
26	07054-63405	DRIVER ATT. A, bearing	1	●	●		Crankshaft seal driving (Seal: 74901) (Use with Item No. 45)
27	07054-63410	DRIVER ATT. B, bearing	1	●	●	●	Hub bearing driver (Use with Item No. 45)
28	07054-63415	DRIVER ATT. C, bearing	1	●	●	●	Clutch release bearing (Use with Item No. 45)

Item No.	Tool No.	Tool Name	Q'ty.	Set Contained			Remarks
				A	B	C	
29	07054-63420	DRIVER ATT. D, bearing	1	●	●	●	Counter and main shaft bearing (Use with Item No. 45)
30	07054-63425	DRIVER ATT. E, bearing	1	●	●	●	Diff. block remover and Diff. bearing (Use with Item No. 45)
31	07093-63401	SOCKET, oil filter	1	●	●	●	Oil filter (Use with Item No. 51)
32	07094-63401	WRENCH, spark plug	1	●	●	●	Spark plug (Used with Item No. 51)
33	07096-63401	WRENCH, lock nut, 30/36	1	●	●	●	Lock nut 36 mm holder (Use with Item No. 34)
34	07096-63405	WRENCH, lock nut, 30/40	1	●	●	●	Lock nut 30 mm holder and steering lock nut, 40 mm
35	07096-63410	WRENCH, lock, 50 mm	1	●	●		Fuel meter unit Dis/assembling

Special Tools (Common with Other Models)

Item No.	Tool No.	Tool Name	Q'ty.	Set Contained			Remarks
				A	B	C	
36	07008-61101	REAMER, valve guide	1	●	●		In. ex. valve guide
37	07009-51202	SHAFT, sliding hammer	1	●	●	●	Front hub Dis/assembling (Use with Item No. 4)
38	07009-57910	WEIGHT, sliding hammer	1	●	●	●	
39	07015-61101	PULLER, timing pulley	1	●	●	●	Timing pulley remover
40	07031-32901	COMPRESSOR, valve spring	1	●	●		Valve spring Dis/assembling
41	07043-61105	GUIDE, ball joint clip	1	●	●		Ball joint clip installer
42	07043-55102	FASTENER, boot band	1	●	●		Model N360 boot band fastener also usable on the model
43	07046-61102	DRIVER, valve guide	1	●	●		(Use with Item Nos. 20 & 21)
44	07047-61105	DRIVER, spring, 8 mm	1	●	●	●	Shift rod, 8 mm pin driver and remover
45	07048-61101	HANDLE, driver	1	●	●	●	(Use with Item Nos. 26 thru 30 & 46)
46	07054-61120	DRIVER ATT., oil seal	1	●	●		Diff. R/D seal driver (Use with Item No. 45)
47	07081-59001	WRENCH, tappet adjust	1	●	●	●	
48	07083-60110	WRENCH, socket, 32 mm	1	●	●	●	Hub nut wrench (Use with Item No. 49)
49	07083-60105	HANDLE, socket wrench	1	●	●	●	Hub nut socket wrench handle
50	07092-55104	PULLER, ball-joint	1	●	●		Dis/assembling Model N360 ball-joint puller also usable on this model
51	07093-61105	WRENCH, ratchet, 14×17 mm	1	●	●	●	Double head (Use with Item No. 31)
52	07097-60505	WRENCH, brake adjust	1	●	●	●	Universal joint type
53	07778-99952	WRENCH, flare nut	1	●	●		
54	07788-99901	WRENCH, hex., 6 mm	1	●	●		Generator Dis/assembling
55	07790-56805	CASE, special tool	1	●	●	●	Special tool set case

Optional Special Tool Kit

Item No.	Tool No.	Tool Name	Q'ty.	Set Contained A	B	C	Remarks
56	07000-63900	KIT, Honda-matic	1	●			A set of special tools for only Honda-matic transmission
56-1	07023-63901	HOLDER, main shaft	(1)	●			16 mm lock nut remover holder (Use with Item No. 56-5)
56-2	07047-61101	DRIVER, spring pin, 5 mm	(1)	●			Parking brake lever Dis/assembling
56-3	07053-61201	COMPRESSOR, spring clutch	(1)	●			D & L clulch Dis/assembling
56-4	07065-63910	JOINT ATT., oil pressure	(2)	●			Oil pressure measurement gauge att.
56-5	07086-63901	WRENCH, lock nut, 16 mm	(1)	●			Main shaft, 16 mm lock nut
56-6	07015-63901	PULLER, mission case	(1)	●			Transmission Dis/assembling
57	07145-63400	KIT, brake power unit	1	●			Brake power unit overhaul, gauge and tool
57-1	07043-57905	CUP GUIDE, master cylinder	(1)	●			Contained in Item No. 57 (07145-63400)
57-2	07043-6ɛ420	HANDLE, power unit Dis/assembly	(2)	●			
57-3	07145-57901	GAUGE, oil pressure	(2)	●			Max., 160 kg/cm²
57-4	07145-63405	JOINT PIPE, pressure gauge	(2)	●			
57-5	07145-63410	VALVE, vacuum check	(1)	●			
57-6	07145-57910	GAUGE, vacuum	(1)	●			
57-7	07145-63415	BAND, tube	(4)	●			
57-8	07145-67920	ATT. C, pressure gauge	(2)	●			
57-9	07145-57955	ATT. I, tube joint	(1)	●			
57-10	07145-63420	GAUGE, adjust bolt	(1)	●			
57-11	07145-63425	TUBE A, vacuum joint	(1)	●			
57-12	07145-63430	TUBE B, vacuum joint	(1)	●			
57-13	07790-63401	CASE, power unit	(1)	●			
58	07146-63400	EMISSION KIT, fuel cap	1	●			Fuel cap inspection tester
58-1	07146-63405	GAUGE, pressure	(1)	●			Contained in Item No. 58 (07146-63400)
58-2	07146-63410	TUBE, by-pass pipe	(1)	●			